# Nutrient Dynamics for Sustainable Crop Production

Ram Swaroop Meena
Editor

# Nutrient Dynamics for Sustainable Crop Production

Springer

*Editor*
Ram Swaroop Meena
Department of Agronomy
Institute of Agricultural Sciences (BHU)
Varanasi, Uttar Pradesh, India

ISBN 978-981-13-8659-6 ISBN 978-981-13-8660-2 (eBook)
https://doi.org/10.1007/978-981-13-8660-2



This Springer imprint is published by the registered company Springer Nature Singapore Pte Ltd.
The registered company address is: 152 Beach Road, #21-01/04 Gateway East, Singapore 189721, Singapore

# Contents

# About the Editor

**Ram Swaroop Meena** was born in a farmer family in VOP, Harsana, Tehsil, Laxmangarh, Alwar District, Rajasthan, India. Dr. Meena had his schooling in the same village and graduated in Agriculture in 2003 from the Sri Karan Narendra Agriculture University, Jobner, Jaipur (Rajasthan). He has obtained his Master's and Doctorate in Agronomy from the Swami Keshwanand Rajasthan Agricultural University, Bikaner (Rajasthan), securing first division in all the classes with triple NET, Junior Research Fellowship (JRF), and Senior Research Fellowship (SRF) from the Indian Council of Agricultural Research, and RGNF Award from the University Grants Commission, Government of India (UGC, GOI). Dr. Meena has been awarded Raman Research Fellowship by the Ministry of Human Resource Development (MHRD), GOI. He has completed his postdoctoral research on soil carbon sequestration under Prof. Rattan Lal, distinguished scientist, and Director, Carbon Management and Sequestration Center (CMASC), Ohio State University, USA. He is working on soil sustainability, crop productivity, and resources use efficiency, under current climatic era. He has supervised 17 postgraduate and 4 Ph.D. students, and he has 9 years of research and teaching experience at the undergraduate/postgraduate/Ph.D. level. He is working on the three externally funded running projects including the Department of Science and Technology (DST), GOI, and involved in many academic and administrative activities going on at the institute/university level. He has published more than 100 research and review papers in peer-reviewed reputed journals and contributed in the edited books with 25 book chapters at national and international levels. He has published 4 books on the national level and another 6 on the international level. He has worked as an expert in the National Council of Educational Research and Training (NCERT), MHRD, GOI, to develop the two books for school education at XI and XII standards. He has been awarded several awards, namely, Young Scientist, Young Faculty, Global Research, Excellence in Research, Honorable Faculty Award, etc. He is a member of 9 reputed national and international societies and is working as a general secretary, editor, and member of the editorial board in 12 national and international peer-reviewed reputed journals and attended several national and international conferences in the country and abroad. He is contributing to the agricultural extension activities on farmers' level as associate coordinator in training, meetings, workshops, and farmers' fair.

# Soil Carbon Sequestration in Crop Production

1

Ram Swaroop Meena, Sandeep Kumar, and Gulab Singh Yadav

**Abstract**

The carbon (C) sequestration potential of global soils are estimated between 0.4 and 1.2 Gt C year$^{-1}$ or 5–15 % (1Pg = 1 × $10^5$ g). The C emission is rising rapidly by 2.3% every year. If the emissions continue to rise, warming could reach the levels that are dangerous for the society, but it looks like global emissions might now be taking a different turn in the last few years. As we know the sustainability of agroecosystem largely depends on its C footprint as the soil organic carbon (SOC) stock; it is an indicator of soil health and quality and plays a key role to soil sustainability. At the same time, continuing unsustainable agricultural approaches under intensive farming have depleted most of the SOC pool of global agricultural lands. Still, the terrestrial ecosystem has enormous potential to store the atmospheric C for a considerable period of time. Therefore, promoting the cultivation of crops sustainably offers multiple advantages, e.g. augmenting crop and soil productivity, adapting climate change resilience, and high turnover of above- and below-ground biomass into the soil system, thus sequestering atmospheric C and dropping concentration of GHGs from the atmosphere. The continuous vegetation on soil surface ensures good soil health and soil C concentration at variable soil depth as per the specific crop. The C sequestration potential and the amount of organic C returned by crop plants rest on specific plant species, depending on the nature of growth, root morphology and physiology, leaf morphology, climatic conditions, soil texture, structure and aggregation, prevailing cropping system, and agronomic interventions during crop

R. S. Meena (✉)
Department of Agronomy, Institute of Agricultural Sciences (BHU), Varanasi, Uttar Pradesh, India

S. Kumar
Department of Agronomy, CCS Haryana Agricultural University, Hisar, Haryana, India

G. S. Yadav
Division of Crop Production, ICAR Research Complex for NEH Region, Lembucherra, India



R. S. Meena (ed.), *Nutrient Dynamics for Sustainable Crop Production*,
https://doi.org/10.1007/978-981-13-8660-2_1

growth period. The above-ground plant biomass, e.g. plant leaves, branches, stem, foliage, fruits, wood, litter-fall, etc., and below-ground plant biomass, e.g. dead roots, released substances from root exudates, rhizospheric deposition, and plant-promoted microbial biomass C, directly contribute to the SOC buildup. Sustainable crop management practice that ensures the increased nitrogen (N) availability accelerates the C input in the soil ecosystem. Farming practices that improve nitrogen and water use efficiency (NUE and WUE) reduce soil disturbance and erosion, increase plant biomass, and together affect N availability and SOC stock. Conservation tillage together with surface residue retention and legume-based sensible crop rotation reduces soil disturbances, surface runoff, and erosion; increases N availability and SOC sequestration; increases soil sustainability by mixed cropping, intercropping, crop rotation, cover cropping, multiple cropping, and relay cropping; and generates and adds greater amount of qualitative plant biomass into the soil. The N addition, especially from bulky organic manure, green manures, leguminous crops, cover crops, biological N-fixing microbes, and farm and kitchen waste materials, is essential for agricultural productivity and SOC sequestration. The C sequestration benefits from addition of chemical nitrogenous fertilizers are compensated by the release of carbon dioxide ($CO_2$) and nitrous oxide ($N_2O$) during manufacturing, transportation, storage, and application of fertilizers. Therefore, approaching integrated nutrient management (INM) encompassing manures and other C-rich resources sustains soil health and increases N availability and SOC sequestration. Moreover, location-specific scientific research is needed to point out the best management practices that enhance NUE, maintain/improve soil health, boost crop production and SOC sequestration, and minimize greenhouse gas (GHG) release in the biosphere. In the view of above, in this chapter, quantifying the C sequestration potential with higher degree of confidence is required in agriculture management. The present book chapter is critically analyses the C sequestration potential of different soil and crop management practices under diverse ecological conditions for sustainable crop productivity.

**Keywords**

Carbon dioxide · Crop production · Soil C sequestration · Sustainable agriculture

## Abbreviations

| | |
|---|---|
| AFS | Agroforestry system |
| C | Carbon |
| $CaCO_3$ | Calcium carbonate |
| $CH_4$ | Methane |
| cm | Centimetre |

| | |
|---|---|
| $CO_2$ | Carbon dioxide |
| $CO_3^{-2}$ | Carbonate |
| FYM | Farmyards manure |
| g | Grams |
| GHGs | Greenhouse gases |
| GPP | Gross primary production |
| Gt | Gigatons |
| ha | Hectare |
| $HCO_3^-$ | Bicarbonate |
| INM | Integrated nutrient management |
| IPCC | Intergovernmental Panel on Climate Change |
| kg | Kilograms |
| Mg | Megagrams |
| Mt. | Metric tons |
| N | Nitrogen |
| $N_2O$ | Nitrous oxide |
| NPP | Net primary productivity |
| NUE | Nitrogen use efficiency |
| OC | Organic carbon |
| OM | Organic matter |
| Pg | Picograms |
| ppm | Parts per million |
| RMPs | Recommended management practices |
| SOC | Soil organic carbon |
| SOM | Soil organic matter |
| Tg | Teragrams |
| WUE | Water use efficiency |

## 1.1 Introduction

Enriching soil organic carbon (SOC) pools in agriculture by encouraging soil C sequestration is an efficient way towards diminishing atmospheric carbon dioxide ($CO_2$) level and inducing soil health (Lal et al. 1999; Post et al. 2004; Bronick and Lal 2005; Lal 2002, 2011; Ashoka et al. 2017). In soil, the C sequestration is characterized by two types: *first*, organic C sequestration – in the form of organic C – which is considered as boon to agriculturalists and, *second*, inorganic C sequestration, in the form of paedogenic calcium carbonate ($CaCO_3$), often called as bane for farmers (Chaudhury et al. 2016; Meena and Meena 2017). The significance of soil as a terrestrial C regulator has been increasingly documented, especially after the Paris Agreement, December 2015, which appeals for action to store and increase the sink capacity of greenhouse gases (GHGs) (FAO 2016). Even after knowing the significance of world's soil as a potential sink and pool of C (Lal 2011), the knowledge about the existing soil C reserves and its capacity of sequestering C is so far incomplete (FAO 2016). However, scientists are trying to optimize the management skills through sustainable crop cultivation so that soils can function as sinks more effectively for C and pay to $CO_2$ diminution strategies (Curtin et al. 2000; Yadav et al. 2018b). After oceans (38,000 gigatons/Gt C), the soil is the second largest C pool of the Earth, and a little change in organic C reserve in soil may cause significant alteration in atmospheric $CO_2$. It is important to understand for the reason that the annual flux of $CO_2$ between soil and atmosphere is big and depends on man-made alterations (Bakker et al. 2007; Kumar et al. 2017b; Dadhich and Meena 2014). The atmosphere holds about 750 Pg (picograms) of C as $CO_2$, whereas globally (excluding permafrost) the upper 100 cm soil holds about 1500 Pg C (1 Pg = 1 Gt = 1015 g) (2500 Pg C in top 200 cm) in the form of SOC and 900–1700 Pg as inorganic C, and this soil exchanges 60 Pg C with the atmosphere every year (Eswaran et al. 1993; Lal 2010; Meena et al. 2015d). It was estimated that global soils hold nearly $1.5 \times 10^{12}$ metric tons of C. In actual, the SOC sequestration potential seems to be between 0.37 and 1.15 Gt C annually (Smith et al. 2008). The rate of soil sequestration in soils under agricultural use varied from 0.1 to 1.0 tons C hectare$^{-1}$ every year (Paustian et al. 2016). Accordingly, there is a huge available gap to reach the potential capacity of soil to sequester C. We should have to manage the billion hectares of land to sequester C so as to touch the annual sequestration rate of 1 Gt C. Moreover, the sequestration level would be comparatively less at the start which would reach at its peak after 20 years and thereafter would decrease gradually (Sommer and Bossio 2014; Yadav et al. 2018a).

The change in organic C content in soil is directly linked with the total amount of Cic substance entered (Buyanovsky and Wagner 2002). The SOC pool is considered as the key indicator of soil fertility and health, and an upmost C pool in terrestrial ecosystem had a very imperative role in global C cycle (Wang et al. 2015). The concentration of SOC in soil is about twice to that of atmosphere and vegetation. So, if the concentration of C is increased, the atmospheric C concentration will get reduced, and it will consequently assuage the problem of global warming and climate change. The soil organic matter (SOM) is linked in a straight line to the SOC; meanwhile, organic C contains 58% of the SOM (Collins et al. 1997). It was projected that 1 ton of SOM is emitted in about 3.667 tons of $CO_2$ into the atmosphere (Meena et al. 2016a). The SOC is the biggest C pool in the terrestrial biosphere, chiefly greater than double of the C accumulated in the atmosphere and vegetative biomass (Jobbagy and Jackson 2000; Liang et al. 2016; Varma et al. 2017a). In top 30 cm soil profile, the average concentration of SOC ranged from 0.30% to 1.05%. It is around 10% of the SOC stocks (140~170 Pg) in agricultural ecosystem and utmost active fragment of the world's terrestrial soil C pool of farmland ecosystem (Liang et al. 2016; Datta et al. 2017a). The farmland harbours of China hold SOC approximately 25–27 Pg and had an imperative contribution in the global C budget (Qin et al. 2013).

The C capturing capacity of soil can be enhanced and improved via improved farming practices that restore soil fertility and health. Promoting sustainable crop cultivation offers multiple advantages: augmenting crop and soil productivity, adapting climate change resilience, sequestering atmospheric C, and dropping concentration of GHGs from the atmosphere (FAO and ITPS 2015). With the purpose to tap the C sequestration potential of soil, the cultivation of plants having higher biomass production capability needs to be endorsed in the agricultural system (FAO and ITPS 2015). Crop residues are one of the chief sources of C in agricultural soils. Agricultural crops produce a considerable quantity of residues, which in turn favours the accumulation of humus in consequent soil C pool upon incorporation into soil (Hajduk et al. 2015; Meena and Yadav 2015). In this chapter, the emphasis is on the magnitude of the potential impacts of agricultural crops that have a capacity to soil C sequestration.

## 1.2 Global Carbon Cycle

It is very important to study the circulation of C on the planet as the C is a major structural component of living organism comprising about 50% of their dry weight, besides its active involvement in the global energy flow and metabolism of natural, human, and industrial systems (Houghton 2003; Dhakal et al. 2015). The C cycle is the biochemical cycle of continuous C exchange among the atmosphere, biosphere, hydrosphere, geosphere, and pedosphere on the planet through the combined process of photosynthesis, respiration, and OM decomposition (Fig. 1.1). The global C cycle is comprised of five major interconnected reservoirs – the atmosphere, terrestrial biosphere, oceans, sediments, and the Earth's interior (David 2010). The C continuously moves through exchange pathways among these reservoirs as a result of numerous physical, chemical, and biological processes (Falkowski et al. 2000; Varma

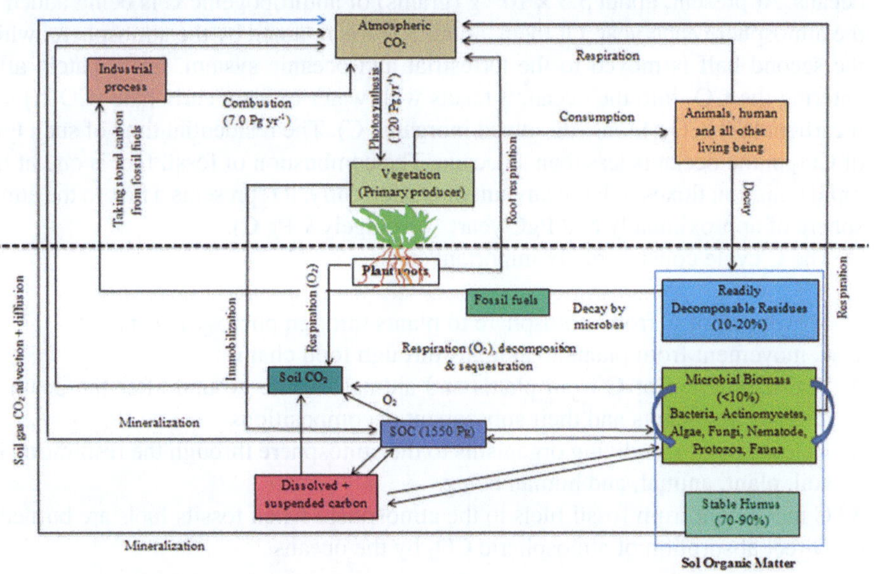


**Fig. 1.1** Schematic diagram of global C cycle. (Data adapted from Lal 2008)

et al. 2017b; Meena and Lal 2018). This cycle starts with the biological C fixation – the conversion of atmospheric $CO_2$ into the living biomass C through the biochemical process of photosynthesis by the more favoured photosynthetic eukaryotes and prokaryotes (Bleam 2012). The photosynthetic process reduces C (+4) in $CO_2$ to C (+1) in the terminal C in glyceraldehyde-3-phosphate, the feedstock for simple sugars, amino acids, and lipids (Bleam 2012). Here, the gross primary production (GPP) is the measure of quantity of atmospheric $CO_2$ removed by photosynthesis every year. According to an estimate, photosynthesis captures 120 Pg C year$^{-1}$ from the atmosphere reservoir and is able to accumulate around 610 Pg C within the living plant at any given time. A part of the photosynthesized biomass C retained by the living plant is directly consumed by the herbivores, while the remaining biomass C becomes the soil residue inviting the diverse soil microbes to attack and decompose, which is known as C mineralization (Bleam 2012; Meena and Yadav 2014). This mineralization of SOC into $CO_2$ occurs through a process called oxidative metabolism in which chemical energy is stored during C-fixation. Respiration (including decomposition of soil biomass) by plant, human, animals, and soil pays back the C into the atmosphere in the form of $CO_2$ and methane ($CH_4$) under anaerobic situations. Forest fires also greatly contribute $CO_2$ and $CH_4$ emission to the atmosphere on annual timescales, but again it is removed by the terrestrial biosphere if vegetation regrows over the decades (IPCC 2007). The plant respiration alone accounts the 50 % of the $CO_2$ (60 Pg C year$^{-1}$) that is returned to the atmosphere in the terrestrial C pool. Similarly, with the decomposition of SOM by the soil microbes, the $CO_2$ is released at the average rate of around 60 Pg C year$^{-1}$. The $CO_2$ released by use of fossil fuel, deforestation, and cement production promoted by human activities accelerates the C exchange chain between atmosphere, terrestrial biosphere, and the oceans. At present, about $5.5 \times 10^{15}$ g (grams) of anthropogenic C is being added in the atmosphere each year. Of them, about 50 % is retained by the atmosphere, while the second half is moved to the terrestrial and oceanic system. Immediately after entering the $CO_2$ into the ocean, it reacts with water to form carbonate ($CO_3^{-2}$) and bicarbonate ($HCO_3^-$) ions (dissolved inorganic C). The residential time of such type of $CO_2$ in the ocean is less than a decade. The combustion of fossil fuel is one of the rapid emission fluxes of large amount of C. Currently, it represents a flux to the atmosphere of approximately 6–8 PgC year$^{-1}$ (averagely 7 Pg C).

The C cycle consists of six important steps:

1. Movement of C from atmosphere to plants through photosynthesis
2. C movement from plants to animals through food chain
3. Transformation of C from plants and animals to the ground after the death of animals and plants and their subsequent decompositions
4. Release of C from living organisms to the atmosphere through the respiration by soil, plant, animal, and human being
5. C movement from fossil fuels to the atmosphere when fossils fuels are burned
6. Direct absorption of atmospheric $CO_2$ by the oceans

## 1.3 Carbon Dioxide Emission Trend and Present Status in Atmosphere

In 1958, Dave Keeling – an American scientist – took the first measurement of $CO_2$ at Mauna Loa Observatory in Hawaii and at Scripps Institution of Oceanography and alerted the globe to the possibility of anthropogenic greenhouse gas effect and global warming. He was the first to register the rise of $CO_2$ in the atmosphere. In 2005, scientists around the world started to keep track of C emissions. Since pre-industrialization time (1750s), the global atmospheric $CO_2$ concentration is continuing to increase from approximately 280 ppm (part per millions) (IPCC 2007) to 406.99 ppm at the end of August 2018 with annual average growth rate of 0.47 ppm $year^{-1}$, although it was 2.7 ppm $year^{-1}$ for the past 2006–2015.The atmospheric $CO_2$ reached the record height of 410.31 ppm in the history for the month of April 2018 as per the report from Mauna Loa Observatory, Hawaii. The increase in annual means from 2015 to 2016, 2.63 ppm, is higher than the increase from 2014 to 2015 and 2013 to 2014 (~2.3 and 2.1 ppm $year^{-1}$, respectively) (WMO 2016). The atmospheric $CO_2$ abundance in 2016 relative to year 1750 was 144.5%. The relative increment from 2015 to 2016 was 0.67%. According to a study, the atmospheric $CO_2$ concentration is now increasing at the rate of 100 times faster over the rate which was at the end of ice age owing to the uncontrolled population growth, rapid industrialization, intensive cultivation, and continuous deforestation promoted by human. Therefore, the release of $CO_2$ into the atmosphere as a result of anthropogenic activities is of great concern. In fact, human activities were responsible for about 110% of observed warming (ranging from 72% to 146%), with natural factors in isolation leading to a slight cooling over the past 50 years as pointed out by IPCC's implied best guess by NASA's Dr. Gavin Schmidt (FAO 2016). In the year 2015, the total $CO_2$ emission from fossil fuel combustion and cement production from industries was 9.9 ± 0.5 Gt C $year^{-1}$, and from land-use pattern mainly deforestation, it was 1.3 ± 0.5 Gt C $year^{-1}$(Le Quéré et al. 2016; WMO 2016). During the last decade (2006–2015), the growth rate of global atmospheric $CO_2$ level, mean ocean $CO_2$ sink, and global residual terrestrial $CO_2$ sink were 4.5 ± 0.1, 2.6 ± 0.5, and 3.1 ± 0.9 Gt C $year^{-1}$, whereas, in 2015, they were 6.3 ± 0.2, 3.0 ± 0.5, and 1.9 ± 0.9, respectively (Le Quéré et al. 2016; Yadav et al. 2017c). The $CO_2$ emitted from the deforestation and land-use change activities was the prime factor behind increased $CO_2$ level in the atmosphere above preindustrial levels (Ciais et al. 2013; Verma et al. 2015c).

Over the globe, the total greenhouse gas $CO_2$ emission in the year 2016 continued to increase at the rate of 0.5 ± 1%, about 53.4 Gt $CO_2$ equivalent (including those from land use and forestry – 4.1 Gt $CO_2$ eq.) (Olivier et al. 2017; Meena et al. 2018a). But, if we look forward, we can find that in the recent 3 years, the amount of $CO_2$ in the atmosphere being released from burning of fossil fuels, gas flaring, and cement manufacturing is consistent. In 2014, the growth in global $CO_2$ emissions was 1.1% (40.3 Gt $CO_2$ eq.); in 2015, it did not grow at all and remains almost stable (39.7 Gt $CO_2$ eq.); and in 2016, they are set to grow very little by just 0.3% (Olivier et al. 2016; Kumar et al. 2018a). This growth in emission trends looks prominently a slowdown over the growth rate of 3.5% in the 2000s and 1.8% in the recent last decade (2006–2015). The main reason behind this slowdown was the change in energy use by the people in China by decreased consumption of coal and fuel and increased use of natural gases and promoting renewable power generation (e.g., wind, solar power, etc.) (Olivier et al. 2017). The leading five emitters China, the United States, India, Russian Federation, and Japan in 2016 covered about 68 % of total global $CO_2$ emissions (Olivier et al. 2017; Meena et al. 2015c). China is the world's top emitter accounting 10,357 metric tons (Mt) (29%) of global $CO_2$ emissions, and the United States is the second biggest emitter, responsible for 5414 Mt. $CO_2$ (15%) of global emissions in 2015. The US emissions since the last decade have been going down because of reduced burning of coal and increased usage of oil and gas; this is why the emissions of the United States fell down by 2.6% in 2015 and also dropped further by 2.0% in 2016 (Olivier et al. 2016, 2017; Yadav et al. 2017b). But it will be a little bit early to say confidently that it has reached its peak as the emissions would increase in the Trump presidency. The emissions across the developing nations are also rising. India is responsible for the 2274 Mt. $CO_2$ (6.3%) of the global $CO_2$ emissions which were increased by 4.7% in 2016. Russia and Japan rank fourth and fifth in global emissions, which account 1617 Mt. (4.5%) and 1237 Mt. $CO_2$ (3.4%), respectively.

C budget is the balance between sink and source of C. The C sources from fossil fuels, industry, and land-use change emissions are balanced by the atmosphere and C sinks on land and in the oceans. The global $CO_2$ emissions and their segregation among the land, ocean, and atmosphere are in balance:

$$E_{FF} + E_{LUE} = G_R + S_O + S_L$$

where $E_{FF}$ is the emissions from fossil fuels and industry, $E_{LUE}$ emissions from land-use change, $G_R$ rate of growth of $CO_2$, $S_O$ mean ocean $CO_2$ sink, and $S_L$ global residual terrestrial $CO_2$ sink.

The growth rate is usually expressed in terms of ppm year$^{-1}$, which can be converted to Gt C year$^{-1}$ (Gt of C year$^{-1}$) using 1 ppm = 2.12 Gt C (Prather et al. 2012; Ballantyne et al. 2012; C. Le Quéré et al. 2016; Dadhich et al. 2015).

However, all $CO_2$ released do not stay in the atmosphere. It is absorbed either by the vegetation on land or in the oceans, minimizing the warming potential which we experience. In 2015, out of the total global $CO_2$ emissions, 44% $CO_2$ remained in the atmosphere (below blue light) and 31% (green) is absorbed by plants and 26% (dark blue) by oceans. The total global $CO_2$ emissions from industrialization time to by the end of 2016 will total 565 billion tons of C which is 92% of the global C budget. Over the last 10 years, the average $CO_2$ released from fossil fuels and industry are responsible for 91% of anthropogenic emissions, whereas the remaining 9% comes from change in land-use pattern. In 2015, 9.9 billion tons of C was emitted in the atmosphere from fossil fuels in the form of $CO_2$, which came from burning of coal (41%), oil (34%), and gas (19%) along with cement production (5.6%) and faring (0.7%) (Meena et al. 2016b; Kumar et al. 2018b).

## 1.4 Soil Carbon Decline Under Intensive Cropping

The intensive cultivation without caring for sustainability of the system resulted in the common problem of reduced SOC stock since long. Most of the global agricultural soils have already lost organic C by 30–75 % from their antecedent SOC flux because of intensive cultivation. It has been projected that the global cultivated soils have already lost 41–55 Pg C (Paustian et al. 1995). Although Smith et al. (2008) stated that the global soils have been experienced as loss of in excess of 40 Pg C due to its cultivation with an average rate of about 1.6 Pg C $year^{-1}$ to the atmosphere in the course of 1990s (Smith et al. 2008; Verma et al. 2015a). However, Lal (2013) reported that the prolonged intensive cultivation is supposed to decrease the soil C stock at the rate of 0.1–1.0 % $year^{-1}$. The soils of India severely depleted the SOC pool which ranged from <1.0 g $kg^{-1}$ (kilograms) to hardly 10–15 Mg (Megagrams) C $ha^{-1}$ (hectare) in upper 40 cm soil horizons (Lal 2015a). The Chinese soils have also lost equal or greater than 30–50 % of the soil C flux (Lal 2013). And in Sweden, nowadays, the C reserve is declining at the annual rate of 1.0 Tg (teragrams) from the total C stock of 270 Tg C in top 25 cm soil surface under agriculture (Andren et al. 2008). The average rate of soil C depletion in soils of England and Wales has been projected to be 0.6% annually (Bellamy et al. 2005). The extent of C loss ranges from 10 to 30 Mg C $ha^{-1}$, reliant on the type of soil and historic land-use pattern, which is higher in soils prone to erosion, salinization, and nutrient diminution than the C loss from least or undegraded soils (Lal 2013). The historical C losses from global soil are estimated to be 78 ± 12 Pg (Lal 2004a, b, c; Buragohain et al. 2017).

Intensive agriculture has a strong capacity to reduce the soil C level in a relatively short time period following initial cultivation, though the degree of reduction varies with the ecosystem and management practices like soil cover, climatic and edaphic characteristics, and farming practices (Poeplau et al. 2011; Powers et al. 2011; Cusack et al. 2013; Meena et al. 2015a). The short-lived impacts are in general dramatic, and agricultural ecosystem may have-long term effects on soil C pool that last for several decades after deserting agriculture (Solomon et al. 2007; Kumar et al. 2017a). The C depletion at the initial time was associated with disruption of soil aggregation, accelerated aeration and decomposition, alteration in plant productivity, biomass production and soil biological properties, and induced soil erosion (Culman et al. 2010; Datta et al. 2017b). The deteriorating soil aggregation as a result of soil cultivation can also lead to increased C loss and consecutive decrement in retention of new C addition (Six et al. 2000). The reduced C status over a long time period was associated with the elongated intensive agricultural practices with less C addition (Solomon et al. 2007). Likewise, the C deposition rate can decrease with time with leftover of C content for longer beneath pre-agricultural levels (Su et al. 2009). These changing trends may expound by increased C losses in the course of cultivation or we can say the lack of ability of agricultural soils to retain the C after crop harvest. The C added by crop plants into the soils is probable to be more liable and susceptible to decomposition than that of the C returned by the woody plants that would be present in the field during the crop growing period (Helfrich et al. 2006; Meena et al. 2017a). Along with these factors, the biomass removal and soil disturbance could result in soil C losses for the duration of cultivation. The lack of strong association of SOC with mineral surfaces is also the reason of reduced soil C retention capacity after crop harvest. To maintain the soil C over long period varies C returns with different practices and the approaches those reduce the C emission from soil. The intensive agriculture can change the C chemistry in the soil through altering plant chemistry, C decomposition rate, etc. (Cusack et al. 2013).

The unsustainable agricultural intensification and change in pattern of land use from natural system to intensive agricultural system management is known to deplete the soil C pool (Guo and Gifford 2002; Söderström et al. 2014; Yadav et al. 2017a). Scientific reports suggested the decreased C stock in permanent cropping system transformed from natural forest land, hastily in the initial years and thereafter at slower rate which reaches at equilibrium after 30–50 years (Nieder and Benbi 2008; Benbi and Brar 2009; Sofi et al. 2018). In the same line, the result of meta-analysis carried out by Guo and Gifford (2002) showed the declined soil C concentration after land-use change from native forest to cropland (−42%) and plantation forest (−13%) and also from pasture to cropland (−52%) and plantation (−10%). This depletion was associated with intensified cultivation practices which have high OM exerting rate, mineralization/oxidation, and soil erosion (Söderström et al. 2014; Ram and Meena 2014). Currently, several agricultural strategies are practiced that expose the agricultural soils to soil erosion. In the last 40 years, about 33 % of global arable land has been lost by erosion or pollution. Soil erosion is the prime factor in substantial removal of SOM and emission of $CO_2$ into the atmosphere. In a experiment on maize diminished SOC level was recorded by 50% in upper 50 cm soil horizons in temperate region at the end of 35 years of intensive cultivation (Arrouays and Pelissier 1994). Liu et al. (2003) also displayed a substantial drop of gross SOC content during the initial 5 years of cultivation with an average annual loss of 2300 kg C $ha^{-1}$ in 0–17 cm soil profile. After 5 years of cultivation till 14 years, the SOC losses also occurred but with decreasing trend with an average annual loss of 950 kg C $ha^{-1}$, and the same decreasing trend still exists between 14 and 50 years of cultivation with a mean loss value of 290 kg C $ha^{-1}$. The overall losses of total SOC in upper 0–43 cm soil profile (0–17 + 18–32 + 33–43 cm) were 17, 28, and 55% after 5, 14, and 50 years, respectively, of intensive cultivation in mollisols of China. The soils of Southern and Central Asia and of sub-Saharan Africa have higher degree of SOC loss. The SOC content in most of South Asian soils ranged from 0.1% to 0.5%. In different regions of India, the SOC concentration significantly decreased after the 1960s (a period of intensive cultivation) as compared to the uncultivated soils prior to the 1960s in top 20 cm soil horizon (Lal 2013). In this line, Jenny and Raychaudhuri (1960) summarized the data of different provinces of India and found the considerable depletion in SOC level (0–20 cm soil) after intensive farming practices. The SOC level in southeastern coast, western coast (per humid), western coast (humid), and Nagpur region of India were decreased from 0.76% to 0.30%, 2.46% to 1.36%, 1.86% to 0.92%, and 1.09% to 0.55%, respectively, when soils were under cultivation. Cusack et al. (2013) examined the potential impact of 200 years of intensive agriculture on soil C level and their chemistry in Hawaii by comparing the reference soil under modem management with intensified pre-European-contact agricultural field system. They reported the declined trend in soil C stocks in Hawaiian agricultural fields (6.1 ± 0.6%) rather than the fallow reference soils (9.3 ± 1.2%). Therefore, the average soil C stock in soil under pre-contact agriculture was reduced by 26 ± 12% relative to the soils of reference sites after intensive 200 years of cultivation.

Globally, the declining C status in soils under agricultural ecosystem is a matter of considerable discussion. As a region of 12 per cent of the total soil C pool is still exists present in cultivated soil (Andren et al. 2008), and the soil under agriculture reside in 35 per cent of the global land surface (Söderström et al. 2014). The technical potential of C sequestration in world soils is 1.2–3.1 Pg year$^{-1}$ for 25–50 years (Lal 2013). By considering the above facts, there is urgent demand of time to rethink about the adoption of sustainable agricultural practices in the twenty-first century. The SOM is not only an indicator of C presence but is also an imperative sink of C sequestration. The SOC represents the largest C pool in terrestrial ecosystems, and is a key factor in deciding the soil quality and input use efficiency (Wiesmeier et al. 2016; Meena et al. 2017b). But the long-term exhaustive farming practices deplete the SOC concentration and result in deterioration of soil structure and consequently the soil productivity (Liu et al. 2013b, c). So, it is a need to improve the critical level of C about 1.1% in the rhizospheric zones (Lal 2013). At present, the intensive agriculture is not sustainable, so the sustainable intensification is a good tactic to save the SOC loss. By changing the land-use pattern following sustainable ways such as through introducing higher biomass-producing crops, shrubs, and tree species in the existing system, the annual C sequestration rate could be increased by 20–75 g C m$^{-2}$ and SOC may reach a new equilibrium in the interior several years (Liu et al. 2013b, c; Kakraliya et al. 2018).

## 1.5 Principles of Soil Carbon Sequestration

Kane (2015) established four pillars for managing soil C dynamics:

1. Reducing soil disturbance through tillage to ensure the physical shelter of C in soil aggregates
2. Enhancing the quantity and quality of plant and animal biomass input in to the soil strata
3. Improving the diversity, abundance and functionaries of beneficial soil microbes
4. Maintaining continuous vegetative cover on soil surface

The capture of atmospheric $CO_2$ and their subsequent storage in the terrestrial ecosystem by a sustainable management of soil and vegetation comprises several agronomic interactions as follows:

- Elimination of mechanical soil disturbance by adopting zero tillage or drastically reduced tillage system (Shaver et al. 2002)

- Continuous surface cover either with living vegetation or crop residue in the form of mulch round the year (Lal 2004a, b, c; 2010; 2016)
- Adoption of agronomic and mechanical measures together to reduce the surface runoff and soil and water erosion by obstructing the velocity of wind and water (Lal 2016)

Accelerating soil health and fertility through practicing INM inclosing organic nutrition sources, biological N fixers/legumes in rotation, mycorrhizae, and organic home wastes promotes in situ OM buildup, potential activities, and diversity of soil bio-organisms and maintains sustainability of soil ecosystem (Liu et al. 2013b, c; Han et al. 2016; Dhakal et al. 2016):

- Maintain adequate soil moisture in crop root zone to increase green water content by improving WUE through introducing drip-cum-fertigation technique and by eliminating or minimizing water loss through evaporation (grey water) and runoff (blue water) (Kumari and Nema 2015).
- Improvements in quality and dietary practices of animal feed to reduce the formation and emission of $CH_4$ through enteric fermentation.
- Follow the system approach rather than an individual crop including livestock and agroforestry along with multiple viable crops in the farming system for efficient resource utilization and biodiversity conservation and to work within the natural ecosystem (Rotenberg and Yakir 2010; Wang et al. 2010, 2015).

## 1.6 Carbon Sequestration Potential of Crop Land

Soil is the major reservoir and a very important sink of C in the terrestrial C cycle because of its capacity to withhold C for relatively a long period of time (Swift 2001). The global soils contain double the amount of C to that of stored in atmosphere plus living vegetation. The C sequestration potential of a soil depends on its capacity to maintain the stock of resistant plant materials to biological decomposition, chemical makeup of SOM, and accumulate the humic fractions more. The amount of C that a soil can sequester rely on the vegetation it supports, soil depth, its drainage capacity, mineral composition, soil temperature, and the relative proportion of soil water and air (Swift 2001). The improved land-use change regulates the budget and transfers of C in terrestrial ecosystem (Lal et al. 2003; Layek et al. 2018). The judicious management of croplands, grasslands, forest, and restored lands are crucial for enhancing the C sequestration potential of soil (Lal 2002), i.e. transforming croplands to grasslands proved in increased soil C. This conversion can be made over the entire field or in confined spots like for shelterbelts, grassed waterway, or field borders. The replacement of conventional agricultural practices by improved land management practices such as introduction of zero tillage or drastically reduced tillage that reduces soil disturbance and incorporation of crop residue into the soil ecosystem has potential to capture the atmospheric C and store in soil as long as these are practiced. The SOC sequestration rate of $570 \pm 140$ kg C $ha^{-1}$ $year^{-1}$ upon conversion of intensive/plow tillage to zero tillage system after

**Table 1.1** Conversion of conventional unscientific farming practices to improved sustainable practices

| S. No. | Conventional practices | Improved sustainable practices |
|---|---|---|
| 1. | Intensive tillage and clean cultivation | Conservation tillage/no-till/drastically reduced tillage |
| 2. | Crop residue burning and removal | Residue retention on soil surface/mulch farming |
| 3. | Summer fallow | Raising cover crops |
| 4. | Synthetic fertilizer use | Site specific nutrient management with compost, biosolids and nutrient cycling |
| 5. | Low input subsistence farming | Judicious use of organic and inorganic nutrient sources |
| 6. | Uncontrolled water use | Water/irrigation conservation/management, water table management |
| 7. | Fence-to-fence cultivation | Marginal agricultural land transformation in to natural conservation/grasslands |
| 8. | Continuous monoculture | Intercropping, mixed cropping, integrated farming system including legumes in rotation |
| 9. | Land use along poverty lines and political boundaries | Integrated watershed management |
| 10. | Draining of wetland | Restoration of wetlands |
| 11. | Deforestation | Afforestation |
| 12. | Naked/barren soil | Soil cover including terrace, vegetative barriers, shelterbelts |
| 13. | Unscientific pasture management | Improved pasture with perennial legume, improved grasses and legume shrubs |
| 14. | Indiscriminate use of pesticides | Integrated pest management |

analysis of 67 long-term experiments in diverse agroecological situations of globe. This figure of SOC pool may reach at new heights in 40–60 years. This conversion of intensive tillage to zero tillage farming on 1500 million ha of cultivated lands besides best recommended management practices (RMPs) could result in sequestration of 0.5–1.0 Pg C year$^{-1}$ by 2050. The conversion of summer fallow by growing of leguminous cover crop permanently is a vital strategy to curtail the depletion in SOC flux. Therefore, the changes in existing land-use pattern towards more ruminative and improved land-use pattern and management practices reduce the soil C depletion, at least partially, and enhance the C sequestration potential of agricultural soils (Table 1.1).

The current rate of C loss due to land-use change (deforestation) and related land-change processes (erosion, tillage operations, biomass burning, excessive fertilizers, residue removal, and drainage of peat lands) is between 0.7 and 2.1 Gt C year$^{-1}$ (World Bank 2012). Presently, the terrestrial sink capacity is increasing at the rate of 1.4 ± 0.7 Pg C annually. Accordingly, terrestrial sink grips nearly 2–4 Pg C year$^{-1}$ whose sink potential could reach at the digit of 5.0 Pg C year$^{-1}$ by 2050 owing to $CO_2$ fertilization effect, sustainable land-use conversion. and viable agronomic management practices. The various improved land conservation practices and their mean soil C sequestration rates across the globe are presented in Table 1.2.

The C sequestration potential of global soil is estimated between 0.4 and 1.2 Gt C year$^{-1}$ or 5–15 % (1Pg = 1 × 10$^{5}$ g) (Lal 2004a, b, c). Similarly, the SOC sequestration

**Table 1.2** Land-use changes and mean soil C sequestration rates (kg C ha$^{-1}$ year$^{-1}$) (World Bank 2012)

| Land-use change | Africa | Asia | Latin America |
|---|---|---|---|
| Crop-to-forest | 1163 | 932 | 528 |
| Crop-to-plantation | – | 878 | 893 |
| Crop-to-grassland | – | 302 | – |
| Crop-to-pasture | – | – | 1116 |
| Pasture-to-forest | – | – | 362 |
| Pasture-to-plantation | – | – | 1169 |
| Pasture improvement | 799 | – | 1687 |
| Grassland-to-plantation | – | – | −406 |
| Annual-to-perennial | – | 1004 | 526 |
| Restoration of wetlands | – | 471 | – |
| Intensive vegetables and specialty crops | – | 2580 | – |
| Exclusion or reduction in grazing | – | 502 | 172 |

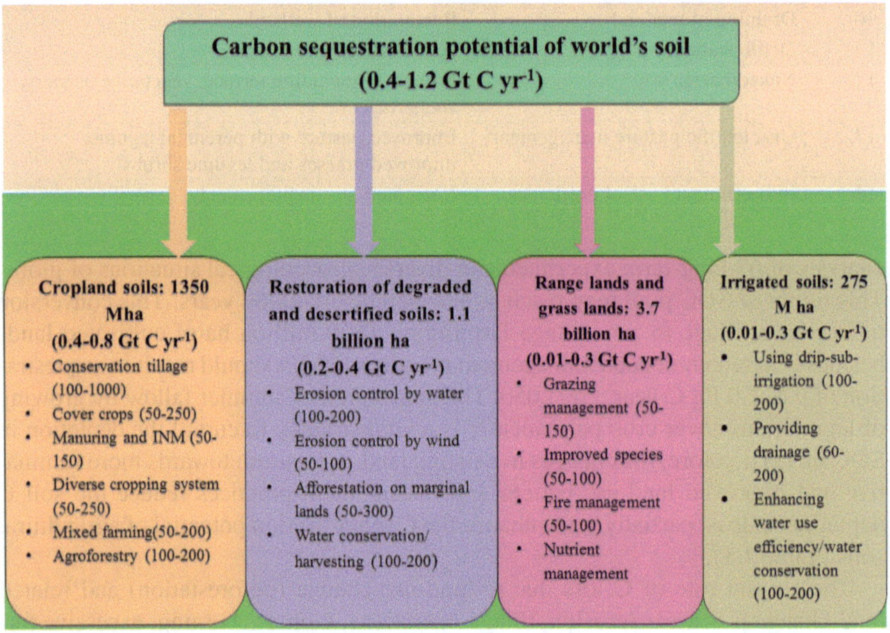


**Fig. 1.2** Carbon sequestration potential of world's soil. (Data adapted from Lal 2004a, b, c)

range of croplands (1350 M ha) varies from 0.4 to 0.8 Gt C year$^{-1}$ in forest and degraded lands (1.1 billion ha) from 0.2 to 0.4 Gt C year$^{-1}$ and 0.01 to 0.3 Gt C year$^{-1}$ in each of rangelands and grasslands (3.7 billion ha), and irrigated soils (275 M ha), respectively (Fig. 1.2).Globally, nearly about 750 million ha of soils is degraded in the tropics with a huge potential of afforestation and soil C restoration. The C sequestration potential of these degraded soils is about 0.5 Mg $ha^{-1}$ $year^{-1}$ as SOC besides additional biomass accumulation rate of 1.0 Mg $ha^{-1}$ $year^{-1}$. Therefore, these soils have the potential to store approximately 1.1 Pg C $ha^{-1}$ $year^{-1}$. According to an estimate (Lal 2002), desertification control in arid and semi-arid regions has the SOC sequestration potential of 0.4–0.7 Pg C $year^{-1}$. According to the Intergovernmental Panel on Climate Change (IPCC) Assessment Report, the global agricultural soils could sequester 400–800 Tg C $year^{-1}$ with the finite capacity saturating after 50–100 years (Verma et al. 2015b). The croplands of Europe have the biological C sequestration potential of 90–120 Tg C annually with best crop and soil management practices when the soil is not disturbed (no/reduced tillage) and efficient utilization of organic amendments. Similarly, the rate of SOC sequestration potential of Chinese soils with improved crop and soil management was estimated to be 2–2.5 Pg C by the 2050s (Sun et al. 2010). Crop and soil management approaches that promote the soil C sequestration take account of the following.

## 1.7 Soil Carbon Pools Improve Sustainability

Sustainability of an agricultural ecosystem strongly hinge on its C footmark. So, the SOC flux is a vital indicator of soil quality and an important driver of agricultural sustainability (Lal 2015b). The changes in land-use system or adaptation of prolonged unsustainable management strategies have already lost the concentration of SOC. The soil C pool is considered as key indicator of soil quality and sustainability of soil ecosystem as a consequence of its influence on soil physical, biological, chemical, and ecological properties (Reeves 1997). Recently, United Kingdom's 'Sustainable Farming and Food Strategy' selected the SOM as the momentous indicator for soil health and quality in the United Kingdom (Anon 2006). The function and significance of SOM is basically associated with its dynamic nature, being constantly synthesized, mineralized, and reorganized (Grego and Lagomarsino 2008). Several researchers documented the improvement in soil physical, biological, chemical, and ecological parameters only because the enrichment of soil by OC is basically based on anecdotal evidence (Bhogal et al. 2009; Meena et al. 2018c). The arable land has been extensively concerned in the worsening of soil health, functionality, and quality through the diminution of soil C stock associated with oxidation next to cultivation. The SOM has long been known as a crucial element in soil quality. The OM has direct effects on the soil available water and indirectly the soil pore distribution. The SOC enhances the stability of soil aggregates and structure because SOM remains physically protected in the core of soil aggregates. The stability of soil aggregation decides the soil water contents, gaseous exchange between soil and atmosphere, soil microbial communities, and nutrient cycling (Sexstone et al. 1985). The soil structure is comprised of primary soil particles and macro- and micro-aggregates acting as physical units of aggregates. The turnover of plant residue in soil is the base of soil aggregation which ensures the availability of C to the soil microbial community as a source of metabolic energy, leading to improvement in soil biological diversity and stimulating biodegradation of harmful soil contaminants (Grego and Lagomarsino 2008; Meena et al. 2015e). These soil microscopic populations and plant-derived carbohydrates are responsible for the creation of soil aggregates by acting as binding force (Six et al. 2000). The turnover rate of SOM influences the biogeochemical transformation of nutrients and associated biochemical processes and thus the agronomic productivity sustainably (Lal 2015b). The increasing SOC stock improves the soil fertility while decreasing the vulnerability of soil to degradation. The plant nutrition is largely owed to the active and water-soluble portions. The dissolved organic fraction has a direct encouraging influence on root growth and nutrient uptake by them (Grego and Lagomarsino 2008). The SOC acts as a buffer counter to immediate change in soil pH filtering agrochemicals and promoting their biodegradation (Grego and Lagomarsino 2008). (Lal 2015a). No doubt, the SOC flux is the utmost reliable pointer of regulating soil degradation, more importantly that caused by androgenic erosions (Rajan et al. 2010). As we know the SOC is a long-lasting component of global C cycle whose concentration in soil is about twice to that of atmosphere and vegetation. So, if the concentration of C is increased, the atmospheric C concentration will get reduced and consequently assuage the problem of global warming and climate change.

## 1.8 Soil Carbon Restoration Options

The SOC sequestration rate ranges between negative to nil in arid and hot climatic regions and 1000 kg C $ha^{-1}$ $year^{-1}$ in temperate and humid regions (Lal 2004a, b, c). But the general mean SOC sequestration rate of agricultural soils ranges between 200 and 250 kg C $ha^{1}$ $year^{-1}$ (Lal 2008). The re-carbonization of the exhausted C flux has need of steady Cic biomass addition which is essential for several functions (Lal 2015a). By looking forward the population explosion and economical emergencies, especially in India, China, Mexico, and Brazil, the significance of innovative agricultural approaches and their impacts on soil and ecological dimensions need to be considered more now than in the ancient. But still, it is needed to critically analyse the biophysical constraints, stabilization mechanisms, relevant economics, and policies with the intension of stabilization of SOC sequestration (Lal 2008). Therefore, implementation of sustainable and viable management practices at ground level in agricultural and forest soils is a vital strategy for soil C sequestration (Lal et al. 2003; Meena et al. 2015b).The practice that can improve the agricultural production in unit area along with a considerable improvement of SOC turnover must be preferred. While, care should be taken when selecting the appropriate farming practice as some approaches are able to accelerate the economical production, but still are C exhaustive in nature, and so increases $CO_2$ emission from soil into the atmosphere. The land improvement practices that accelerate C addition through increasing net primary productivity (NPP) should be enhance to the C sequestration close to their potential mark. However, it is assumed that by the implementation of sustainable management practices only 50–66% of their capacity is attainable.

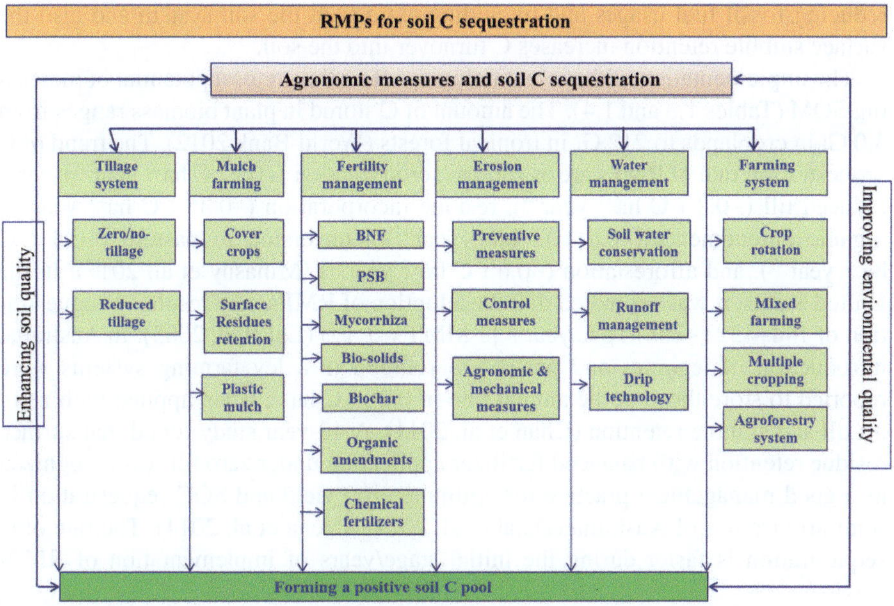


**Fig. 1.3** Recommended management practices (RMPs) for soil carbon sequestration. (Modified Lal 2004a, b, c)

In agricultural ecosystem, the rate of soil C sequestration can be regulated through change in existing land-use pattern, farming system, tillage, soil fertility maintenance, and pest management methods. Practically, there are numerous improved sustainable agricultural practices to be followed instead of non-scientific traditional approaches in C-depleted soils for ensuring good soil C build-up (Fig. 1.3). The sustainable management practices improve the soil, need based nutrient to sustain the soil health, and efficient water management to improve water use efficiency, sustainable pest management with minimal possible use of agro-chemicals, conservation tillage, surface residue retention, mulching, crop rotation, mixed farming, intercropping, cover cropping, strip cropping, and vegetative barriers enlarges C accumulation in soil. Besides this, agricultural strategies also include rescheduling of farm management practices such as irrigation and nutrient application to better match critical growth stages and introducing and implementing efficient technologies that conserve water and soil. Appropriate land uses through intensifying the prime agricultural lands, multiple cropping, improved pasture with low stocking rate, and restoring wetlands and by converting marginal agricultural land to grassland are more desirable options for soil C enrichment. The improved farming practices via adapting ecologically sustained strategy with high diversity, mixed farming, sensible crop rotation while inclosing legume, agroforestry system (AFS), and adding of shrubs in silvipastoral system are found to be good in terms of sustainable soil C sequestration. Reduced or no-tillage reduces the C losses by reducing fossil fuel usages and by adding extra C in the soil system and also the surface stubble retention increases C turnover into the soil.

The implementation of these technologies offers the greatest potential of increasing SOM (Tables 1.3 and 1.4). The amount of C stored in plant biomass ranges from 3.0 Gt in croplands to 212 Gt in tropical forests (World Bank 2012). The trend of C sequestration rate of RMPs are as follows: crop rotation (~0.2 t C $ha^{-1}$ $year^{-1}$), zero/reduced till (~0.3 t C $ha^{-1}$ $year^{-1}$), residue incorporation (~0.35 t C $ha^{-1}$ $year^{-1}$), organic amendments (~0.5 t C $ha^{-1}$ $year^{-1}$), conversion to pasture (~0.5 t C $ha^{-1}$ $year^{-1}$), and afforestation (~0.6 t C $ha^{-1}$ $year^{-1}$) (Minasny et al. 2017). In the United States, it was estimated that the adoption of RMPs may results in sequestration of 144–432 (~288) Tg C $year^{-1}$ [1 MMT = 1 Tg] (Lal et al. 2003). In Australia, introduction of legumes and pastures a rotation in a ley farming systems were reported to store the C at the annual rate of 0.26 t C $ha^{-1}$, when applied with zero/no-till and stubble retention (Chan et al. 2011). A 40-year study found that surface residue retention with balanced fertilizer application under zero till was recognized as a good management practice for optimum crop yield and SOC sequestration in semi-arid tropics of Australia (Dalal et al. 2011; Meena et al. 2014). The rate of C sequestration is faster during the initial stage/years of implementation of RMPs

**Table 1.3** Soil carbon sequestration rates under USDA Natural Resources Conservation Service (NRCS) conservation practices for cropland (Lal et al. 1998; Swan et al. 2015; Chambers et al. 2016)

| Conservation practices | C sequestration rate in soil (Mg C $ha^{-1}$ $year^{-1}$) |
|---|---|
| Conservation agriculture | 0.10–0.40 |
| Conservation cover – retiring marginal soils | 0.42–0.94 |
| Crop rotation | 0.15–0.17 |
| Forage-based rotation | 0.05–0.20 |
| Elimination of summer fallow | 0.05–0.20 |
| Cover crop | 0.15–0.22 |
| Residue management cum zero till | 0.15–0.27 |
| Residue management cum reduced till | 0.02–0.15 |
| Mulch till | 0.07–0.18 |
| Strip till | 0.07–0.17 |
| Strip cropping | 0.02–0.17 |
| Filter strips | 0.42–0.95 |
| Contour buffer strips | 0.42–0.94 |
| Field border | 0.42–0.94 |
| Vegetative wind barriers | 0.42–0.95 |
| Vegetative barriers | 0.42–0.94 |
| Grassed waterways | 0.42–0.96 |
| Organic amendments | 0.20–0.30 |
| Water table management/irrigation | 0.05–0.10 |
| Use of improved varieties | 0.05–0.10 |
| Soil fertility management | 0.05–0.10 |
| Lawns and turfs | 0.50–1.00 |
| Mined soil reclamation | 0.50–1.00 |

**Table 1.4** Effect of land-use change RMPs on soil carbon sequestration potential of drylands (Lal et al. 1998)

| Practice | C sequestration potential (t C $ha^{-1}$ $year^{-1}$) |
|---|---|
| Water management and conservation | 0.10–0.30 |
| Conservation agriculture | 0.15–0.30 |
| Conservation tillage | 0.10–0.20 |
| Compost (20 mg $ha^{-1}$ $year^{-1}$) | 0.10–0.20 |
| Integrated nutrient management | 0.10–0.20 |
| Restoration of eroded soils | 0.10–0.20 |
| Agricultural intensification | 0.10–0.20 |
| Mulching or cover cropping (4–6 mg $ha^{-1}$ $year^{-1}$) | 0.05–0.10 |
| Elimination of summer fallow | 0.05–0.10 |
| Restoration of salt-affected soils | 0.05–0.10 |
| Afforestation | 0.05–0.10 |
| Grassland and pastures | 0.05–0.10 |

which declines with time as soil attains equilibrium (Minasny et al. 2017). The actual/net quantity of C sequestered in the different soil horizons with the different soil management or farming practices highly varies with the countries, climatic situations, ecosystem, soil texture, and initial C level of that site.

### 1.8.1 Conservation Tillage

The increase in SOC flux is one of the key objects of sustainable soil resource management (Lal and Kimble 1997). Conventional tillage may negatively affect the soil C pool due to increased soil erosion and breakdown of soil structure. Conservation tillage is a basic term that encompasses all the tillage practices that reduce surface runoff and soil and water erosion over the conventional practices and provide protection from the falling raindrop impacts. As the soil under zero tillage system remains without interruption, soil aggregates remain intact, physically protecting C. Soil management and conservation tillage practices also endorse the availability of N and SOC sequestration. The enhancement of soil micro-aggregation, deeper placement of SOC in lower horizons, and reversal of soil-degrading processes are the prime tools of C sequestration with conservation tillage system (Lal and Kimble 1997) (Fig. 1.4). Consequently, soil can uphold the C content upon replacing the conventional intensive tillage by zero or drastically reduced or conservation tillage instead by way of decreasing fallow period, plummeting soil disturbance, and incorporation of crop residue in soil strata in the rotation cycle (Fig. 1.5). Avoiding summer fallowing in dry ecosystems and implementing zero till system with surface residue retention as mulch improve the soil structure, infiltration rate, and C accumulation and thus lower the bulk density (Shaver et al. 2002; Meena et al. 2018b). According to Han et al. (2010), zero till + straw returning and rotary tillage + straw

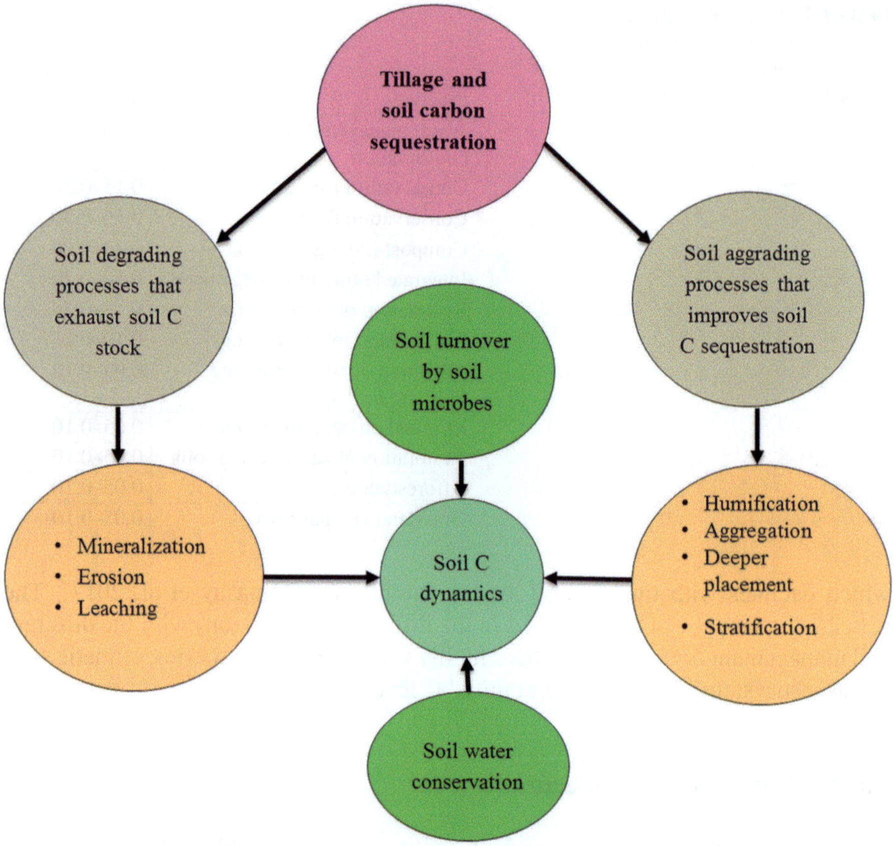


**Fig. 1.4** Tillage and soil carbon dynamics. (Adapted from Lal and Kimble 1997)

returning increased the SOC accumulation by 18.0 and 17.6% in top 5.0 cm surface soil over the conventional tillage practice. The mean soil C sequestration rate with adaptation of zero tillage, crop residue management, mulch farming, and cover cropping in Asia, Africa, and Latin America is presented in Fig. 1.5 (World Bank 2012). The adoption of conservation tillage has a great potential to sequester about 43 Tg C in wider Europe including Soviet Union or 23 Tg C in European Union annually (Smith et al. 1998). By 2020, conversing conventional tillage to conservation tillage may cause to a global C sequestration of $1.5 \times 10^{15}$ to $4.9 \times 10^{15}$ g C (Lal 1997). According to Lee et al. (1993), transforming the corn and soybean farms in the corn belt of the United States from conventional tillage to no-tillage could sequester $3.3 \times 106$ tons C year$^{-1}$ over the next 100 years. Besides, as soil is not manipulated and pulverized in conservation tillage, it reduces the rapid microbial breakdown of SOM and plant residues and can therefore reduce the $CO_2$ evaluation in the biosphere. The tillage and C sequestration rates under diverse cropping system of world are presented in Table 1.5.

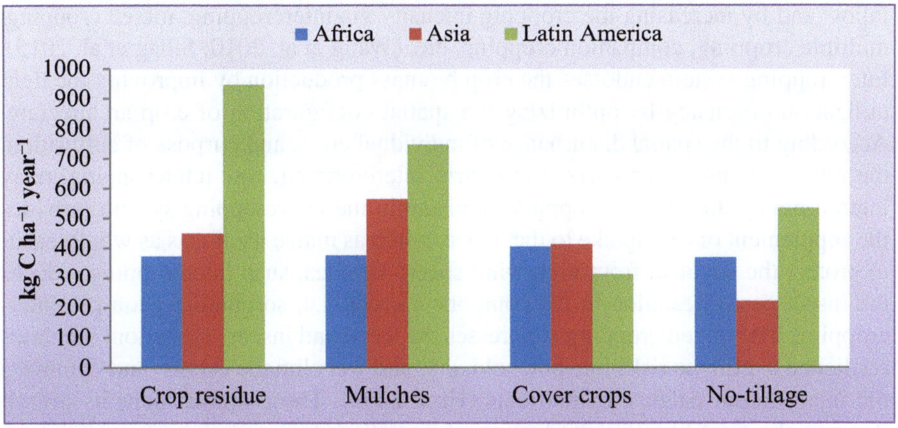


**Fig. 1.5** Tillage, crop residue management, and mean soil carbon sequestration rates (World Bank 2012)

**Table 1.5** Tillage and carbon sequestration rate under diverse cropping systems of world

| Cropping system | Location | Tillage system | C sequestration (Mg C $ha^{-1}$ $year^{-1}$) | Reference |
|---|---|---|---|---|
| Wheat-corn (6) | Gto, Mexico | Conventional tillage | 1.05 | Follett et al. (2005) |
| Wheat-corn (6) | Gto, Mexico | Zero tillage | −0.03 | Follett et al. (2005) |
| Wheat-fallow (27) | Nebraska, USA | Zero tillage | 0.18 | Kettler et al. (2000) |
| Wheat-fallow (27) | Nebraska, USA | Conventional tillage | −0.007 | Kettler et al. (2000) |
| Various crops (6) | Georgia, USA | Conventional tillage, zero tillage, minimum tillage | 0.02 | Sainju et al. (2002) |
| Rye-corn (20) | Kentucky, USA | Zero tillage | 0.37 | Ismail et al. (1994) |
| Rye-corn (20) | Kentucky, USA | Conventional tillage | 0.15 | Ismail et al. (1994) |

### 1.8.2 Cropping System

The field experiments suggested the increased SOC content by increasing cropping intensity over the monoculture owing to higher biomass and residue production in diverse cropping system (Wang et al. 2010, 2015). The deposition of organic C largely depends on the cumulative input of crop residue on soil surface and their subsequent incorporation in soil strata (Kuo and Jellum 2002). Hence, it is important to increase the total crop biomass input in soil to upsurge the SOC concentration. The biomass addition in soil can be enhanced by eliminating the summer fallow and by increasing the cropping intensity via intercropping, mixed cropping, multiple cropping, companion cropping, etc. (Wang et al. 2010; Sihag et al. 2015). Intercropping system endorses the crop biomass production by improving the light utilization efficiency by optimizing the spatial configuration of crop architecture. According to the spatial disturbance of individual crops and purpose of cultivation, the intercropping is categorized into strip intercropping, row intercropping, relay intercropping, and mixed cropping. Soybean in the intercropping system provides the supplement of (N) uptake to the maize, whereas maize itself acts as windbreaker to protect the soybean from high wind speed. Besides, strip intercropping reduces the insect-pest infestation in the component crops, i.e. sorghum-pigeon pea intercropping. The mixed cropping suppresses the weed and insect infestation; increases resilience to climate risks like hot, cold, dry, and wet climatic events; and optimizes the input-output balance of nutrients (Hirst 2009). These mutual benefits overall improve the total biomass production of overall system and show a potential for biomass return and SOC sequestration. Wang et al. (2010) showed the improved soil C in intercropping depending upon the component crops. The accelerated nutrient removal in intercropping system over the natural ecosystem is the critical logic for enhanced C sequestration. The SOC accumulation rate ranged with a modest value of about 1.0 Mg C ha$^{-1}$ (Nair et al. 2009; Mitran et al. 2018).

### 1.8.3 Legume-Based Crop Rotation

The SOC can be enriched by the use of apposite crop rotations (Lal 2010). Crop rotation can improve biomass production and thereafter the soil C sequestration, principally the rotations of legumes with non-legumes. This was because of the higher conversation efficiency from residue C to soil C by legumes in rotation over the monoculture wheat crop. The legume-based rotations are more efficient in converting biomass C in to SOC in compression to the grass-based rotation. Inclusion of legumes in rotation has the potential of guaranteeing the in situ availability of N which in turn played a vital role in generating higher biomass C. It also promotes the release of C via root exudation in to the rhizospheric zone (Hajduk et al. 2015). N fixed by the root nodules of legumes also accelerates the C sequestration potential of succeeding crop in the rotation, more likely because of the improved microbial functionaries and biomass production by successive crop. The provided by the legumes enhances the NUE and produces more root biomass and thus C inputs in soil. Lal (2010) in their research advocated that the legumes based rotation endorsed the accumulation of liable C pool in soil ecosystem considerably greater than C returned from the contentious wheat and uncultivated fallow period. The effect of leguminous crop species on SOC sequestration is more pronounced for green manure, cover crops, and forage which give back a large quantity of C and N in soil system. The GHG abatements of crop rotation were 0.7–1.5 t $CO_2$ equivalent ha$^{-1}$ year$^{-1}$ (World Bank 2012).

### 1.8.4 Cover Crops

Inclusion of the cover crops in the cropping system is a promising way of C sequestration in cultivated soils. Raising leguminous crops enriches biological diversity, the crop residue quality, and soil C flux (Lal 2004a, b, c). The higher the biodiversity of an ecosystem, the more will be the capturing and sequestration capacity over the system exhibiting low biodiversity. The unique advantage of cover crops over the other management options is that they not only enhance the SOC stock but also reduce the C loss, unlike organic manures. The prime object of soil C sequestration through cover crops, and its coming back to the soil ecosystem in such a way that some of the biomass C is not escape back into the biosphere. The improved biomass C below and above the soil surface due to cover cropping can build a C-rich zone through offsetting mineralization and plummeting losses by erosion (Lal 2016), because the soil erosion alone is responsible for the loss of 1.1 Pg C $year^{-1}$ in paedologic pool. Since the entry of cover crops in the cropping system, the change in SOC stock ($R^2 = 0.19$) was tracked for a period of 54 years in a meta-analysis by Poeplau and Don (2015) and reported the annual change rate of 0.32 ± 0.08 Mg $ha^{-1}$ $year^{-1}$ in mean 22 cm soil depth. The predicted new steady state was reached after 155 years of cover crop cultivation with a total mean SOC stock accumulation of 16.7 ± 1.5 Mg $ha^{-1}year^{-1}$for a soil depth of 22 cm. The cover cropping generated the abatement rates of 1.7–2.4 t $CO_2$ equivalent $ha^{-1}$ $year^{-1}$ (World Bank 2012). Legume-based cropping systems improve SOC (Sainju et al. 2002) and decrease the C and N evaluation (Drinkwater et al. 1998). Hence, cover cropping improves the soil quality by enriching SOC through their biomass and they also promote soil aggregation, and protect the surface soil from runoff and erosion. The biomass production and the subsequent turnover rate of organic materials in soil depend on the growing environments of cover crop. Therefore, the rate of C sequestration hinge on selection of suitable cover crop, agronomic management practices, climatic zone, and soil texture (Lal 2016).

### 1.8.5 Integrated Nutrient Management

The C sequestration potential of agricultural soil is being reduced continuously in the presence of imbalanced nutrient management. The balanced application of organic and inorganic fertilizer in agricultural soils for crop production is crucial for soil C sequestration. Several scientific studies advocated that judicious and balanced application of synthetic fertilizers and organic manure for long term can enhance the soil productivity and SOC pool (Johnston et al. 2009; Nayak et al. 2012; Liu et al. 2013b, c; Han et al. 2016). The plots treated with higher rate of N exhibits improved rate of C sequestration with a mean value of 1.0–1.4 Mg C $ha^{-1}$ over the non-fertilized plots. The influence of fertilization on rate of SOC sequestration will be greater when the soil is deficient in nutrient. In such conditions, the practices which improve N use efficiency are critical for SOC accumulation (Fig. 1.6). These should be based on the principle of 5Rs (right time, right method,

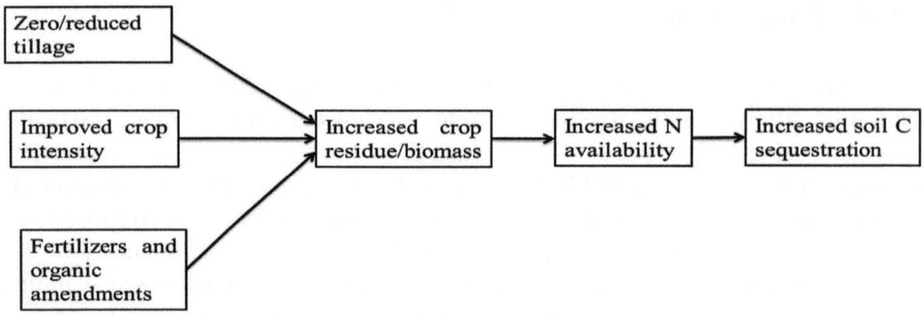


**Fig. 1.6** RMPs that increase N availability and soil carbon sequestration

right source, right amount, right place). The sequestration rate can be increased either by increasing the content of crop biomass C or by reducing the $CO_2$ emission from the soil or by both. The fertilizer management strategies in cultivated soils, e.g. synthetic fertilizers, organic manures (e.g., farm yard manure (FYM), compost, vermicompost, biosolids, and biochar), surface residue retention, and green manuring, have been documented as promising way to enhance SOC accumulation and to reduce $CO_2$ evaluation from the soil. Adequate availability of nutrient elements from these sources improves the crop yield, biomass-C generation, and, so, crop residue and root input in soil (Kätterer et al. 2011).

In general, the supply of same amount of nutrient through organic manures and compost in soil considerably enhanced the accumulation of SOC, particulate OC, microbial biomass, and, thus, the rate of C sequestration as compared to the inorganic fertilizers. Organic amendments and surface stubble retention are recognized as prominent practices for bringing the change in SOC levels (Maillard and Angers 2014). Their effect on soil C sequestration becomes worthier when it is adapted with conservation tillage and organic farming (Han et al. 2016). A field trial with application of FYM increased SOC concentration by 200% over a period of 100 years at Rothamsted, UK (Johnston et al. 2009). The continuous straw retention of surface soil improved the soil C sequestration in Ultuna, Sweden, at the end of 54 years of experiment (Kätterer et al. 2011; Meena et al. 2017c). It describes the importance of long-term application of organic amendments in building the C reserve in soil strata. Han et al. (2016) carried out a metal-analysis on relation of different nutrient management practices on change in rate of SOC content over a wide range of climatic and ecological regions. The outcome of this analysis was the increased level of SOC by 3.2–3.8 (~3.5 or 36.2%), 1.9–2.2 (~2.0 or 19.5%), 1.2–2.3 (~1.7 or 15.4%), and 0.7–1.0 g $kg^{-1}$ (~0.9 or 10.0%) at 95% confidence interval in topsoil with application of synthetic fertilizer + organic manure (FM), synthetic fertilizer + straw (FS), balanced synthetic fertilizer (BF), and unbalanced synthetic fertilizers (UF), respectively. This estimation of C sequestration under FM and FS was over duration of 26–117 and 28–73 years, respectively, over highly variable ecological conditions. Table 1.6 clearly shows the effects of increasing N availability on soil C sequestration rate in different regions by adapting INM strategy under irrigated and rain-fed conditions.

**Table 1.6** Effect of N availability increasing RMPs on soil carbon sequestration rate (kg C $ha^{-1}$ $year^{-1}$) and potential (Tg C $year^{-1}$) in the United States, Canada, and Mexico

| | Dryland (irrigated) | | | | | | Humid cropland (non-irrigated) | | | | | |
|---|---|---|---|---|---|---|---|---|---|---|---|---|
| Agroecosystem | Canada | | United States | | Mexico | | Canada | | United States | | Mexico | |
| Fertilizer | 50–150 | 0.04–0.12 | 30–390 | 0.7–9 | 105–117 | 0.7 | 70–680 | 4–35 | 20–1690 | 3–260 | 50–150 | 1–3 |
| Fertilizer + manure | 50–150 | 0.04–0.12 | 111–1545 | 2–34 | 50–150 | 0.3–0.9 | 50–150 | 3–8 | 719–927 | 111–143 | 50–150 | 1–3 |
| Fertilizer + legume as cover crop | 100–300 | 0.08–0.2 | 500–900 | 11–20 | 30–320 | 0.2–2.0 | 61–132 | 3–7 | 18–224 | 3–35 | 100–300 | 2–6 |
| Fertilizer + manure + legume as cover crop | 100–300 | 0.08–0.2 | 557 | 12 | 100–300 | 0.6–2 | 100–300 | 5–15 | 656 | 101 | 100–300 | 2–6 |

Adapted from Christopher and Lal (2007)

### 1.8.6 Irrigation Management

The application of irrigation water has a large potential to enhance the rate of soil C sequestration. As a result, judicious application of irrigation water in arid and semi-arid ecosystem accelerates the biomass production, improves the above- and below-ground plant parts returned to the soil, and therefore increases the SOC stock. Besides, appropriate water table management, including drip/sprinkler irrigation methods, and effective water recycling are required for SOC sequestration. The experimental results showed the annual C sequestration range of 0.05–0.15 t C $ha^{-1}$ SOC (Conant et al. 2001) and 0.05–0.10 t C $ha^{-1}$ SIC (Nordt et al. 2000) in soil.

Crop production and quantity of organic residues returned in soil is the function of availability of irrigation water for the crop plants. Soil moisture has substantial impacts on soil-atmosphere C exchange mechanisms and SOM decomposition by microbes. Availability of moisture in soil governs vegetative growth and NPP and thus affects C addition to the soil ecosystem (Yuste et al. 2007). Irrigation to the cropland has both positive and negative impacts on SOC accumulation in soils over long time. The improved water supply promotes plant biomass production and increase C input to the soil in the forms of root exudates, rhizo-deposition, dead roots and other vegetative parts (Kochsiek et al. 2009). In contrast, irrigation endorses the soil moisture build-up and associated microbial activities. This results in increased SOM decomposition and $CO_2$ emanations into the free atmosphere (Trost et al. 2013; Gogoi et al. 2018). This may lead to reduction in SOC reservoir. Lack of adequate soil moisture in drought-prone areas can inhibit the performance of soil fauna and flora and can therefore cut the SOM decomposition which results in decreases in loss in soil C (Lai et al. 2013). Trost et al. (2013) in their investigation in different dryland ecosystems reported an increase in 90% to more than 500% of SOC owing to application of irrigation in cultivated desert soils. Irrigation increases SOC concentration by 11–35% in semi-arid regions but not in humid regions. Although this relationship between irrigation and SOC build-up is not independent, this also depends on other factors like fertilizer, tillage, etc. This process is simplified by a diagrammatic representation in Fig. 1.7. At last we can conclude that irrigation application leads to upsurge SOC concentration in arid and desert cultivated soils as compared to the non-irrigated soils. Whereas in humid and in soils already rich in SOC content, irrigation has no considerable effects on SOC build-up. In dryland ecosystem, life-saving irrigation and water harvesting minimize the risk in crop production and sequester the atmospheric C in to the soil (Table 1.7). The improved irrigation produced low to medium moderately high abatement rates of 0.2–3.4 t $CO_2$ equivalent $ha^{-1}$ $year^{-1}$ (World Bank 2012).

### 1.8.7 Agroforestry System

Agroforestry system (AF) consists of mixture of trees, agricultural crops, and livestock to exploit the economic and ecological benefits of agroecosystem. It is a crucial leader of terrestrial C sequestration containing about 12% of the global terrestrial

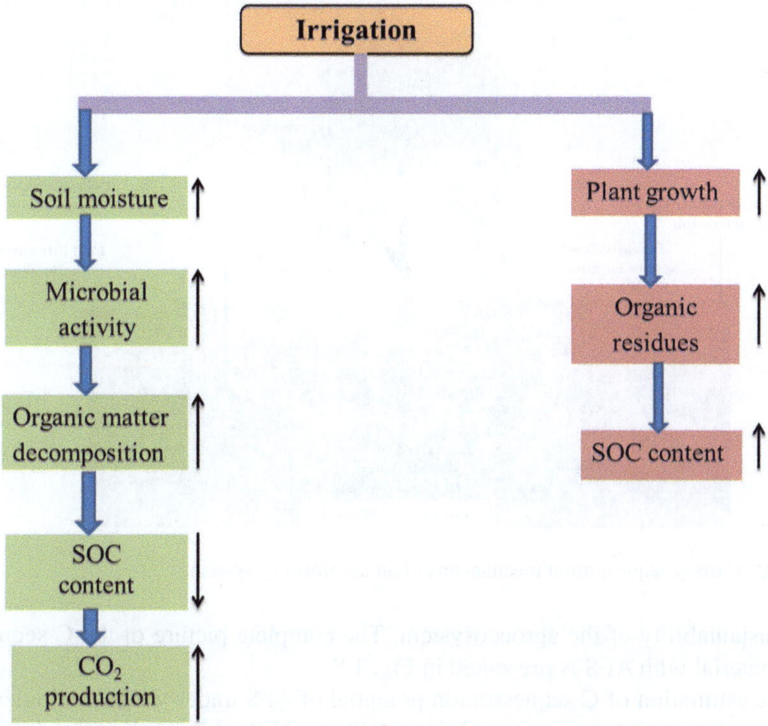


**Fig. 1.7** Diagrammatic representation of basic effects of irrigation on SOC (Trost et al. 2013)

**Table 1.7** Water management and mean soil carbon sequestration rates (kg C ha$^{-1}$ year$^{-1}$) (World Bank 2012)

| Practice | Africa | Asia | Latin America |
|---|---|---|---|
| Rainwater harvesting | 839 | 1086 | – |
| Improved irrigation | – | 1428 | 571 |
| Cross-slope barriers | 1193 | – | – |

C (Dixon 1995). The trees capture and store C by tumbling respiration rate and by growing rapidly by exploring the benefits of favourable temperature at early growth stage (Rotenberg and Yakir 2010). The roots of forest tress and perennial crops penetrate deeper subsurface horizons, thus placing SOC at deeper horizons far away from the range of tillage implements (Lorenz and Lal 2014). Therefore, the SOC pool do not remains for a longer time as a permanent C pool. The acts as mulch and covers the land surface of cultivated field that decompose with passage of time and form the part of SOC pool. Besides, this obstructs the speed of blowing wind and flowing water and reduces soil runoff which is a crucial process of soil C dynamics. It also moderates the soil moisture loss from soil surface as evaporation. Thereby, the increased C content in AFS ensures the better agricultural productivity

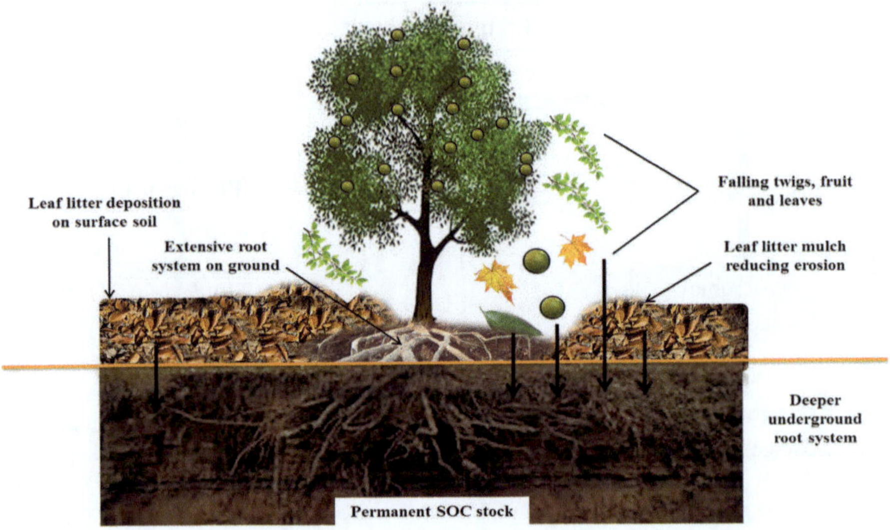


**Fig. 1.8** Carbon sequestration mechanisms of an agroforestry system

and sustainability of the agroecosystem. The complete picture of the C sequestration material with AFS is presented in Fig. 1.8.

The estimation of C sequestration potential of AFS under varied ecological and management environment ranged from 0.29 to 15.21 Mg ha$^{-1}$ year$^{-1}$ in above-ground plant biomass and 30–300 Mg ha$^{-1}$ year$^{-1}$ in below-ground plant parts up to a depth of 1.0 m (Nair et al. 2010). Above-ground biomass is a direct measure of C sequestration, assuming that 50% of the biomass is made up by C (Nair et al. 2010). The cumulative C sequestrat

ion including above- and below ground parts under AFS is considerably greater as compared to the treeless croplands in the same ecological and management conditions. Some of the agroforestry practices are silvipastoral, ally cropping, forest farming, windbreakers, home gardens, riparian buffers, woodlots, etc.

The annual accumulation rate of C in soil is expected to increase at the rate of 1.3 Mg ha$^{-1}$ in the next two decades; after that it would decelerate by 0.20 Mg ha$^{-1}$ year$^{-1}$ in the next eight decades (Silver et al. 2000). So, it is very crucial to highlight the significance of AFS in capturing and storing C in soil for the duration of first 2–5 decades. Along with the food, feed, fibre, fuel, and fodder, AFS are also important in relation to the soil fertility and soil C sequestration (Abberton 2010) (Fig. 1.9). It was found that the forest system is supposed to capture equal to 3 Pg Cs yearly (Ibrahim et al. 2010) and also estimated that the global forest system contributes on behalf of about 90% of annual C pool between soil and atmospheric C (Wani and Qaisar 2014). Agroforestry has been recognized as having the greatest potential for C sequestration of all the land-use system (Minasny et al. 2017).Their

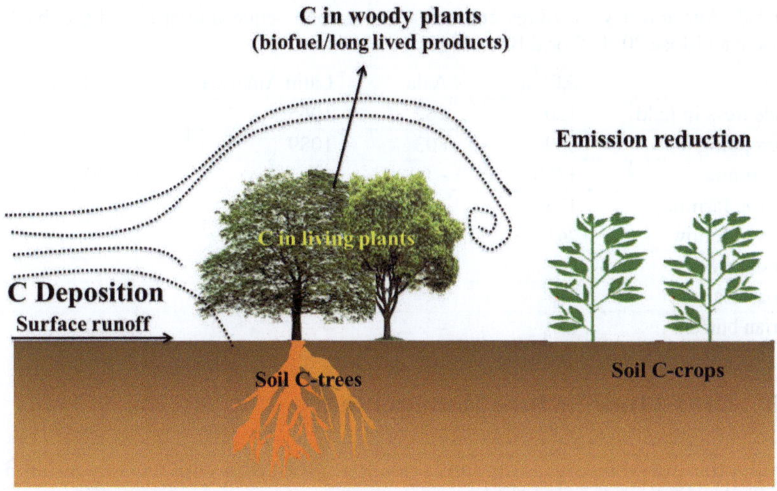


**Fig. 1.9** Agroforestry for reducing wind velocity, surface water runoff, and soil C loss

sequestration potential depends on the $CO_2$ capturing capacity from atmosphere or photosynthetic rate and transformation of $CO_2$ into long-lived C material as such. Up to 2.2 Pg C (1 Pg C = 1 picogram of C – $10^{15}$gC = 1 gigatons C = 1 Gt C = 1 billion metric tons of C) could be stored below- and above-ground over 50 years in AFS (Lorenz and Lal 2014). The SOC sequestration in AFS is uncertain and may reach up to 300 Mg C $ha^{-1}$ to 100 cm depth. Nair et al. (2009) estimated the C sequestration range of 5–10 kg C $ha^{-1}$ in 25 years in AFS of arid and semi-arid ecosystem and 100–250 kg C $ha^{-1}$ in humid environment in 10 years. According to a report of IPCC (2007), agroforestry has the potential of 1.1–2.2 Pg C sequestrations in terrestrial ecosystem in the next 50 years (Jose 2009). According to Oelbermann et al. (2004), the C storage capacity in above-ground plant parts in AFS to be estimated is $1.9 \times 10^9$ and $2.1 \times 10^9$ Mg C $year^{-1}$ in temperate and tropical ecosystem. The C storage capacity of agri-silviculture system varies 68–81 and 12–228 Mg C $ha^{-1}$ in dry lowland and humid tropical lands of Southern Asia. The potential of silvipastoral systems in North America is highest with a storage value of 90–198 Mg C $ha^{-1}$ (Murthy et al. 2013). In accordance with Richards and Stokes (2004), the forest lands can sequester up to 250 million metric tons of C $year^{-1}$ which shares about 12% of the $CO_2$ emissions in the United States. The advanced plantation of *Cassia siamea* increases the SOC concentration at the rate of 50 kg $ha^{-1}year^{-1}$in upper 10 cm soil profile due to its capacity of higher litter-fall (5–7 Mg $ha^{-1}$ $year^{-1}$) that helps to sustain the higher SOC content (Lal et al. 1998). The mean C sequestration rate of different agroforestry measures in different ecological conditions is presented in Table 1.8.

**Table 1.8** Agroforestry measures and mean soil carbon sequestration rate (kg C $ha^{-1}$ $year^{-1}$) (Udawatta and Jose 2011; World Bank 2012)

| Practice | Africa | Asia | Latin America | North America |
|---|---|---|---|---|
| Include tress in field | 1204 | 562 | 1065 | – |
| Intercropping | 629 | 803 | 1089 | – |
| Ally farming | 1458 | – | – | 3400 |
| Tree crop farming | 1359 | – | – | – |
| Improved fellow | 2413 | – | – | – |
| Diversify trees | – | – | 1365 | – |
| Silvipastoral | – | – | – | 6100 |
| Riparian buffers | – | – | – | 2600 |

### 1.8.8 Grassland/Pasture Management

Globally, the grasslands/grazing lands occupy 3460 Mha which cover about 31% of the Earth's land surface (Lal 2004a, b, c). They are grouped into three categories based on their relative soil C sequestration potential. First are the are *natural grasslands*, which are not protected and are not under livestock, agriculture, and other usages and, therefore, remain undisturbed in natural state. Second are the *degraded grasslands* are poorly managed where no improvement can be expected in short term. Third are the grasslands which are prone to management improvements. There is a wide scope to enhance the SOC and SIC storage of degraded grasslands through restoration and implementation of sustainable soil conservation approaches. Moreover, transforming marginal croplands to more ruminative pastures also confiscate C.

Globally, grassland ecosystem shares more than 10% of the cumulative C storage among all the vegetation (Nosberger et al. 2000). In grassland ecosystem, up to 98% of the total C can be found sequestered below-ground, that is why the soil is the largest C storing body of the terrestrial C pool (Jones and Donnelly 2004). Grassland management mainly affects the soil C sequestration by altering C inputs in soil via root turnover and exudation, root and shoot biomass, and NPP (Schuman et al. 2002). Beside the root biomass and their decomposition, root exudation, rhizodeposition, mucilage production, and sloughing from living roots also contribute to soil C. In most of the grassland ecosystem, about 75–80% of the cumulative root biomass remains in top 30 cm soil profile, but accurate determination of C transfer from different sources is difficult because the root growth, death, and subsequent decomposition occur concurrently and at varied rates as per the species and climatic conditions. In temperate grasslands, an extensive stock of accumulated C is situated in soil profile in roots and soil. The measured and modelled rate of C sequestration ranges from 0 to >8 Mg C $ha^{-1}$ $year^{-1}$ (Jones and Donnelly 2004).

The agronomic management approaches strongly influence the C sequestration rates and the future C stocks in grasslands. Grasslands have potential for building the C stock in the soil strata which can be substantially enhanced by change in management environments. According to an estimation of NRSC, on adopting the C-rich conservation agricultural practices on grasslands (grazing and pasture lands),

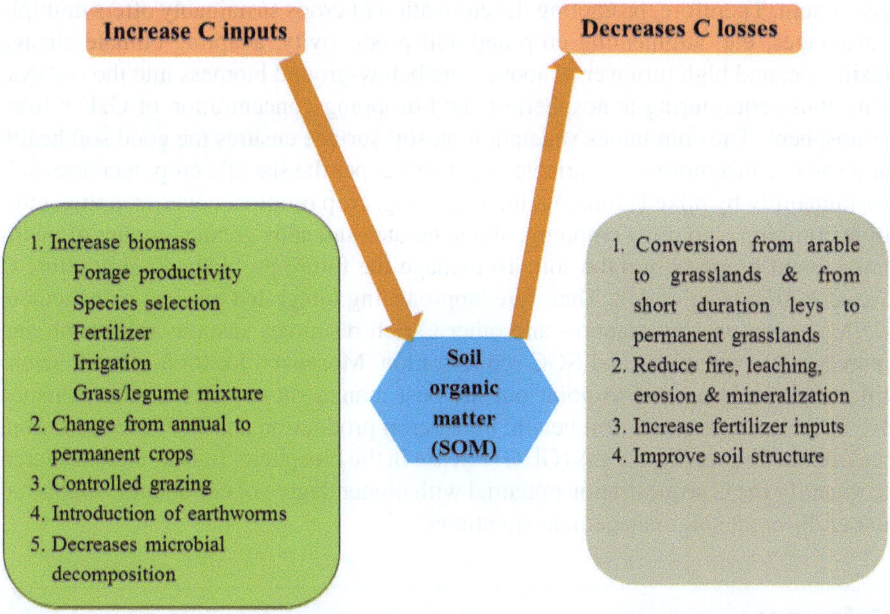


**Fig. 1.10** Schematic illustration of management options to increase SOM in grassland ecosystems

0.020.44 Mg C $ha^{-1}$ $year^{-1}$ can be accumulated in soil in the coming decades (Chambers et al. 2016). The implementation of sustainable agronomic practices in 40.5 Mha grasslands over the next decades could result in sequestration of 18 Tg C $year^{-1}$. These management options include judicious use of organic and inorganic sources of nutrition, controlled grazing, appropriate mixture of grasses and legumes as per the climatic conditions, expansion of soil microbial diversity, and irrigation (Lal 2004a, b, c). The improved pasture management results in SOC sequestration of 0.11–3.04 Mg C $ha^{-1}$ $year^{-1}$ at the annual building rate of 0.54 Mg C $ha^{-1}$ (Conant et al. 2001). In the United Kingdom, SOC content increased at the annual rate of 0.02% for 12 years by adapting grass leys. However, the amount of SOC retained or sequestered by soil depends on the input-output balance of C by different strategies under grassland ecosystem (Fig. 1.10).

## 1.9 Conclusion and Future Outlook

The amount of C that a soil can sequester rely on the vegetation it supports, soil depth, its drainage capacity, mineral composition, soil temperature, and the relative proportion of soil water and air. The C sequestration potential of a soil depends on its capacity to maintain the stock of resistant plant materials to biological decomposition, chemical makeup of SOM, and accumulate the humic fractions more. The improved land-use change regulates the budget and transfers of C in terrestrial ecosystem. Therefore, promoting the cultivation of crops sustainably offers multiple advantages, e.g. augmenting crop and soil productivity, adapting climate change resilience, and high turnover of above- and below-ground biomass into the soil system, thus sequestering atmospheric C and dropping concentration of GHGs from atmosphere. The continuous vegetation on soil surface ensures the good soil health and soil C concentration at variable soil depth as per the specific crop; increases soil sustainability by mixed cropping, intercropping, crop rotation, cover cropping, multiple cropping, and relay cropping; and generates and adds greater amount of qualitative plant biomass into the soil. To manage the future problems in agriculture C sequestration is an option. Therefore, approaching integrated nutrient management (INM) encompassing manures and other C-rich resources sustains soil health and increases N availability and SOC sequestration. Moreover, location-specific scientific research is needed to point out the best management practices that enhance NUE, maintain/improve soil health, boost crop production and SOC sequestration, and minimize greenhouse gas (GHG) release in the biosphere. In fact, more research to quantify the C sequestration potential with higher degree of confidence is required under different soil management situations.

## References

Abberton (2010) Enhancing the role of legumes: potential and obstacles. In: Grassland C sequestration: management, policy and economics. Proceedings of the workshop on the role of grassland C sequestration in the mitigation of climate change. Rome, April 2009

Andren O, Katterer T, Karlsson T, Eriksson J (2008) Soil C balances in Swedish agricultural soils 1990-2004, with preliminary projections. Nutr Cycl Agroecosyst 81:129–144

Anon (2006) Techniques for measuring soil physical properties, MAFF reference book 441. HMSO, London

Arrouays D, Pelissier P (1994) Changes in C storage in temperate humic loamy soils after forest clearing and continuous corn cropping in France. Plant Soil 160:215–223

Ashoka P, Meena RS, Kumar S, Yadav GS, Layek J (2017) Green nanotechnology is a key for eco-friendly agriculture. J Clean Prod 142:4440–4441

Bakker MM, Govers G, Jones RA, Rounsevell MDA (2007) The effect of soil erosion on Europe's crop yields. Ecosystems 10:1209–1219. https://doi.org/10.1007/s10021-007-9090-3

Ballantyne AP, Alden CB, Miller JB, Tans PP, White JWC (2012) Increase in observed net C dioxide uptake by land and oceans during the last 50 years. Nature 488:70–72

Bellamy PH, Loveland PJ, Bradley RI, Lark RM, Kirk GJD (2005) C losses from all soils across England and Wales 1978–2003. Nature 437:245–248

Benbi DK, Brar JS (2009) A 25-year record of C sequestration and soil properties in intensive agriculture. Agron Sustain Dev 29(2):257–265

Bhogal A, Nicholson FA, Chambers BJ (2009) Organic C additions: effects on soil bio-physical and physico-chemical properties. Eur J Soil Sci 60:276–286. https://doi.org/10.1111/j.1365-2389.2008.01105.x

Bleam WF (2012) Natural organic matter and humic colloids. In: Soil and environmental chemistry, pp 209–256. https://doi.org/10.1016/B978-0-12-415797-2.00006-6

Bronick CJ, Lal R (2005) Soil structure and management: a review. Geoderma 124:3–22

Buragohain S, Sharma B, Nath JD, Gogaoi N, Meena RS, Lal R (2017) Impact of ten years of bio-fertilizer use on soil quality and rice yield on an inceptisol in Assam, India. Soil Res. https://doi.org/10.1071/SR17001

Buyanovsky GA, Wagner GH (2002) C cycling in cultivated land and its global significance. Glob Chang Biol 4(2):131–141. https://doi.org/10.1046/j.1365-2486.1998.00130.x

Chambers A, Lal R, Paustian K (2016) Soil C sequestration potential of US croplands and grasslands: implementing the 4 per thousand initiative. J Soil Water Conserv 71(3):68–74. https://doi.org/10.2489/jswc.71.3.68A

Chan KY, Conyers MK, Li GD, Helyar KR, Poile G, Oates A, Barchia IM (2011) Soil C dynamics under different cropping and pasture management in temperate Australia: results of three long-term experiments. Soil Res 49:320–328

Chaudhury ST, Bhattacharyya SPW, Pal DK, SahrawatKL NA, Chandran P, VenugopalanMV TB (2016) Use and cropping effects on C in black soils of semi-arid tropical India. Curr Sci 110(9):1652–1698

Christopher SF, Lal R (2007) Nitrogen management affects C sequestration in North American cropland soils. Crit Rev Plant Sci 26(1):45–64. https://doi.org/10.1080/07352680601174830

Ciais P et al (2013) C and other biogeochemical cycles. In: Stocker T, Qin D, Platner GK (eds) Climate change the physical science basis. Cambridge University Press, Cambridge

Collins HP, Paul EA, Paustian K, Elliot ET (1997) Characterization of soil organic C relative to its stability and turnover. In: Paul IEA et al (eds) Soil organic matter in temperate agroecosystems. CRC Press, Boca Raton, pp 51–72

Conant RT, Paustian K, Elliott ET (2001) Grassland management and conversion into grassland: effects on soil C. Appl Ecol 11:343–355

Culman SW, DuPont ST, Glover JD, Buckley DH, Fick GW, Ferris H, Crews TE (2010) Long-term impacts of high input annual cropping and unfertilized perennial grass production on soil properties and belowground food webs in Kansas, USA. Agric Ecosyst Environ 137:13–24

Curtin D, Wang H, Selles F, Zentner RP, Biederbeck VO, Campbell CA (2000) Legume green manure as partial fallow replacement in semiarid Saskatchewan: effect on C fluxes. Can J Soil Sci 80:499–505

Cusack DF, Chadwick OA, Ladefoged T, Vitouse PM (2013) Long-term effects of agriculture on soil C pools and C chemistry along a Hawaiian environmental gradient. Biogeochemistry 112:229–243. https://doi.org/10.1007/s10533-012-9718-z

Dadhich RK, Meena RS (2014) Performance of Indian mustard (*Brassica juncea* L.) in response to foliar spray of thiourea and thioglycolic acid under different irrigation levels. Indian J Ecol 41(2):376–378

Dadhich RK, Meena RS, Reager ML, Kansotia BC (2015) Response of bio-regulators to yield and quality of Indian mustard (*Brassica juncea* L. Czernj. and Cosson) under different irrigation environments. J App Nat Sci 7(1):52–57

Dalal RC, Allen DE, Wang WJ, Reeves S, Gibson I (2011) Organic C and total nitrogen stocks in a vertisol following 40 years of no-tillage, crop residue retention and nitrogen fertilisation. Soil Tillage Res 112:133–139

Datta R, Baraniya D, Wang YF, Kelkar A, Moulick A, Meena RS, Yadav GS, Ceccherini MT, Formanek P (2017a) Multi-function role as nutrient and scavenger of free radical in soil. Sustain MDPI 9:402. https://doi.org/10.3390/su9081402

Datta R, Kelkar A, Baraniya D, Molaei A, Moulick A, Meena RS, Formanek P (2017b) Enzymatic degradation of lignin in soil: a review. Sustain MDPI 1163(9):1–18. https://doi.org/10.3390/su9071163

David A (2010) The global C cycle. Princeton University Press, Princeton. ISBN 9781400837076

Dhakal Y, Meena RS, De N, Verma SK, Singh A (2015) Growth, yield and nutrient content of mungbean (*Vigna radiata* L.) in response to INM in eastern Uttar Pradesh, India. Bangladesh J Bot 44(3):479–482

Dhakal Y, Meena RS, Kumar S (2016) Effect of INM on nodulation, yield, quality and available nutrient status in soil after harvest of green gram. Legum res 39(4):590–594

Dixon RK (1995) Agroforestry systems: sources or sinks of greenhouse gases? Agrofor Syst 31:99–116

Drinkwater LE, Wagoner P, Sarrantonio M (1998) Legume based cropping systems have reduced C and nitrogen losses. Nature (London) 396:262–265

Eswaran H, Vandenberg E, Reich P (1993) Organic C in soils of the world. Soil Sci Soc Am J 57:192–194
Falkowski P, Scholes RJ, Boyle E, Canadell J, Canfield D et al (2000) The global C cycle: a test of our knowledge of earth as a system. Science 290(5490):291–296
FAO (2016) Hunger, poverty and climate change: the challenges today and tomorrow
FAO and ITPS (Intergovernmental Technical Panel on Soils) (2015) Status of the world's soil resources (SWSR) – main report. FAO, Rome
Follett RF, Castellanos JZ, Buenger ED (2005) C dynamics and sequestration in an irrigates vertisol in Central Mexico. Soil Tillage Res 83:148–158
Gogoi N, Baruah KK, Meena RS (2018) Grain legumes: impact on soil health and agroecosystem. In: Meena et al (eds) Legumes for soil health and sustainable management. Springer, Singapore. https://doi.org/10.1007/978-981-13-0253-4_16
Grego S, Lagomarsino A (2008) Soil organic matter in the sustainable agriculture: source or sink of C. In: Soil C sequestration under organic farming in the Mediterranean environment by Sara Marinari and Fabio Caporali, pp 39–51. ISBN: 978-81-7895-327-4
Guo LB, Gifford RM (2002) Soil C stocks and land use change: a meta-analysis. Glob Chang Biol 8:345–360
Hajduk E, Wlaoeniewski S, Szpunar-Krok E (2015) Influence of legume crops on content of organic C in sandy soil. Soil Sci Annu 66(2):52–56. https://doi.org/10.1515/ssa-2015-0019
Han B, Kong FL, Zhang HL, Chen F (2010) Effects of tillage conversion on C sequestration capability of farmland soil doubled cropped with wheat and corn. Ying Yong Sheng Tai Xue Bao 21(1):91–98
Han P, Zhang W, Wang G, Sun W, Huang Y (2016) Changes in soil organic C in croplands subjected to fertilizer management: a global meta-analysis. Sci Rep 6:271–299. https://doi.org/10.1038/srep27199
Helfrich M, Ludwig B, Buurman P, Flessa H (2006) Effect of land use on the composition of soil organic matter in density and aggregate fractions as revealed by solid state 13C NMR spectroscopy. Geoderma 136:331–341
Hirst KK (2009) Mixed cropping, agricultural technique known as mixed cropping. http://archaeology.about.com/od/historyofagriculture/qt/mixed_cropping.html
Houghton RA (2003) The contemporary carbo cycle. Treatise Geochem 8:473–513. https://doi.org/10.1016/B0-08-043751-6/08168-8
Ibrahim M, Guerra L, Casasola F, Neely C (2010) Importance of Silvopastoral systems for mitigation of climate change and harnessing of environmental benefits. In: Proceedings of the Workshop on the role of grassland C sequestration in the mitigation of climate change. Rome, April 2009
IPCC (2007) Intergovernmental Panel on Climate Change. Assessment report
Ismail I, Blevins RL, Frye WW (1994) Long-term no-tillage effects on soil properties and continuous corn yields. Soil Sci Am J 58:193–198
Jenny H, Raychaudhuri SP (1960) Effect of climate and cultivation on nitrogen and organic matter reserves in Indian soils. ICAR, New Delhi
Jobbagy EG, Jackson RB (2000) The vertical distribution of soil organic C and its relation to climate and vegetation. Ecol Appl 10:423–436
Johnston AE, Poulton PR, Coleman K (2009) Soil organic matter: its importance in sustainable agriculture and C dioxide fluxes. Adv Agron 101:1–57
Jones MB, Donnelly A (2004) C sequestration in temperate grassland ecosystems and the influence of management, climate and elevated $CO_2$. New Phytol 164(3):423–439
Jose S (2009) Agroforestry for ecosystem services and environmental benefits: an overview. Agrofor Syst 76:1–10
Kakraliya SK, Singh U, Bohra A, Choudhary KK, Kumar S, Meena RS, Jat ML (2018) Nitrogen and legumes: a meta-analysis. In: Meena RS et al (eds) Legumes for soil health and sustainable management. Springer, Singapore. https://doi.org/10.1007/978-981-13-0253-4_9
Kane D (2015) C sequestration potential on agricultural lands: a review of current science and available practices. Scholes RJ & Noble IR 2001. Storing C on land. Science 294:1012–1013

Kätterer T, Bolinder MA, Andrén O, Kirchmann H, Menichetti L (2011) Roots contribute more to refractory soil organic matter than above-ground crop residues, as revealed by a long-term field experiment. Agric Ecosyst Environ 141:184–192

Kettler TA, Lyon DJ, Doran JW, Powers WL, Stroup WW (2000) Soil quality assessment after weed-control tillage in a no-till wheat-fallow cropping system. Soil Sci Soc Am J 64:339–346

Kochsiek AE, Knops JMH, Walters DT, Arkebauer TJ (2009) Impacts of management on decomposition balance in irrigated and rainfed no till agricultural systems. Agric For Meteorol 149:1983–1993

Kumar S, Meena RS, Pandey A, Seema (2017a) Soil acidity management and an economics response of lime and sulfur on sesame in an alley cropping system. Int J Curr Microb Appl Sci 6(3):2566–2573

Kumar S, Meena RS, Yadav GS, Pandey A (2017b) Response of sesame (*Sesamum indicum* L.) to Sulphur and lime application under soil acidity. Int J Plant Soil Sci 14(4):1–9

Kumar S, Meena RS, Bohra JS (2018a) Interactive effect of sowing dates and nutrient sources on dry matter accumulation of Indian mustard (*Brassica juncea* L.). J Oilseed Brassica 9(1):72–76

Kumar S, Meena RS, Lal R, Yadav GS, Mitran T, Meena BL, Dotaniya ML, EL-Sabagh A (2018b) Role of legumes in soil carbon sequestration. In: Meena RS et al (eds) Legumes for soil health and sustainable management. Springer, Singapore. https://doi.org/10.1007/978-981-13-0253-4_4

Kumari P, Nema AK (2015) Soil C sequestration enhancement techniques: an emergent technology to mitigate climate change. Climate Change 1(4):463–468

Kuo S, Jellum E (2002) Influence of winter cover crop and residue management on soil nitrogen availability and corn. Agron J 94(3):501–508

Lai L, Li Y, Tian Y, Jiang L, Zhao X, Zhu L, Chen X, Gao Y, Wang S, Zheng Y, Rimmington GM (2013) Effects of added organic matter and water on soil C sequestration in an arid region. PLoS One 8(7):e70224. https://doi.org/10.1371/journal.pone.0070224

Lal R (1997) Residue management, conservation tillage and soil restoration for mitigating greenhouse effect by CO,-enrichment. Soil Tillage Res 43:81–107

Lal R (2002) Soil C dynamics in cropland and rangeland. Environ Pollut 116:353–362

Lal R (2004a) Soil C sequestration impacts on global climate change and food security. Science 304:1623–1627

Lal R (2004b) Soil C sequestration in India. Clim Chang 65:277–296

Lal R (2004c) Soil C sequestration to mitigate climate change. Geoderma 123:1–22

Lal R (2008) C sequestration. Philos Trans R Soc Lond Ser B Biol Sci 363(1492):815–830

Lal R (2010) Enhancing eco-efficiency in agro-ecosystems through soil C sequestration. Crop Sci 50(Suppl 1):S–120

Lal R (2011) Sequestering C in soils of agro-ecosystems. Food Policy 36:S33–S39

Lal R (2013) Intensive agriculture and the soil C pool. J Crop Improv 27:735–751. https://doi.org/10.1080/15427528.2013.845053

Lal R (2015a) Soil C sequestration in agroecosystems of India. J Indian Soc Soil Sci 63(2):125–143. https://doi.org/10.5958/0974-0228.2015.00018.3

Lal R (2015b) Restoring soil quality to mitigate soil degradation. Sustainability 7:5875–5895. https://doi.org/10.3390/su7055875

Lal R (2016) Soil C sequestration and aggregation by cover cropping. J Soil Water Conserv 70(6):329–339. https://doi.org/10.2489/jswc.70.6.329

Lal R, Kimble JM (1997) Conservation tillage for C sequestration. Nutr Cycl Agroecosyst 49:243–253

Lal R, Kimble JM, Follett RF, Cole CV (1998) The potential of U.S. cropland to sequester C and mitigate the greenhouse effect. Ann Arbor Sci Publ Chelsea, MI 128

Lal R, Follettr F, Kimblej M, Colec V (1999) Managing U.S. cropland to sequester C in soil. J Soil Water Conserv 54:374–381

Lal R, Follett RF, Kimble JM (2003) Achieving soil C sequestration in the United States: a challenge to the policy makers. Soil Sci 168:827–845

Layek J, Das A, Mitran T, Nath C, Meena RS, Singh GS, Shivakumar BG, Kumar S, Lal R (2018) Cereal+Legume intercropping: an option for improving productivity. In: Meena RS et al (eds) Legumes for soil health and sustainable management. Springer, Singapore. https://doi.org/10.1007/978-981-13-0253-4_11

Le Quéré C et al (2016) C budget. Earth Syst Sci Data 8:1–45. www.earth-syst-sci-data.net/8/1/2016/. https://doi.org/10.5194/essd-8-1-2016
Lee JJ, Phillips DL, Liu R (1993) The effect of trends in tillage practices on erosion and C content of soils in the U.S. corn belt. Water Air Soil Pollut 70:389–401
Liang F et al (2016) Three-decade long fertilization-induced soil organic C sequestration depends on edaphic characteristics in six typical croplands. Sci Rep 6:30350. https://doi.org/10.1038/srep30350
Liu X, Han X, Song C, Herbert SJ, Xing B (2003) Soil organic C dynamics in black soils of China under different agricultural management systems. Commun Soil Sci Plant Anal 34(7–8):973–984. https://doi.org/10.1081/CSS-120019103
Liu E, Yan C, Mei X, Zhang Y, Fan T (2013b) Long-term effect of manure and fertilizer on soil organic C pools in dryland farming in Northwest China. PLoS One 8(2):e56536. https://doi.org/10.1371/journal.pone.0056536
Liu JB, Zhang YQ, Wu B, Qin SG, Jia X, Feng W (2013c) Changes in soil C, nitrogen and phosphorus along a chrono sequence of *Caragana microphylla* plantation, North western China. Pol J Environ Stud 23(2):385–391
Lorenz K, Lal R (2014) Soil organic C sequestration in agroforestry systems. A review. Agron Sustain Dev 34:443–454. https://doi.org/10.1007/s13593-014-0212-y
Maillard É, Angers DA (2014) Animal manure application and soil organic C stocks: a meta-analysis. Glob Chang Biol 20:666–679
Meena RS, Lal R (2018) Legumes and sustainable use of soils. In: Meena RS et al (eds) Legumes for soil health and sustainable management. Springer. https://doi.org/10.1007/978-981-13-0253-4_1
Meena H, Meena RS (2017) Assessment of sowing environments and bio-regulators as adaptation choice for clusterbean productivity in response to current climatic scenario. Bangladesh J Bot 46(1):241–244
Meena RS, Yadav RS (2014) Phonological performance of groundnut varieties under sowing environments in hyper arid zone of Rajasthan, India. J App Nat Sci 6(2):344–348
Meena RS, Yadav RS (2015) Yield and profitability of groundnut *(Arachis hypogaea* L) as influenced by sowing dates and nutrient levels with different varieties. Legum Res 38(6):791–797
Meena RS, Yadav RS, Meena VS (2014) Response of groundnut (*Arachis hypogaea* L.) varieties to sowing dates and NP fertilizers under Western dry zone of India. Bangladesh J Bot 43(2):169–173
Meena RS, Dhakal Y, Bohra JS, Singh SP, Singh MK, Sanodiya P (2015a) Influence of bioinorganic combinations on yield, quality and economics of Mungbean. Am J Exp Agric 8(3):159–166
Meena RS, Meena VS, Meena SK, Verma JP (2015b) The needs of healthy soils for a healthy world. J Clean Prod 102:560–561
Meena RS, Meena VS, Meena SK, Verma JP (2015c) Towards the plant stress mitigate the agricultural productivity: a book review. J Clean Prod 102:552–553
Meena RS, Yadav RS, Meena H, Kumar S, Meena YK, Singh A (2015d) Towards the current need to enhance legume productivity and soil sustainability worldwide: a book review. J Clean Prod 104:513–515
Meena RS, Yadav RS, Reager ML, De N, Meena VS, Verma JP, Verma SK, Kansotia BC (2015e) Temperature use efficiency and yield of groundnut varieties in response to sowing dates and fertility levels in Western dry zone of India. Am J Exp Agric 7(3):170–177
Meena H, Meena RS, Singh B, Kumar S (2016a) Response of bio-regulators to morphology and yield of clusterbean [*Cyamopsis tetragonoloba* (L.) Taub.] under different sowing environments. J App Nat Sci 8(2):715–718
Meena RS, Bohra JS, Singh SP, Meena VS, Verma JP, Verma SK, Shiiag SK (2016b) Towards the prime response of manure to enhance nutrient use efficiency and soil sustainability a current need: a book review. J Clean Prod 112:1258–1260
Meena RS, Gogaoi N, Kumar S (2017a) Alarming issues on agricultural crop production and environmental stresses. J Clean Prod 142:3357–3359

Meena RS, Kumar S, Pandey A (2017b) Response of sulfur and lime levels on productivity, nutrient content and uptake of sesame under guava (*Psidium guajava* L.) based agri-horti system in an acidic soil of eastern Uttar Pradesh, India. J Crop Weed 13(2):222–227

Meena RS, Meena PD, Yadav GS, Yadav SS (2017c) Phosphate solubilizing microorganisms, principles and application of microphos technology. J Clean Prod 145:157–158

Meena H, Meena RS, Lal R, Singh GS, Mitran T, Layek J, Patil SB, Kumar S, Verma T (2018a) Response of sowing dates and bio regulators on yield of clusterbean under current climate in alley cropping system in eastern U.P., India. Legum Res 41(4):563–571

Meena RS, Kumar V, Yadav GS, Mitran T (2018b) Response and interaction of *Bradyrhizobium japonicum* and *Arbuscular mycorrhizal* fungi in the soybean rhizosphere: a review. Plant Growth Regul 84:207–223

Meena BL, Fagodiya RK, Prajapat K, Dotaniya ML, Kaledhonkar MJ, Sharma PC, Meena RS, Mitran T, Kumar S (2018c) Legume green manuring: an option for soil sustainability. In: Meena et al. (ed) Legumes for Soil Health and Sustainable Management. Springer, Singapore. https://doi.org/10.1007/978-981-13-0253-4_12

Minasny B, Malone BP, McBratney AB et al (2017) Soil C 4 per mille. Geoderma 292:59–86. https://doi.org/10.1016/j.geoderma.2017.01.002

Mitran T, Meena RS, Lal R, Layek J, Kumar S, Datta R (2018) Role of soil phosphorus on legume production. In: Meena et al (eds) Legumes for soil health and sustainable management. Springer, Singapore. https://doi.org/10.1007/978-981-13-0253-4_15

Murthy IK, Gupta M, Tomar S, Munsi M, Tiwari R et al (2013) C sequestration potential of agroforestry systems in India. J Earth Sci Clim Chang 4:131. https://doi.org/10.4172/2157-7617.1000131

Nair PKR, Nair VD, Kumar BM, Haile GS (2009) Soil C sequestration in tropical agroforestry systems: a feasibility appraisal. Environ Sci Pol 12(8):1099–1111

Nair PKR, Nair VD, Kumar BM, Showalter JM (2010) C sequestration in agroforestry systems. Adv Agron 108:237–307

Nayak AK, Gangwar B, Shukla AK, Mazumdar SP, Kumar A, Raja R, Kumar A, Kumar V, Rai PK, Mohan U (2012) Long-term effect of different integrated nutrient management on soil organic C and its fractions and sustainability of rice–wheat system in Indo Gangetic Plains of India. http://hdl.handle.net/10919/69922

Nieder R, Benbi DK (2008) C and nitrogen in the terrestrial environment. Springer, New York, p 430

Nordt LC, Wilding LP, Drees LR (2000) Pedogenic Cate transformations in leaching soil systems. In: Lal R, Kimble JM, Stewart BA (eds) Global climate change and tropical ecosystems. CRC/Lewis Publishers, Boca Raton, pp 43–63

Nosberger J, Blum H, Fuhrer J (2000) Crop ecosystem responses to climatic change: productive grasslands. In: Hodges HF (ed) Climate change and global crop productivity. CAB International, Wallingford, pp 271–291

Oelbermann M, Voroney RP, Gordon AM (2004) C sequestration in tropical and temperate agroforestry systems: a review with examples from Costa Rica and southern Canada agriculture. Ecosyst Environ 104:359–377

Olivier JGJ, Janssens-Maenhout G, Muntean M, Peters JAHW (2016) Trends in global $CO_2$ emissions (2016): report. PBL Netherlands Environmental Assessment Agency/European Commission, Joint Research Centre, The Hague/Ispra

Olivier JGJ, Schure KM, Peters JAHW (2017) Trends in global $CO_2$ and total greenhouse gas emissions: 2017 report. PBL Netherlands Environmental Assessment Agency, The Hague

Paustian K, Robertson GP, Elliott ET (1995) Management impacts on C storage and gas fluxes ($CO_2$, $CH_4$) in mid-latitude cropland. In: Lal R et al (eds) Soil management and the greenhouse effect. Lewis Publishers, Boca Raton, pp 69–83

Paustian K, Lehman J, Ogle S, Reay D, Robertson GP, Smith P (2016) Climate-smart soils. Nature 532:49–57

Poeplau C, Don A (2015) C sequestration in agricultural soils via cultivation of cover crops – a meta-analysis. Agric Ecosyst Environ 200:33–41. https://doi.org/10.1016/j.agee.2014.10.024

Poeplau C, Don A, Vesterdal L, Leifeld J, Van Wesemael B, Schumacher J, Gensior A (2011) Temporal dynamics of soil organic C after land-use change in the temperate zone – C response functions as a model approach. Glob Change Biol 17:2415–2427

Post WM, Izaurralde RC, Jastrow JD, McCarl BA, Amonette JE, Bailey VL, Jardine PM, West TO, Zhou J (2004) Enhancement of C sequestration in US soils. Bioscience 54:895–908. https://doi.org/10.1641/0006-3568

Powers JS, Corre MD, Twine TE, Veldkamp E (2011) Geographic bias of field observations of soil C stocks with tropical land-use changes precludes spatial extrapolation. Proc Natl Acad Sci U S A 108:6318–6322

Prather MJ, Holmes CD, Hsu J (2012) Reactive greenhouse 30 gas scenarios: systematic exploration of uncertainties and the role of atmospheric chemistry. Geophys Res Lett 39:L09803

Qin ZC, Huang Y, Zhuang QL (2013) Soil organic C sequestration potential of cropland in China. Glob Biogeochem Cycles 27:711–722

Rajan K, Natarajan A, Kumar K, Badrinath M, Gowda R (2010) Soil organic C—the most reliable indicator for monitoring land degradation by soil erosion. Curr Sci 99:823–827

Ram K, Meena RS (2014) Evaluation of pearl millet and mungbean intercropping systems in arid region of Rajasthan (India). Bangladesh J Bot 43(3):367–370

Reeves DW (1997) The role of soil organic matter in maintaining soil quality in continuous cropping systems. Soil Tillage Res 43:131–167

Richards KR, Stokes C (2004) A review of forest C sequestration cost studies: a dozen years of research. Clim Chang 63:1–48

Rotenberg R, Yakir D (2010) Contribution of semi-arid forests to the climate system. Science 327:451–454. https://doi.org/10.1126/science.1179998

Sainju UM, Singh BP, Yaffa S (2002) Soil organic matter and tomato yield following tillage, cover cropping and nitrogen fertilization. Agric J 94:594–602

Schuman GE, Janzen HH, Herrick JE (2002) Soil C dynamics and potential C sequestration by rangelands. Environ Pollut 116(3):391–396

Sexstone AJ, Revsbech NP, Parkin TB, Tiedje JM (1985) Direct measurement of oxygen profiles and denitrification rates in soil aggregates. Soil Sci Soc Am J 49(3):645–651

Shaver TM, Peterson GA, Ahuja LR, Westfall DG, Sherrod LA, Dunn G (2002) Surface soil physical properties after twelve years of dryland no till management. Soil Sci Soc Am J 66:1296–1303

Sihag SK, Singh MK, Meena RS, Naga S, Bahadur SR, Gaurav YRS (2015) Influences of spacing on growth and yield potential of dry direct seeded rice (*Oryza sativa* L) cultivars. Ecoscan 9(1–2):517–519

Silver WL, Ostertag R, Lugo AE (2000) The potential for C sequestration through reforestation of abandoned tropical agricultural and pasture lands. Restor Ecol 8:394–407

Six JP, Elliott KET, Combrink C (2000) Soil structure and organic matter distribution of aggregate-size classes and aggregate-associated C. Soil Sci Soc Am J 64:681–689

Smith P, Powlson DS, Glendining MJ, Smith JU (1998) Preliminary estimates of the potential for C mitigation in European soils through no-till farming. Glob Chang Biol 4:679–685

Smith P et al (2008) Greenhouse gas mitigation in agriculture. Philos Trans R Soc B 363:789–813

Söderström et al (2014) What are the effects of agricultural management on soil organic C (SOC) stocks? Environ Evid 3:2. https://doi.org/10.1186/2047-2382-3-2

Sofi PA, Baba ZA, Hamid B, Meena RS (2018) Harnessing soil rhizobacteria for improving drought resilience in legumes. In: Meena RS et al (eds) Legumes for soil health and sustainable management. Springer, Singapore. https://doi.org/10.1007/978-981-13-0253-4_8

Solomon D et al (2007) Long-term impacts of anthropogenic perturbations on dynamics and speciation of organic C in tropical forest and subtropical grassland ecosystems. Glob Change Biol 13:511–530

Sommer R, Bossio D (2014) Dynamics and climate change mitigation potential of soil organic C sequestration. J Environ Manag 144:83–87

Su YZ, Liu WJ, Yang R, Chang XX (2009) Changes in soil aggregate, C, and nitrogen storages following the conversion of cropland to alfalfa forage land in the marginal oasis of Northwest China. Environ Manag 43:1061–1070

Sun W, Huang Y, Zhang W, Yu Y (2010) C sequestration and its potential in agricultural soils of China. Glob Biogeochem Cycles 24:GB3001. https://doi.org/10.1029/2009GB003484

Swan A, Williams SA, Brown K, Chambers A, Creque J, Wick J, Paustian K (2015) COMET Planner. C and greenhouse gas evaluation for NRCS conservation practice planning. A companion report to www.comet-planner.com

Swift RS (2001) Sequestration of C by soil. Soil Sci 166(11):858–871

Trost B, Prochnow A, Drastig K, Meyer-Aurich A, Ellmer F, Baumecker M (2013) Irrigation, soil organic C and N2O emissions. A review. Agron Sustain Dev 33:733–749. https://doi.org/10.1007/s13593-013-0134-0

Udawatta RP, Jose S (2011) C sequestration potential of agroforestry practices in temperate North America. In: Mohan Kumar B, Ramachandran Nair PK (eds) C sequestration potential of agroforestry systems opportunities and challenges. Springer, Dordrecht/Heidelberg/London/New York. https://doi.org/10.1007/978-94-007-1630-8

Varma D, Meena RS, Kumar S (2017a) Response of mungbean to fertility and lime levels under soil acidity in an alley cropping system in Vindhyan region, India. Int J Chem Stud 5(2):384–389

Varma D, Meena RS, Kumar S, Kumar E (2017b) Response of mungbean to NPK and lime under the conditions of Vindhyan region of Uttar Pradesh. Leg Res 40(3):542–545

Verma JP, Jaiswal DK, Meena VS, Meena RS (2015a) Current need of organic farming for enhancing sustainable agriculture. J Clean Prod 102:545–547

Verma JP, Meena VS, Kumar A, Meena RS (2015b) Issues and challenges about sustainable agriculture production for management of natural resources to sustain soil fertility and health: a book review. J Clean Prod 107:793–794

Verma SK, Singh SB, Prasad SK, Meena RN, Meena RS (2015c) Influence of irrigation regimes and weed management practices on water use and nutrient uptake in wheat (*Triticum aestivum* L. Emend. Fiori and Paol.). Bangladesh J Bot 44(3):437–442

Wang Q, Li Y, Alva A (2010) Cropping systems to improve C sequestration for mitigation of climate change. J Environ Prot 1:207–215. https://doi.org/10.4236/jep.2010.13025

Wang J et al (2015) Contributions of wheat and maize residues to soil organic C under long-term rotation in North China. Sci Rep 5:11409. https://doi.org/10.1038/srep11409

Wani NR, Qaisar KN (2014) C per cent in different components of tree species and soil organic C pool under these tree species in Kashmir valley. Curr World Environ 9(1):174–181

Wiesmeier M et al (2016) Projected loss of soil organic C in temperate agricultural soils in the 21st century: effects of climate change and C input trends. Sci Rep 6:32525. https://doi.org/10.1038/srep32525

WMO (2016) Greenhouse gas bulletin The state of greenhouse gases in the atmosphere based on global observations through 2015. No. 12: ISSN 2078-0796

World Bank (2012) C sequestration in agricultural soils, Report No. 67395-GLB

Yadav GS, Babu S, Meena RS, Debnath C, Saha P, Debbaram C, Datta M (2017a) Effects of godawariphosgold and single supper phosphate on groundnut (*Arachis hypogaea*) productivity, phosphorus uptake, phosphorus use efficiency and economics. Indian J Agric Sci 87(9):1165–1169

Yadav GS, Lal R, Meena RS, Babu S, Das A, Bhomik SN, Datta M, Layak J, Saha P (2017b) Conservation tillage and nutrient management effects on productivity and soil carbon sequestration under double cropping of rice in North Eastern region of India. Ecol Indic. http://www.sciencedirect.com/science/article/pii/S1470160X17305617

Yadav GS, Lal R, Meena RS, Datta M, Babu S, Das LJ, Saha P (2017c) Energy budgeting for designing sustainable and environmentally clean/safer cropping systems for rainfed rice fallow lands in India. J Clean Prod 158:29–37

Yadav GS, Das A, Lal R, Babu S, Meena RS, Saha P, Singh R, Datta M (2018a) Energy budget and carbon footprint in a no-till and mulch based rice–mustard cropping system. J Clean Prod 191:144–157

Yadav GS, Das A, Lal R, Babu S, Meena RS, Patil SB, Saha P, Datta M (2018b) Conservation tillage and mulching effects on the adaptive capacity of direct-seeded upland rice (*Oryza sativa* L.) to alleviate weed and moisture stresses in the North Eastern Himalayan region of India. Arch Agron Soil Sci. https://doi.org/10.1080/03650340.2018.1423555

Yuste JC, Baldocchi DD, Gershenson A, Goldstein A, Misson L et al (2007) Microbial soil respiration and its dependency on C inputs, soil temperature and moisture. Glob Change Biol 13:2018–2035

# Soil Quality for Sustainable Agriculture

# 2

Duraisamy Vasu, Pramod Tiwary, Padikkal Chandran,
and Surendra Kumar Singh

**Abstract**

Being integral to all functions of terrestrial ecosystem, soil is intended to produce food for feeding the growing population of the world. However, food security is facing threat from soil degradation occurring worldwide. Soils degrade due to the exerting pressure from various sectors of the society including urbanization and industrialization. The major driving forces of soil degradation are deforestation, change in land use, soil erosion, uncontrolled grazing, waste disposal, and unscientific land management. Globally, 24% (350 lakh km$^2$) of the land has degraded which is increasing at the rate of 50–100 lakh ha year$^{-1}$ and poses threat to the livelihood of more than 1500 million people. In this scenario, sustaining soil quality (SQ) is the major challenge to meet the increasing food demand. Hence, evaluating and monitoring SQ is crucial to sustain agricultural production and to overcome the vagaries of climate change on soil functions. However, soil quality *per se* is complex and site-specific because of the larger variety of soil usage, and its evaluation is difficult due to the subjectivity. Nonetheless, soil quality can be quantified in the form of an index for temporal and spatial comparison of various land use and management systems. In this chapter, we discuss the concept and importance of SQ, indicators of SQ, minimum data set (MDS) for evaluating SQ, methods of MDS selection, and indexing of the soil quality. It will bring out the effect of soil and crop management practices such as tillage, cropping systems, cover crops, and nutrient management on soil quality and crop production focusing in tropical environments. We conclude that principal component analysis is an effective method to select MDS from a large set of soil properties and weighted index method of quantifying SQ proved to be efficient in predicting changes in SQ under various crop production systems. Conservation tillage methods coupled with integrated nutrient management sustains or aggrades the soil quality in different agroecosystems.

D. Vasu (✉) · P. Tiwary · P. Chandran · S. K. Singh
ICAR – National Bureau of Soil Survey and Land-Use Planning, Nagpur, India





R. S. Meena (ed.), *Nutrient Dynamics for Sustainable Crop Production*,
https://doi.org/10.1007/978-981-13-8660-2_2

**Keywords**
Conservation agriculture · Earthworm population · Factor analysis · Indexing soil quality · Soil organic matter

## Abbreviations

| | |
|---|---|
| BD | Bulk density |
| C | Carbon |
| CEC | Cation exchange capacity |
| $CO_2$ | Carbon dioxide |
| CT | Conservation tillage |
| DT | Decision trees |
| EC | Electrical conductivity |
| ESP | Exchangeable sodium percentage |
| FAO | Food and Agriculture Organization |
| Fe | Iron |
| FYM | Farmyard manure |
| I | Iodine |
| IGP | Indo-Gangetic plains |
| INM | Integrated nutrient management |
| K | Potassium |
| MBC | Microbial biomass carbon |
| MDS | Minimum data set |
| M-SQR | Muencheberg soil quality rating |
| MT | Minimum tillage |
| N | Nitrogen |
| NRCS | Natural resources conservation service |
| NT | No tillage |
| P | Phosphorus |
| PC | Principal components |
| PCA | Principal component analysis |
| QBS-ar | Soil biological quality-arthropod |
| S | Sulfur |
| SMT | Stubble-mulch tillage |
| SOC | Soil organic carbon |
| SOM | Soil organic matter |
| SQ | Soil quality |
| SQI | Soil quality index |
| Zn | Zinc |

## 2.1 Introduction

The soil is a heterogeneous natural resource which supports life on terrestrial earth. Being integral to all functions of terrestrial ecosystems, soil is intended to produce food for feeding the growing population of the world (Paustian et al. 2016). Globally, agriculture intensification in the last century by the use of high-yield crop varieties, irrigation expansion, and chemical amendments such as high analysis fertilizers and lime led to green revolution and self-sufficiency in food production in most of the countries. Primarily, food and timber production increased by 170% and 60%, respectively, in the past four decades (Foley et al. 2011), showing the important role of soil in sustaining the food security and other ecosystem services (Fig. 2.1).

Nutrients and water in the soil, solar radiation, and carbon dioxide ($CO_2$) are used by the plants for the photosynthesis process and to produce food for the humans and animals. Moreover, soils store water received from irrigation and rainfall and then release it for sustaining plant growth and reproduction. Soils also act as a filter of nonhazardous and toxic metals through various mechanisms such as clay surface adsorption and precipitation which balances the composition of soil chemical environment. Most of the above functions of the soil benefit the humans and animals (Palm et al. 2007). Soils affect human life directly and indirectly by the quality of food produced from agriculture. If pathogens and toxic metals are absorbed by the

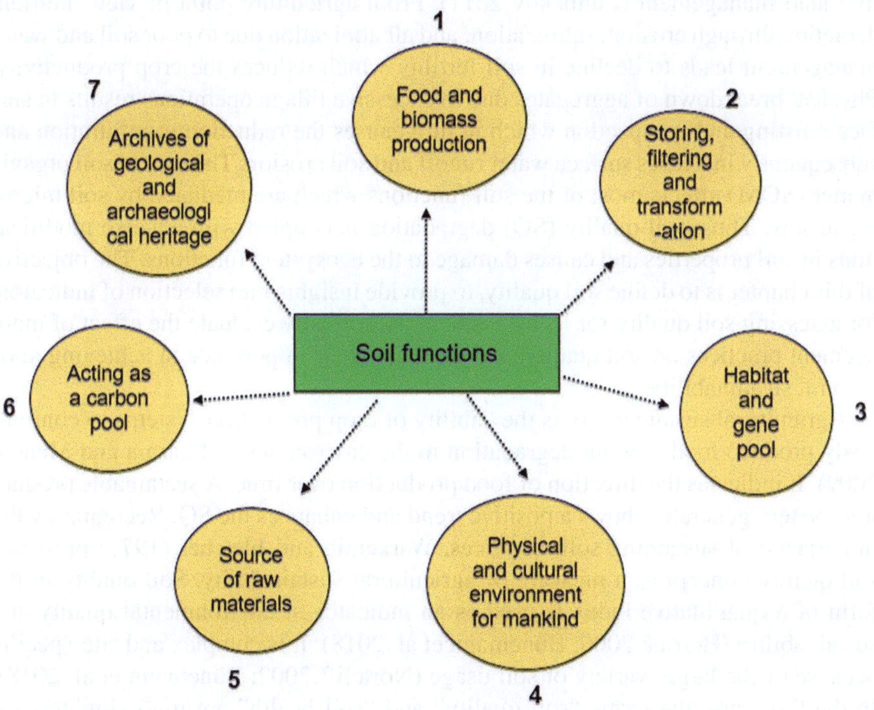


**Fig. 2.1** Seven functions of soil. (Adapted from European Commission 2006)

human body via food chain, it is a threat to human health and nutritional security (Oliver and Gregory 2015). The poor nutritional quality of the food is attributed to agriculture, social, and political causes. Moreover, the food quality is influenced by soil fertility. For example, the inherently poor fertile soils in most of the areas of Asia and Africa support only low crop yields because they require additional fertilizers. Therefore, the food consumed by two-thirds of the global population is deficient in essential nutrients, viz., 60% in iron (Fe), 30% in zinc (Zn), and iodine (I). As human body cannot produce these trace elements, their lack of presence in the soil causes nutrient deficiency diseases like anemia in humans. Hence, soil fertility is indispensable for food security.

In recent decades, global soils are under prodigious pressure due to competing demands from various sectors of the society in general and diversion of prime arable lands to nonarable uses like urbanization and industrialization, in particular (Foley et al. 2011). Owing to the exerting pressure, 24% (350 lakh $km^2$) of the global land area is degraded (Lal 2012), which is increasing at the rate of 50–100 lakh ha $year^{-1}$ and poses threat to the livelihood of more than 1500 million people (Bai et al. 2013; Stavi and Lal 2015). India is behind only to China with respect to rural people affected by soil degradation (Bai et al. 2013). The cultivated lands produce only 1.5 t $ha^{-1}$ of food grain in these regions due to poor distribution of rainfall (Srinivasarao et al. 2014). Major causes for the soil degradation are deforestation, change in land use, soil erosion, uncontrolled grazing, waste disposal, and unscientific land management (Zalibekov 2011). From agriculture point of view, nutrient depletion through erosion, salinization, and alkalinization due to poor soil and water management leads to decline in soil fertility which reduces the crop productivity. Physical breakdown of aggregates due to excessive tillage operations results in surface crusting and compaction which in turn causes the reduction in infiltration and subsequently increases surface water runoff and soil erosion. The loss of soil organic matter (SOM) affects most of the soil functions which are mediated by soil microorganisms. Thus, soil quality (SQ) degradation is coupled with adverse modifications in soil properties and causes damage to the ecosystem functions. The objective of this chapter is to define soil quality, to provide insights into selection of indicators for assessing soil quality for its assessment, to critically evaluate the effect of management practices on soil quality, and establish their importance in achieving agricultural sustainability.

Agricultural sustainability is the "ability of crop production systems to continuously produce food without degradation to the environment" (Sharma and Mandal 2009). It indicates the direction of food production over time. A sustainable production system generally shows a positive trend and enhances the SQ. Recognizing the importance of sustaining soil resources, Warkentin and Fletcher (1977) proposed soil quality concept as a measure of agricultural sustainability. Soil quality in the form of a quantitative index is used as an indicator of environmental quality and sustainability (Herrick 2000; Bünemann et al. 2018). It is complex and site-specific because of the larger variety of soil usage (Nortcliff 2002; Bünemann et al. 2018). In the literature, the terms "soil quality" and "soil health" are used similarly for nearly half a century (More 2010; Bünemann et al. 2018). It is difficult to distinguish between them going by their definition. However, they may be differentiated in terms of timescale that the term "soil health" indicates condition of soil in a short period and "soil quality" over a longer period much analogous to the condition of a human at a particular time (health) and long time period (quality of life) (Acton and Gregorich 1995). In general, these terms are being used as an indicator of current soil status, and their assessment is basically aimed at measuring the impact of past and present land use on future agricultural sustainability. In this chapter, the concept and importance of soil quality, measuring and periodic monitoring of soil quality, and the influence of various crop and land management practices on soil quality in the tropical environment are discussed.

## 2.2 Soil Quality: Concept and Importance

### 2.2.1 The Concept of Soil Quality

Warkentin and Fletcher (1977) introduced the concept of soil quality for appropriate input allocation to increase the production of food and fiber. During the 1970s and 1980s, soil quality was synonymously used in the context of land evaluation, focusing on inherent soil properties which are linked to soil genesis and pedo-environment. Many researchers proposed several definitions for soil quality from the early 1990s (Table 2.1) which mainly focused on linking soil with agricultural productivity. Later, by expanding the linkage of soil functions, Karlen et al. (1997) defined soil quality as "the capacity of specific kind of soil to function, within natural or managed ecosystem boundaries, to sustain plant and animal productivity, maintain or enhance water and air quality, and support human health and habitation." Further, Bouma et al. (2017) provided a broader view to soil quality by defining soil quality

**Table 2.1** Various definitions of soil quality

| Definition | References |
|---|---|
| Inherent attributes of soil inferred from soil characteristics | SSSA (1987) |
| The capacity of soil to function within ecosystem boundaries | Larson and Pierce (1991) |
| The capacity of soil to function within ecosystem boundaries to sustain biological productivity, maintain environmental quality, and promote plant and animal health | Doran and Parkin (1996) |
| Soil's capacity or fitness to support growth without resulting in soil degradation or otherwise harming the environment | Acton and Gregorich (1995) |
| Soil quality is the capacity of a specific kind of soil to function within natural or managed ecosystem boundaries, sustain plant and animal productivity, maintain or enhance water and air quality, and support human health and habitation | SSSA (1995) |
| Productivity and environment moderation capacity | Lal (1997) |
| The intrinsic capacity of a soil to contribute to ecosystem services, including biomass production | Bouma et al. (2017) |

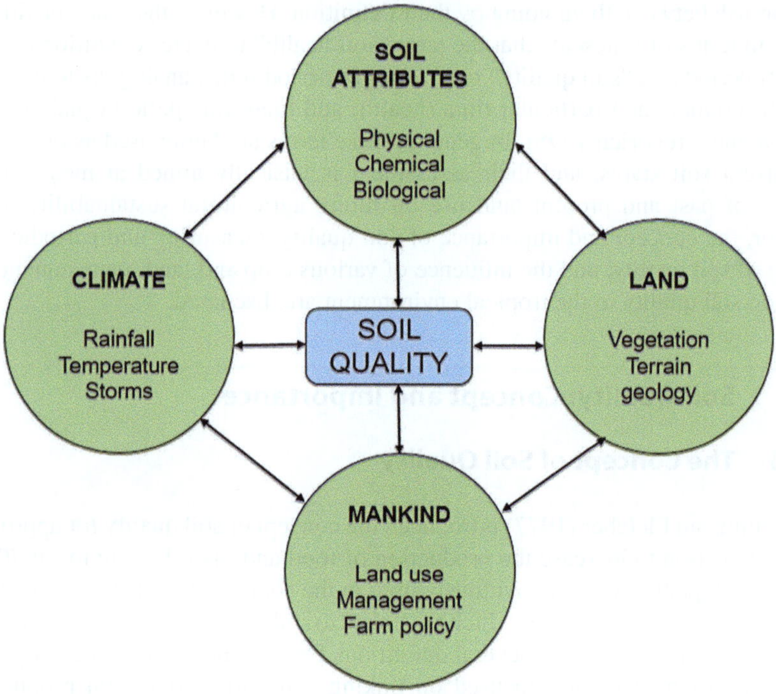


**Fig. 2.2** Factors influencing soil quality. (Modified from Arshad and Coen 1992)

as "the intrinsic capacity of a soil to contribute to ecosystem services, including biomass production."

Generally, soil quality is conceptualized as "inherent soil quality" and "dynamic soil quality." The first one is influenced by soil's inherent properties and the second involves changes in soil properties determined by human use. The inherent soil quality exhibits minimum change, but the dynamic soil quality rapidly responds to agriculture management practices (Seybold et al. 1999; Vasu et al. 2016; Biswas et al. 2017). The changes in soil properties may occur within hours to a period of decades with respect to the response level of soil properties. However, the limits to which the dynamic soil properties can change are determined by inherent properties. The changes in SQ are also influenced by management systems, agroecology, hydrogeology, and cropping systems (Fig. 2.2).

At the initial stage, SQ assessments focused on measuring dynamic soil properties mostly from surface soil at 0–25 cm depth (Karlen et al. 2003). Subsequently, a range of soil parameters that represent the soil functions is identified using landscape characteristics and knowledge of pedology to understand how the soil is functioning and to select appropriate indicators for evaluation (Norfleet et al. 2003). Recently, the importance of soil control section or soil profile is recognized. As a result, soil profile characteristics are now being increasingly used in SQ evaluation. Many studies also proved that the evaluation of soil quality warrants inclusion of both inherent and dynamic properties (Merril et al. 2013; Ray et al. 2014;

Moncada et al. 2014; Vasu et al. 2016). The difference between inherent and dynamic soil properties are context-dependent, and care should be employed in relating the soil properties with specific soil functions (Schwilch et al. 2016). In recent literature, the soil quality concept is incorporated with land evaluation, soil management, environmental monitoring, land degradation, and land restoration (Bünemann et al. 2018).

### 2.2.2 Importance of Soil Quality

Measuring and monitoring soil quality is essential because of its multiple functions. Degradation of the soil resources is among the most serious and widespread threat to life in the terrestrial earth. Karlen et al. (2003) outlined the following reasons why measuring and monitoring soil quality is important:

(a) Many stakeholders are interested in soil resources.
(b) Change in priorities and demand for soil resources.
(c) Land-use decisions are made more in a human or institutional context.

From crop production point of view, soil quality evaluation can help the farmers, farm managers, extension workers, and policymakers to identify the sustainability of a given land use. Soil quality evaluation can be used for the following purposes (Andrews et al. 2002; Aparicio and Costa 2007; Vasu et al. 2016; Bouma et al. 2017; Biswas et al. 2017):

(a) Evaluation of the sustainability of long-term cropping systems
(b) Identification of soil degradation by crop and soil management practices
(c) Evaluation of the effect of farm machinery on soil properties
(d) Finding the suitability of soil for introducing new crops
(e) Assessing the capacity of the soil to produce more food
(f) Evaluating the effect of natural hazards on soil properties
(g) Assessment of the effect of forest fire on soil properties and soil biota
(h) Impact assessment of the effect of deforestation on ecosystem services

The quantification and comparison of soil quality among different land use, crop production systems, and management practices facilitate better land-use planning for sustainable utilization of the nonrenewable soil resources (Norfleet et al. 2003). The SQ concept and its implementation can subsequently address the issues of both productivity and sustainability (USDA-NRCS 1999). Moreover, in a broader view, SQ assessment is important to (i) target conservation efforts for the improvement of soil, (ii) assess soil and crop management practices, (iii) correlate soil quality with other natural resources, (iv) quantification and determination of soil quality trends, and (v) help in decision-making (More 2010).

## 2.3 Soil Quality Assessment

The basic need for SQ assessment is because of the deleterious changes in soil functions caused by inappropriate management and other natural factors. Soil quality assessment is an exercise in measuring the changes in soil properties due to management, change in land use, deforestation, etc. However, SQ uses soil taxonomy as the foundation, and most of the inherent soil properties, a product of the five soil-forming factors, are given due importance (Andrews et al. 2004). Soil quality per se is a series of threshold value of selected soil properties as indicators of SQ. The indicators are defined as "the soil properties and processes which are most sensitive to changes in soil function" (Doran and Parkin 1996). For considering soil properties as SQ indicators, the following criteria are generally followed:

(a) Correlate well with ecosystem processes and defined management goal.
(b) Be attainable to most users.
(c) Responds rapidly to management practices.
(d) Uncomplicated to determine.

Soil quality is generally evaluated by two approaches. The first one is a comparative method in which SQ of a given land use or management is evaluated in a given time. The second one is a dynamic approach where the SQ is assessed using temporal data (Shukla et al. 2006). Essentially, threshold values of indicator properties are necessary to draw comparisons and identify whether soil quality is degrading or improving after the imposed management in both the short and long term (Biswas et al. 2017). Benchmark sites were successfully used to assess the changes in soil quality over time (Acton and Gregorich 1995) (Table 2.2). Karlen et al. (1997) proposed a conceptual framework for SQ evaluation and then Andrews et al. (2004) developed a quantitative formula, and they suggested that the SQ must be monitored by focusing on soil functions. In recent times, SQ is used as a method to evaluate land-use systems at various scales from regional to the national level (Mukherjee and Lal 2014; Vasu et al. 2016).

### 2.3.1 Soil Quality Indicators

Soil properties are influenced by changes in land use, management, and other external factors. The modifications in soil properties are the reaction of soil to the changes in land use. However, these changes are generally slow. Hence, it is difficult to assess the change in soil quality unless there is an irreversible change in any of the soil properties (Nortcliff 2002). Because of this reason, it became necessary to identify a few soil properties as soil quality indicators which can reflect the changes in soil quality. Soil quality indicators are properties that are sensitive to soil functions and should be easy to measure (Dumanski and Pieri 2000; Aparicio and Costa 2007). The soil quality indicators are generally classified into four categories (More 2010):

**Table 2.2** Soil quality assessment at various scales and methods

| Scale | Country | Methodology | References |
|---|---|---|---|
| Global | Germany, Russia, China, New Zealand, Canada, UK, Denmark, etc. | Index method using both inherent and dynamic soil properties with a weighing factor | Mueller et al. (2012) |
| Regional | Iran | Index method using MDS and TDS; digital soil mapping using random forest model | Nabiollahi et al. (2018) |
| Regional | China | Index method using minimum data set | Liu et al. (2015) |
| 23 benchmark sites (national) | Canada | Trend analysis | Acton and Gregorich (1995) |
| National | UK | Trigger values | Loveland and Thompson (2002) and Merrington (2006) |
| Block (18 soil profiles) | India | Index method using minimum data set | Vasu et al. (2016) |
| Plot scale | India | Index method using minimum data set | Biswas et al. (2017) |
| 200 locations | Netherlands | Target values | Wattel-Koekkoek et al. (2012) |
| 511 sites (national) | New Zealand | Comparative values | Sparling et al. (2004) |
| National and regional | Australia | Target values | Gonzalez-Quiñones et al. (2015) |
| Plot scale | USA | Index method using scoring curves | Karlen et al. (2001) and Andrews et al. (2004) |

1. Visual indicators
2. Physical indicators
3. Chemical indicators
4. Biological indicators

#### 2.3.1.1 Visual Indicators

The visual indicators are field observations of mostly qualitative soil properties, viz., soil depth, color, erosion, gully formation, salt deposition, drainage, surface ponding, soil structure, consistence, mottles, rooting depth, root development, earthworm population, rodent activity, etc. These indicators are assessed in the field and interpreted by both experts and farmers. The main advantage of visual SQ indicators is that they are immediately interpreted without time-consuming laboratory analysis (Bünemann et al. 2018). Among the many visual soil quality indicators, soil structure is given importance in the recent literature (Emmet-Booth et al. 2016), and various methods such as Peerlkamp test, SOILpak, profile cultural method, visual evaluation of soil structure, and Muencheberg Soil Quality Rating (M-SQR) are widely used to evaluate soil quality (Guimaraes et al. 2011; Mueller et al. 2012; Ball et al. 2013).

Visual indicators are easy to measure and interpret and can be used for replicating the SQ assessment in various locations. Mueller et al. (2012) developed a framework for SQ evaluation using M-SQR which is based on visual indicators such as surface horizon depth, topsoil and subsoil structure, rooting depth, slope, and relief. They are rated and quantified to assess soil quality in various scales. The Natural Resources Conservation Service (NRCS) proposed a set of indicators and their ranking for SQ evaluation (Table 2.3). These indicators are qualitative and farmers can use them for on-farm assessment of soil quality (Adeyolanu and Ogunkunle 2016).

#### 2.3.1.2 Physical Indicators

Physical properties such as texture, structure, hydraulic conductivity, infiltration, porosity, bulk density, and aggregate stability are used as physical SQ indicators (Table 2.4). They are used to evaluate physical SQ and linked with seedling emergence, root growth, water movement, water holding capacity, penetration resistance, etc. Physical properties play a vital role in determining the soil erodibility and soil-plant-water-atmosphere relationships (More 2010). More recently, Dexter (2004) proposed the "S-value" as an indicator to measure soil physical quality. The "S-value" is related to hydraulic conductivity, compaction, water content, penetration resistance, and aggregate stability (Dexter and Czyz 2007).

#### 2.3.1.3 Chemical Indicators

Important soil chemical processes are ionic diffusion, leaching, acidification, alkalinization, salinization, mineralization, etc. Maintaining a favorable nutrient content is critical to soil chemical quality. Both long-term use of subsistence agricultural practices without proper fertilization and heavy usage of chemical fertilizers in intensive high productive agricultural systems rapidly decline soil chemical quality. The chemical indicators of SQ are pH, EC, salinity, sodicity, organic carbon, nitrogen fractions, phosphorus concentration, cation exchange capacity (CEC), and heavy metal concentrations. Among the chemical indicators, P concentration, cation exchange capacity, exchangeable sodium and magnesium, and hydraulic conductivity (which are interrelated) are considered important in rainfed agriculture production systems, and they are also used to assess chemical and physical degradation (Vasu et al. 2016; Vasu et al. 2018). Soil pH and available P are the most used chemical indicators in SQ assessment as they indicate most of the nutrient-related transformations in soil.

#### 2.3.1.4 Biological Indicators

The microorganisms play an important role in organic matter decomposition and recycling of nutrients. The microbes have the capacity to alleviate the consequences of disturbances on soil ecosystem services, due to their resistance, resilience, and/or functional redundancy (Allison and Martiny 2008). The soil microbes reciprocate

**Table 2.3** Qualitative visual soil quality indicators and their ranking

| Soil quality indicators | Low | Moderate | High | Method of assessment |
|---|---|---|---|---|
| Earthworm | Few worms (1–4) per shovel, no casts or holes | More worms (5–8) per shovel, some casts and holes | Many worms (>8) per shovel, many casts and holes | Use of quadrant and counting the number of earthworms or casts (five quadrant throws per site) |
| Organic matter | No visible roots or residues | Some plant residues and roots | Lots of roots/residues in many stages of decomposition | Presence and abundance of visible residues or roots, color |
| Subsurface compaction | Hard layers, tight soil, restrict wire penetration, obvious hardpan, roots turned awkwardly | Firm soil, moderate shovel resistance, penetration beyond tillage layer | Loose soil, unrestricted wire penetration, no hardpan, mostly vertical root plant growth | Degree of resistance to a stick (100 cm × 1 cm in diameter) when inserted into the soil |
| Erosion | Obvious soil deposition, large gullies joined, obvious soil drifting | Some deposition, few gullies, some colored runoff, some evidence of soil drifting | No visible soil movement, no gullies, clear or no runoff, no obvious soil drifting | Presence of gullies, rills, or any evidence of runoff |
| Water holding capacity | Plant stress immediately following rain or irrigation, soil has limited capacity to hold water, soil requires frequent irrigation | Crops did not easily suffer from dry spell in the area, soil requires moderate irrigation | Soil holds water well for long time, thick topsoil for water storage, crops do well in dry spells, soil requires little irrigation | Rate at which water runs out after a good rain, with or without puddling |
| Drainage | Excessive wet spots on the field, ponding, root disease | Some wet spots on the field and profile, some root diseases | Water is evenly drained through field and soil profile, no evidence of root disease | Degree of wetness or dryness, ponding, or runoff |
| Crop condition | Stunted growth, uneven stand, discoloration, low yield | Some uneven or stunted growth, slight discoloration, signs of stress | Healthy, vigorous, and uniform stand | Leaf color and rate of crop growth throughout season |

Modified from USDA-NRCS (1999)

**Table 2.4** Some of the commonly used soil quality indicators to measure soil functions

| Category | Soil properties |
|---|---|
| Physical indicators | Bulk density |
| | Soil texture |
| | Aggregate stability |
| | Water storage |
| | Soil compaction |
| | Soil depth |
| | Penetration depth |
| | Porosity |
| | Hydraulic conductivity |
| | Infiltration |
| | Penetration resistance |
| Chemical indicators | pH |
| | EC |
| | CEC |
| | Organic matter |
| | Labile C and N |
| | Total and available N |
| | Available K |
| | Available P |
| | Sodicity and salinity |
| | Heavy metals |
| Biological indicators | Soil respiration |
| | Enzyme activities |
| | Microbial biomass |
| | Earthworms |
| | Micro-arthropods |
| | N-mineralization |

Adapted from Andrews et al. (2004) and Bünemann et al. (2018)

rapidly to changes in soil and indicate the factors and processes modifying the soil quality. The high sensitivity of microbes to the changes in the soil processes is an advantage as they can be used to monitor the short-term changes in the soil effectively (de La Rosa 2005).

Population of micro- and macroorganisms, earthworms, nematodes, termites, and their actions are important indicators of soil quality. The microbial biomass is an important part of the active ingredient in soil responsible for nutrient circulation and degradation of organic pollutants (Stenberg et al. 1998). Respiration rate and microbial biomass carbon (MBC) are used to measure microbial activity, more specifically microbial decomposition of organic matter in the soil. For example, ergosterol (fungal by-product) is used to measure the activity of the organisms that govern the formation and stability of soil aggregates.

Enzymes in soil are produced by microbes, plant roots, and fauna, and they have indispensable role in nutrient cycling. Enzymes such as dehydrogenase, urease, phosphatases, and glucosidase are used to measure nutrient mineralization in soil, and they can provide an early warning to the potential threats to soil quality (Comino et al. 2018). Recent studies also used enzyme activity to assess the effect of tillage practices on soil quality (Raiesi and Kabiri 2016).

Soil organic carbon (SOC) is a vital SQ indicator and ubiquitously used in almost every SQ assessment. The SOC is closely related to other soil properties, including soil structure, nutrient availability, water holding capacity, and erosion resistance (Zuber et al. 2017). However, as the SOC pool is large, it is difficult to detect changes in total soil organic matter in response to management (Haynes 2005). Nonetheless, the fractions of carbon can provide a clear picture of SOC changes. Particularly, labile fraction of SOC which makes up approximately 15% of the total SOC is easily decomposed and determines the magnitude of microbial activity in soil (Hajek et al. 1990). Since the labile carbon is the primary source of energy for microbes in the soil, they change more rapidly and to a higher extent than the total SOC. This effectively connects SOC between soil chemical and biological properties and makes it an effective SQ indicator for most of the agroecosystems.

Earthworm population is the best indicator of the structural, microclimatic, nutritive, and toxic status of soil. Earthworms play an important role in conserving and improving soil structure, recycling soil nutrients, promoting the gradual mixing of the soil layers, and creating a better aeration and drainage system in the soil. The earthworms are indicators of both water and nutrient cycling. Apart from earthworms, Italian scientists proposed the activity of micro-arthropods as SQ indicators (Parisi et al. 2005). The index QBS-ar is a consolidated form of micro-arthropods community in the soil. The QBS-ar takes into account soil micro-arthropods, invertebrates belonging to the Arthropoda phylum, having a range size between 0.2 and 2 mm (mesofauna). Based on the adaptation mechanisms of these organisms, QBS-ar considers that the number of micro-arthropod groups well adapted to the soil is high in soil with good quality (Menta et al. 2018). Recently, soil cover and dung beetle are used as biological SQ indicators (Chaves et al. 2017). Soil cover protects the soil against erosion, adds nutrients, and maintains the soil fauna (Van Elsas et al. 2007). The dung beetle is highly sensitive to changes in soil conditions, contributes to nutrient cycling, and facilitates water infiltration (Nortcliff 2002).

### 2.3.2 Selection of Soil Quality Indicators

The definition of soil quality reflects complexity due to the soil composition which has solid, liquid, and gaseous phases. Hence, the quantitative determination of soil quality is a difficult process. Moreover, analyzing all the soil properties increases the cost of SQ assessment especially in large-scale attempts. Researchers developed various methods to identify minimum soil data set (MDS) as indicators to determine soil quality. Principal component analysis (PCA) (Andrews and Carroll 2001), expert opinion (Andrews et al. 2002), pedotransfer functions, linear and multiple regression, decision trees (Moncada et al. 2014), and factor analysis (Shukla et al.

2006) are some of the methods used commonly for selection of indicators. Four steps are followed to quantify SQ in the form of soil quality index (SQI):

(a) Identification of production or sustainability goal
(b) Selection of SQ indicators
(c) Transformation of soil properties
(d) Computation of soil quality index

#### 2.3.2.1 Selection of Minimum Data Set (MDS)

Principal component analysis (PCA) is a commonly used and widely accepted method for selection of MDS. The primary use of PCA is to reduce the dimension of data without losing the message conveyed by the data (Armenise et al. 2013). Components with high "eigenvalues" explain the variability among the data set, and varimax rotation is used to maximize the variability (Andrews et al. 2002; Waswa et al. 2013). Principal components (PCs) with eigenvalues ≥1 (Kaiser 1960) are retained for identifying highly weighed soil properties. In each PC, variables with high weight are selected as indicators. If the selected soil properties are correlated ($r > 0.70$), then the property with the highest weight is retained as an indicator (Andrews and Carroll 2001).

In recent times, decision trees (DT) models are frequently used in the selection of MDS. The DTs help to quantify the data which has both ordinal and categorical soil properties (Moncada et al. 2014). Generally, classification trees and model trees are employed to identify the relations of many soil properties with soil quality. The classification trees predict the values of individual property with a set of nominal values, whereas model trees use linear functions (Debeljak and Dzeroski 2011). The trees are built as per the splitting rule, which performs the splitting of a learning sample into smaller parts. Tree-based models are fitted by successively splitting a data set into increasingly homogeneous subsets. The attributes are selected and rules are generated to relate the explanatory soil properties with soil quality class (Debeljak and Dzeroski 2011).

#### 2.3.2.2 Indexing Soil Quality

Soil quality can be evaluated using appropriate indicators which represent soil functions linked to the defined management goal. However, a quantitative value is necessary to make comparisons to management systems. The SQI is calculated by aggregating the indicators after converting the indicators into dimensionless units. The absolute values are used to derive SQI by various methods, viz., additive (Andrews and Carroll 2001), weighted, and hierarchical decision support system (Andrews et al. 2002). These methods of SQI calculation are indirect approaches and widely accepted because of their advantages in identifying the systematic complexity of soil under natural or managed ecosystems. Opportunities to compare different index methods are rare since it is not common to have more than one SQI for a particular area (Qi et al. 2009).

Selected SQ indicators in MDS have transformed scored into values ranging from 0 to 1 using linear and/or nonlinear scoring methods (Liebig et al. 2001). In the nonlinear scoring method, the indicators are grouped into (i) more is better; (ii) less is better, and (iii) optimum range and standard nonlinear curves are established. Indicators are arranged in increasing or decreasing order based on whether a higher value is "good" or "bad" with respect to soil function. For "higher is better" category, each value of the selected property is divided by the highest value so as the highest value is scored as 1. For "less is better" category, the lowest value is divided by each data value so as the lowest value is scored as 1. For indicators such as pH, "optimum" threshold function is used. They are scored as "higher is better" up to a threshold value (e.g., pH 7.5) then scored as "lower is better" above the threshold (Andrews et al. 2002).

Two methods are used to calculate SQI, viz., additive and weighted index methods. In additive method, the index is calculated by adding the transformed scores of the indicators from MDS. In weighted method, index is calculated by the following procedure: the transformed indicator data is the assigned weightage based on the variability explained by the PCs. The fraction of variability accounted by each PC as a part of total variability is used as weight factor for indicators selected from the respective PCs (Ray et al. 2014). The transformed scores are multiplied by the weight factors and then added to derive SQI. In expert opinion method, the weight factor is determined by the relative importance of selected indicators in influencing changes in soil function.

## 2.4 Soil Quality in Intensive Agriculture

In general, it is inferred that high soil quality corresponds to high productivity without environmental concerns which are not always the case. Apart from SQ, crop yield is influenced by many external factors such as solar radiation, temperature, evapotranspiration, and precipitation. Hence, it is necessary to establish a correlation between soil quality and defined management goal (Karlen et al. 2003) in intensively cultivated areas to have a complete understanding of the factors determining the management goal such as sustainability and productivity.

Agricultural intensification involves the adoption of high-yielding varieties with high cropping intensity and employing management practices to sustain productivity without causing environmental pollution. Continuous intensive cultivation improves soil fertility; however, in the recent decades, inappropriate management practices caused adverse consequences for the overall environment. Application of high amount of chemical fertilizers, especially P, and reduction in the use of organic manures leads to eutrophication at the surrounding water bodies. Excessive application of nitrogen decreases C fixation capacity due to C release from the soil. Carbon release and N leaching contribute to the greenhouse effect and threaten safety of ground and surface water (Qi et al. 2009). Unbalanced use of N and P fertilizers decreases the yield under intensive cultivation of rice-wheat-jute, soybean-wheat, and sorghum-wheat systems by unfavorably modifying soil properties, whereas the application of organic manure and integrated nutrient management sustained the soil quality and positively influenced the crop yield (Manna et al. 2005).

In intensive crop production systems, relating one particular soil property to crop yield is irrational because soil is a complex system. Here, SQI can be successfully used to interpret crop performance. For example, de Paul Obade and Lal (2016) evaluated crop performance using SQI developed by partial least square regression method by correlating with crop yields. Mueller et al. (2012) used M-SQR to successfully relate with crop yield in a wheat-maize system. Moreover, SQI approach is advantageous to evaluate sustainability of crop production in different agroecological regions as region-specific SQI can be developed using suitable SQ indicators (Sinha et al. 2014; Vasu et al. 2016).

The intensively cultivated lands are susceptible to degradation due to various reasons including inappropriate crop and soil management. It is important to understand the degradation patterns through appropriate indicators to design and implement site-specific management practices. Being site-specific in nature, SQI is an effective method to assess and map potential land degradation threats. For example, Waswa et al. (2013) established SOC, pH, CEC, clay content, and available P as best SQ indicators to identify soil degradation in western districts of Kenya. Sewage water irrigation is a common practice in some intensively cultivated areas of India and it deteriorates soil physical properties. A study by Masto et al. (2008) identified CEC, ESP, available P, and dehydrogenase activity as indicators of soil degradation by sewage water irrigation in Nalgonda district. Sharma et al. (2005) used available N, P, S, microbial biomass carbon, and hydraulic conductivity as key indicators to evaluate the effect of long-term soil management on crop yields and soil quality in dryland Alfisols. The SQI method is also employed to differentiate production potential of agricultural soil and generate an integrated index of soil at national level and used for agrarian planning (Vilček and Koco 2018).

In intensive cultivation, crop rotation influences soil quality through inputs related to the crop species included in the rotation. The quantity and substrate quality differ among the crop residues added to the soil in the crop rotation. This is commonly evaluated by the C and N cycles. Therefore, SQ parameters related to C and N mineralization can be used to quantify the effect of agricultural intensification on soil quality. For example, Zuber et al. (2017) compared the effect of maize-soybean rotation under contrasting tillage management in the intensively cultivated Illinois soil and showed that management significantly alters soil quality.

In countries like China, conversion of natural forest into plantations for timber production is a common practice (Selvalakshmi et al. 2017). Similarly, in Northeast India, shifting cultivation is a practice of cutting and burning of the natural forest trees and using the land for growing annual crops for a period before allowing the natural vegetation to regenerate (Mishra et al. 2017). These changes in land use affect soil quality either positively or negatively. Yu et al. (2018) showed that conversion of grasslands under alkaline soil to cropland degraded the soil quality by using N:P ratio, invertase enzyme, water-extractable organic C, and labile C as SQ indicators. Similarly, continuous cultivation of Chinese fir led to degradation of soil quality because of nutrient depletion by intensive monoculture (Zhijun et al. 2018;

Selvalakshmi et al. 2018). Also, many studies showed that the shifting cultivation practice decreased soil quality due to loss of organic matter, topsoil erosion, and poor vegetation cover (Mishra et al. 2017).

## 2.5 Management Effects on Soil Quality

Crop management practices have the potential to either enhance or degrade soil quality based on the intensity, suitability, and resilience of soil under use. Tillage, water and fertilizer management, cropping systems, and land-use conversion are some of the management practices which have pronounced impact on soil quality. Moreover, the effect of a set of management practices on soil quality is influenced by the inherent properties like texture, depth, CEC, clay mineralogy, and water holding capacity. In this section, the effect of tillage, cropping systems, and nutrient management are discussed.

### 2.5.1 Effect of Tillage

Tillage is one of the main land management practices integral to crop production. Tillage is generally carried out to create favorable soil environment, i.e., increase soil aeration and infiltration rate, seedbed preparation, soil and moisture conservation, expose the soil-borne pathogens and insects to light, and weed control. Tillage also has negative effect on soil properties such as subsoil compaction, erosion, enhanced mineralization, and decomposition of soil organic matter, etc. Tillage operations using double disk, chisel, and moldboard reduce the topsoil thickness by an average of 10 cm (Mendoza et al. 2008). It influences most of the soil quality indicators such as organic carbon, bulk density, porosity, hydraulic conductivity, water holding capacity, microbial community, and earthworms, either positively or negatively. To minimize the negative effects of tillage on soil quality, modern tillage concepts such as no tillage (NT), minimum tillage (MT), stubble-mulch tillage (SMT), and conservation tillage (CT) practices are introduced.

Generally, tillage accelerates the loss of soil organic carbon by increasing biological oxidation and soil erosion. Hence, tillage-based farming systems in drylands and arid ecosystems are not sustainable (Fageria et al. 2002). The NT or CT methods have now become popular across the dryland ecosystems of the world and implemented in various crop production systems. They considerably improve many soil properties which in turn sustains the soil quality and crop productivity (Table 2.5). Specifically, in soil characterized with low organic matter, poor structure, and impaired soil physical properties, the NT system has greater positive effects such as soil protection from erosion, soil water conservation by decreased evaporation, and increase in organic matter by reduced mineralization. The NT system increases soil organic carbon by lowering the temperature at the soil surface, increasing soil water content, and lack of residue incorporation and mixing in the soil. It preserves soil biological component by the means of little damage to soil

**Table 2.5** Influence of different tillage methods on some soil quality indicators

| Soil quality indicator | No tillage/ conservation tillage | Conventional tillage | Deep ploughing | References |
|---|---|---|---|---|
| Organic carbon | Increase in surface soil | Reduced | Reduced | Motta et al. (2002), Shukla et al. (2006), Mendoza et al. (2008), Fortuna et al. (2008) and Schmidt et al. (2018) |
| Bulk density | Decrease | No effect | No effect/ increase | Shukla et al. (2006) and Idowu and Kircher (2016) |
| Soil compaction | Reduce | Increase | Increase | Idowu and Kircher (2016) and Schmidt et al. (2018) |
| pH | No effect | Slight decrease | Slight decrease | Motta et al. (2002) |
| Total N | Increase | decrease | Decrease | Fortuna et al. (2008) |
| Aggregate stability | High | Low | Low | Shukla et al. (2006), Mendoza et al. (2008) and Idowu and Kircher (2016) |

aggregates or structure and increase in soil organic matter content. Long-term SMT and NT with good residue management improve the physical, chemical, and biological properties of soil (Seybold et al. 2002). However, care should be employed before implementing NT system that any plow pan present in the subsoil layer due to CT systems needs to be disrupted to make the NT effective (Schmidt et al. 2018).

The CT systems increase the yield (8–11%) in cereal-pulse-based systems than conventional tillage systems. Zero tillage or CT systems combined with crop residue incorporation increases the possibility of sustaining crop production especially under rainfed conditions in semiarid regions (Sharma et al. 2005). However, primary tillage is essential for continuous successful cultivation in semiarid tropical soil such as Alfisols to break the hard layer formed in the dry season so that root penetration and infiltration are not hindered. Soil quality parameters considerably improved under minimum tillage under rainfed conditions in an Inceptisol under pearl millet-based cropping system (Sharma et al. 2014). In humid and subhumid climatic conditions, claypan soil with poor drainage are prone to erosion due to restricted compacted subsurface layer. The NT and CT systems with maize-soybean rotation have the potential to increase SOC, total N, and aggregate stability by >30, 35, and 40%, respectively (Jung et al. 2008). Though NT and CT systems lead to temporary immobilization of N, it increases the N availability during the later stages and following crops. Legumes as cover crops and rotations with cereals increase SOC stocks in the soil (Fortuna et al. 2008).

### 2.5.2 Effect of Cropping Systems

Cropping system refers to temporal and spatial arrangements of crops and management of soil, water, and vegetation in order to optimize the biomass/agronomic production per unit area, per unit time, and per unit input (Lal 2003). Globally, cereal-based cropping systems are dominant and they cover 61% of the cultivated land. Wheat, maize, rice, barley, and millets occupy more than two-thirds of the cropland. The Caribbean and Central Africa are exceptions where sugarcane and cassava dominates, respectively (Leff et al. 2004). The major area with wheat-based systems is under temperate soil except in the Indo-Gangetic Plains, where tropical alluvial soils are common. Maize and other millets are prevalent in tropical and subtropical parts of the world. Next to cereals, oilseeds and pulses cover 14% of the total cultivated land. In general, agroforestry systems and tree plantations sustain soil quality than annual crop-based systems. The cereal cropping systems with pulses improve the soil quality than cereals without pulses (Wienhold et al. 2006). Sugarcane is one of the major sugar crops cultivated under irrigated conditions. Though management varies with respect to different locations, the sugarcane trash incorporation with CT system considerably improves aggregate stability, SOC content, and soil permeability thus improving overall soil quality (Cairo-Cairo et al. 2017).

In rice-based cropping systems, ploughing (puddling) leads to the breakdown of capillary pores, reduced void ratio, poor soil aggregates, dispersed fine clay particles and low soil strength, surface crust formation, and cracks after drying which ultimately degrades the physical quality of the soil (Masto et al. 2008). In the commonly followed rice-wheat system in the IGP of India, nitrogen losses due to leaching and denitrification are common due to alternate wetting and drying cycles. The productivity of the rice-wheat system decreased over the years and became unsustainable (Chaudhury et al. 2005). Also, in the rainfed cultivation, rice-based systems lead to decrease in soil quality than fruit trees and vegetable-based systems (Mandal et al. 2011). However, cereal-based systems have high potential to maintain SOC stock in soil because of the quantity and quality of their residues that are returned to the soil (Mandal et al. 2007). Legume crop inclusion and stubble incorporation in a cereal-based cropping system improve soil quality by increasing SOC and soil physical properties in the long term. Results from Armenise et al. (2013) indicate that wheat-bean double cropping system with stubble incorporation improved soil quality over continuous wheat monocropping with stubble burning. Soybean-based crop rotation increased SOC (29%), MBC (27%), mean weight diameter (9%), and dehydrogenase activity (5%) in Vertisols of Central India and produced high SQI (Sharma et al. 2016).

Pulse crop-based cropping systems and cover crops protect the soil from erosion and nutrient loss (Weerasekara et al. 2017). The added advantages of growing cover crops in crop rotation are suppression of weeds, carbon sequestration, soil moisture conservation, and reduced nonpoint source pollution. Legume cover crops increase soil N by fixing atmospheric nitrogen and thus reduce the quantity of external N fertilizer requirement. Agroforestry systems with perennial trees as primary components interact with soil components in long term. They modify soil properties through biomass addition and root and microbe's interaction mechanisms (Udawatta et al. 2008). Alley crops are the annual crops grown in rows between trees in the agroforestry system, and these crops enhance carbon sequestration (Udawatta et al. 2014).

### 2.5.3 Nutrient Management and Soil Quality

Concerns are growing over the long-term sustainability of agriculture because both over- and underapplication of fertilizer nutrients have damaged the environment (Armenise et al. 2013). In developed countries, overapplication of inorganic fertilizers resulted in degradation of soil and contamination of water sources. In developing countries, vagaries of climate change, population pressure, land constraints, and the decline of traditional soil management practices have often reduced soil fertility (Bai et al. 2013). Additionally, soil erosion, nutrient imbalance, and low SOM are the major soil fertility-related issues resulting in degradation of soil quality. Future increase in crop yields in developing countries will have to come from crop intensification for which integrated approach toward the soil nutrients management is the potential solution advocated by FAO, and it has been used successfully under various crop production systems for the past two decades (Bouma et al. 2017). Integrated nutrient management (INM) is composed of fertilizers, organic manures, legumes, crop residues, and biofertilizers as main components. INM enhances the availability of applied as well as native soil nutrients (Lal 2003). It synchronizes the nutrient demand of the crop with nutrient supply from native and applied sources.

The INM provides balanced nutrition to crops and minimizes the antagonistic effects resulting from hidden deficiencies and nutrient imbalance (Mandal et al. 2007). Long-term INM practice sustained the yield of maize-wheat system and substantially improved the soil aggregation stability and physical quality (Dutta et al. 2015). The application of recommended dose of N, P, and K with farmyard manure increased dehydrogenase activity and microbial biomass carbon and improved the soil quality under rice-wheat-jute system in the IGP region of India (Table 2.6) (Chaudhury et al. 2005). The NPK with FYM at 15 t $ha^{-1}$ sustained rice and wheat yield and aggraded the soil quality by increasing SOC and Zn availability in Mollisols (Ram et al. 2016).

The effect of INM is more pronounced in the rainfed production conditions than irrigated conditions. Long-term INM practice increased the aggregation stability, labile carbon, and dehydrogenase activity in the Inceptisols under pearl millet cropping system in tropical soil of India (Sharma et al. 2014). Application of organic manures such as cattle dung manure, vermicompost, and poultry manure along with mineral fertilizers based on N equivalents and nutrient requirements of crops improved the SOC content, available nutrients, MBC, and dehydrogenase and alkaline phosphatase in the topsoil under soybean-wheat cropping system in Vertisols (Ramesh et al. 2009). Farm compost application increases SOC content, earthworm population, and MBC and reduces BD in soil with light texture, thus creating favorable environment for crop growth and increasing the yield of maize-based cropping system (D'Hose et al. 2012).

**Table 2.6** Average yield (t ha$^{-1}$) and sustainable yield index (SYI) of jute, rice, and wheat from a 30-year long-term fertilizer experiment in a sandy loam alluvial soil

| | Jute | | Rice | | Wheat | | |
|---|---|---|---|---|---|---|---|
| Treatment | Yield | SYI | Yield | SYI | Yield | SYI | Overall SYI |
| 100% N | 1.627 | 0.479 | 3.094 | 0.379 | 1.9 | 0.4 | 0.419 |
| 100% NP | 1.721 | 0.528 | 3.501 | 0.433 | 2.175 | 0.478 | 0.480 |
| 100% NPK | 1.957 | 0.641 | 3.797 | 0.439 | 2.257 | 0.529 | 0.536 |
| 100% NPK + FYM | 2.063 | 0.662 | 3.856 | 0.538 | 2.34 | 0.537 | 0.579 |
| Control | 0.874 | 0.255 | 1.564 | 0.363 | 0.736 | 0.312 | 0.310 |

Adapted from Chaudhury et al. (2005)

## 2.6 Conclusion and Future Perspectives

Changes in land use are rapid in the last century and more large-scale changes are expected in the future due to competing demands from various sectors of the society. In agriculture, conversion from conventional to organic farming is promoted by policy decisions, and shift from intensive tillage to conservation tillage methods is encouraged to prevent soil erosion and minimize threats like eutrophication. Soil quality has now become an integral part of sustainable agriculture by means of an effective tool in monitoring both short- and long-term changes in soil caused by management practices. Many physical and chemical indicators are used to evaluate soil quality, but the biological indicators are only recently given importance. The selection of biological indicators for SQ evaluation is not an easy task because of the lack of direct link with the soil functions. Moreover, determining the active part of the population of organisms in soil is difficult. However, it is now established that direct linkages such as microbial richness affect C and N cycles, and decrease in the diversity of soil microbes reduces C sequestration and N turnover. Instead of using total SOC, its labile fraction can be used for detecting short-term changes in SQ. The recent developments in molecular biology made the rapid assessment of soil biological properties easier and the potential for many biological properties for inclusion in SQ assessment is increased. The inclusion of site-specific appropriate biological indicators in SQ assessment will make SQI a more comprehensive and stronger tool for evaluating various production systems.

Moreover, factors such as climate, site characteristics, crop yield, and plant nutrient content are seldom linked with soil quality. This indicates that the SQ evaluation is not linked with soil threats and other ecosystem services. The expert knowledge with respect to the study area might add value to SQ assessment, and hence consideration of expert opinion in indicator selection could improve the validity and applicability of SQI in the context of a defined management goal. For rapid assessment of the chemical and biological indicators, remote sensing and spectroscopic techniques may be employed. Recent developments in the visual evaluation techniques can be explored for SQ assessment in the field, and it could be advantageous as land owners and farmers can participate in visual assessment of soil quality.

## References

Acton DF, Gregorich LJ (1995) Understanding soil health. In: Acton DF, Gregorich LJ (eds) The health of our soil: towards sustainable agriculture in Canada. Centre for Land and Biological Resources Research Branch, Agriculture and Agri-Food Canada, Ottawa. 138pp

Adeyolanu OD, Ogunkunle AO (2016) Comparison of qualitative and quantitative approaches to soil quality assessment for agricultural purposes in South-western Nigeria. Cogent Food Agric 2:1149914

Allison SD, Martiny JBH (2008) Resistance resilience and redundancy in microbial communities. Proc Natl Acad Sci U S A 105:11512–11519

Andrews SS, Carroll CR (2001) Designing a soil quality assessment tool for sustainable agroecosystem management. Ecol Appl 11:1573–1585

Andrews SS, Karlen DL, Mitchell JP (2002) A comparison of soil quality indexing methods for vegetable production systems in northern California. Agric Ecosyst Environ 90:25–45

Andrews SS, Karlen DL, Cambardella CA (2004) The soil management assessment framework. Soil Sci Soc Am J 68:1945–1962

Aparicio V, Costa JL (2007) Soil quality indicators under continuous cropping systems in the Argentinean Pampas. Soil Tillage Res 96:155–165

Armenise E, Redmile-Gordon MA, Stellacci AM, Ciccarese A, Rubino P (2013) Developing a soil quality index to compare soil fitness for agricultural use under different managements in the Mediterranean environment. Soil Tillage Res 130:91–98

Arshad MA, Coen GM (1992) Characterization of soil quality: Physical and chemical criteria. Am J Altern Agric 7(1–2):25–31

Bai Z, Dent D, Wu Y, de Jong R (2013) Land degradation and ecosystem services. In: Rattan et al (eds) Ecosystem services and C sequestration in the biosphere. Springer, Dordrecht/Heidelberg/New York/London. https://doi.org/10.1007/978-94-007-6455-2_15

Ball BC, Munkholm LJ, Batey T (2013) Applications of visual soil evaluation. Soil Tillage Res 127:1–2

Biswas S, Hazra GC, Purakayastha TJ, Saha N, Mitran T, Roy SS, Basak N, Mandal B (2017) Establishment of critical limits of indicators and indices of soil quality in rice-rice cropping systems under different soil orders. Geoderma 292:34–48

Bouma J, van Ittersum MK, Stoorvogel JJ, Batjes NH, Droogers P, Pulleman MM (2017) Soil capability: exploring the functional potentials of soil. In: Field D et al (eds) Global soil security. Springer, Cham, pp 27–44

Bünemann EK, Bongiorno G, Bai Z, Creamer RE et al (2018) Soil quality – a critical review. Soil Biol Biochem 120:105–125

Cairo-Cairo P, Diaz-Martin B, Rodriguez-Urrutia A (2017) Soil quality indicators in vertisols under sugarcane. Arch Agron Soil Sci 63:1477–1488

Chaudhury J, Mandal UK, Sharma KL, Ghosh H, Mandal B (2005) Assessing soil quality under long-term rice-based cropping system. Commun Soil Sci Plant Anal 36:1141–1161

Chaves HML, Lozada CMC, Gaspar RO (2017) Soil quality index of an oxisol under different land uses in the Brazilian savannah. Geoderma Reg 10:183–190

Comino F, Aranda V, García-Ruiz R, Ayora-Cañada MJ, Domínguez-Vidal A (2018) Infrared spectroscopy as a tool for the assessment of soil biological quality in agricultural soil under contrasting management practices. Ecol Indic 87:117–126

D'Hose T, Cougnon M, Vliegher AD, Bockstaele EV, Reheul D (2012) Influence of farm compost on soil quality and crop yields. Arch Agron Soil Sci 58(sup1):S71–S75

de la Rosa D (2005) Soil quality evaluation and monitoring based on land evaluation. Land Degrad Dev 16:551–559

de Paul Obade V, Lal R (2016) Towards a standard technique for soil quality assessment. Geoderma 265:96–102

Debeljak M, Dzeroski S (2011) Decision trees in ecological modelling. In: Jopp F, Reuter H, Breckling B (eds) Modelling complex ecological dynamics. Springer, Berlin/Heidelberg

Dexter AR (2004) Soil physical quality Part I. Theory effects of soil texture density and organic matter and effects on root growth. Geoderma 120:201–214
Dexter AR, Czyz EA (2007) Applications of S-theory in the study of soil physical degradation and its consequences. Land Degrad Dev 18:369–381
Doran JW, Parkin TB (1996) Quantitative indicators of soil quality: a minimum data set. In: Doran JW, Jones AJ (eds) Methods for assessing soil quality SSSA special publication no 49. SSSA, Madison, pp 25–37
Dumanski J, Pieri C (2000) Land quality indicators (LQI): research plan. Agric Ecosyst Environ 81:93–102
Dutta J, Sharma SP, Sanjay K, Sharma GD, Sharma N, Sankhyan K (2015) Indexing soil quality under long-term maize-wheat cropping system in an acidic alfisol. Commun Soil Sci Plant Anal 46:1841–1862
Emmet-Booth JP, Forristal PD, Fenton O, Ball BC, Holden NM (2016) A review of visual soil evaluation techniques for soil structure. Soil Use Manag 32:623–634
European Commission (2006) Thematic strategy for soil protection commission of the European communities, p 12
Fageria NK, Baligar VC, Clark RB (2002) Micronutrients in crop production. Adv Agron 77:185–268
Foley JN, Ramankutty KA, Brauman ES, Cassidy JS et al (2011) Solutions for a cultivated planet. Nature 478:337–342
Fortuna ARL, Blevins W, Frye W, Grove J, Cornelius P (2008) Sustaining soil quality with legumes in no-tillage systems. Commun Soil Sci Plant Anal 39:1680–1699
Gonzalez-Quiñones V, Murphy DV, Bowles RW, Mele PM (2015) A national soil quality monitoring framework. GRDC soil biology initiative II, final report. UWA000138, p 258
Guimaraes RML, Ball BC, Tormena CA (2011) Improvements in the visual evaluation of soil structure. Soil Use Manag 27:395–403
Hajek BF, Karlen DL, Lowery B, Power JF, Schumaker TE, Skidmore EL, Sajoka RE (1990) Erosion and soil properties. In: Larson WE, Foster GR, Allmaras RR, Smith CM (eds) Research Issues in Soil Erosion Productivity. University of Minnesota, Saint Paul, pp 23–40
Haynes RJ (2005) Labile organic matter fractions as central components of the quality of agricultural soil: an overview. Adv Agron 85:221–268
Herrick JE (2000) Soil quality: an indicator of sustainable land management? Appl Soil Ecol 15:75–83
Idowu OJ, Kircher P (2016) Soil quality of semi-arid conservation reserve program lands in eastern New Mexico. Arid Land Res Manag 30:153–165
Jung W, Newell R, Kenneth K, Sudduth A, Kremer RJ (2008) Contrasting grain crop and grassland management effects on soil quality properties for a north-central Missouri claypan soil landscape. Soil Sci Plant Nutr 54:960–971
Kaiser HF (1960) The application of electronic computers to factor analysis. Educ Psychol Meas 29:141–151
Karlen DL, Mausbach MJ, Doan JW, Cline RG, Harris RF, Schuman GE (1997) Soil quality: a concept definition and framework for evaluation. Soil Sci Soc Am J 61:4–10
Karlen DL, Andrews SS, Doran JW (2001) Soil quality: current concepts and applications. Adv Agron 74:1–40
Karlen DL, Andrews SS, Doran JW, Wienhold BJ (2003) Soil quality – humankind's foundation for survival. J Soil Water Conserv 58:171–179
Lal R (1997) Degradation and resilience of soil. Philos Trans R Soc (Biology) London 352:997–1010
Lal R (2003) Cropping Systems and Soil Quality. J Crop Prod 8:33–52. https://doi.org/10.1300/J144v08n01_03
Lal R (2012) Restoring degraded lands and the flow of its provisioning services In: Proceedings of the 4th international conference on drylands deserts and desertification. Ben Gurion University of the Negev Sde-Boqer, p 65

Larson WE, Pierce FJ (1991) Conservation and enhancement of soil quality evaluation of sustainable land management in the developing world. International Board for Soil Research and Management, Bangkok

Leff B, Ramankutty N, Foley JA (2004) Geographic distribution of major crops across the world. Glob Biogeochem Cycles 18:GB1009. https://doi.org/10.1029/2003GB002108200

Liebig MA, Varvel G, Doran JW (2001) A simple performance based index for assessing multiple ecosystem functions. Agronomy and Horticulture Faculty Publications. Paper 336

Liu Z, Zhou W, Li S, He P, Liang G, Lv J, Jin H (2015) Assessing soil quality of gleyed paddy soil with different productivities in subtropical China. Catena 133:293–302

Loveland PJ, Thompson TRE (2002) Identification and development of a set of national indicators for soil quality, R&D Project, Record P5–053/PR/02, Environment Agency UK

Mandal M, Patra AK, Singh D, Masto RE (2007) Effect of long-term application of manure and fertilizer on biological and biochemical activities in soil during crop development stages. Bioresour Technol 98:3585–3592

Mandal UK, Ramachandran K, Sharma KL et al (2011) Assessing soil quality in a semiarid tropical watershed using a geographic information system. Soil Sci Soc Am J 75:1114–1160

Manna MC, Swarup A, Wanjari RH, Ravankar HN, Mishra B, Saha MN, Singh YV, Sahi DK, Sarap PA (2005) Long-term effect of fertilizer and manure application on soil organic carbon storage soil quality and yield sustainability under sub-humid and semi-arid tropical India. Field Crop Res 93:264–280

Masto RE, Chhonkar PK, Singh D, Patra AK (2008) Alternative soil quality indices for evaluating the effect of intensive cropping fertilization and manuring for 31 years in the semi-arid soil of India. Environ Monit Assess 136:419–435

Mendoza RB, Franti TG, Doran JW, Powers TO, Zanner CW (2008) Tillage effects on soil quality indicators and nematode abundance in loess soil under long-term no-till production. Commun Soil Sci Plant Anal 39:2169–2190

Menta C, Conti FD, Pinto S, Bodini A (2018) Soil biological quality index (QBS-ar): 15 years of application at global scale. Ecol Indic 85:773–780

Merril SD, Liebig MA, Tanaka DL, Krupinsky JM, Hanson JD (2013) Comparison of soil quality and productivity at two sites differing in profile structure and topsoil properties. Agric Ecosyst Environ 179:53–61

Merrington G (2006) The development and use of soil quality indicators for assessing the role of soil in environmental interactions. The Environment Agency Science Report SC030265

Mishra G, Marzaioli R, Giri K, Borah R, Dutta A, Jayaraj RSC (2017) Soil quality assessment under shifting cultivation and forests in north-eastern Himalaya of India. Arch Agron Soil Sci 63:1355–1368

Moncada MP, Gabriels D, Cornelis WM (2014) Data-driven analysis of soil quality indicators using limited data. Geoderma 235–236:271–278

More SD (2010) Soil quality indicators for sustainable crop productivity. J Indian Soc Soil Sci 58:5–11

Motta ACV, Reeves DW, Touchton JT (2002) Tillage intensity effects on chemical indicators of soil quality in two coastal plain soil. Commun Soil Sci Plant Anal 33(5–6):913–932

Mueller L, Schindler UT, Shepherd G, Ball BC et al (2012) A framework for assessing agricultural soil quality on a global scale. Arch Agron Soil Sci 58(sup1):S76–S82

Mukherjee A, Lal R (2014) Comparison of soil quality index using three methods. PLoS One 9:1–15

Nabiollahi K, Taghizadeh-Mehrjardi R, Eskandari S (2018) Assessing and monitoring the soil quality of forested and agricultural areas using soil-quality indices and digital soil-mapping in a semi-arid environment. Arch Agron Soil Sci 64:696–707

Norfleet ML, Ditzler CA, Puckett WE, Grossman RB, Shaw JN (2003) Soil quality and its relationship to pedology. Soil Sci 168(3):149–155

Nortcliff S (2002) Standardisation of soil quality attributes. Agric Ecosyst Environ 88:161–168

Oliver MZ, Gregory PJ (2015) Soil, food security and human health: a review. Eur J Soil Sci 66:257–276

Palm C, Sanchez P, Ahamed S, Awiti A (2007) Soil: a contemporary perspective. Annu Rev Environ Resour 32:99–129

Parisi V, Menta C, Gardi C, Jacomini C, Mozzanica E (2005) Microarthropod communities as a tool to assess soil quality and biodiversity: a new approach in Italy. Agric Ecosyst Environ 105:323–333

Paustian K, Lehmann J, Ogle S, Reay D, Robertson GP, Smith P (2016) Climate smart soil. Nature 532:49–57

Qi Y, Darilek JL, Huang B, Zhao Y, Sun W, Gu Z (2009) Evaluating soil quality indices in an agricultural region of Jiangsu Province China. Geoderma 149:325–334

Raiesi F, Kabiri V (2016) Identification of soil quality indicators for assessing the effect of different tillage practices through a soil quality index in a semi-arid environment. Ecol Indic 71:198–207

Ram S, Singh V, Sirari P (2016) Effects of 41 years of application of inorganic fertilizers and farm yard manure on crop yields soil quality and sustainable yield index under a rice-wheat cropping system on Mollisols of north India. Commun Soil Sci Plant Anal 47:179–193

Ramesh P, Panwar NR, Singh AB, Ramana S, Rao AS (2009) Impact of organic-manure combinations on the productivity and soil quality in different cropping systems in Central India. J Plant Nutr Soil Sci 172:577–585

Ray SK, Bhattacharyya T, Reddy KR, Pal DK, Chandran P, Tiwary P et al (2014) Soil and land quality indicators of the Indo-Gangetic Plains of India. Curr Sci 107:1470–1486

Schmidt ES, Villamil MB, Amiotti NM (2018) Soil quality under conservation practices on farm operations of the southern semiarid pampas region of Argentina. Soil Tillage Res 176:85–94

Schwilch G, Bernet L, Fleskens L et al (2016) Operationalizing ecosystem services for the mitigation of soil threats: a proposed framework. Ecol Indic 67:586–597

Selvalakshmi S, Vasu D, Zhijun H, Guo F, Ma XQ (2017) Influence of long-term successive rotations and stand age of Chinese fir plantations on soil properties. Geoderma 306:127–134

Selvalakshmi S, Vasu D, Zhijun H, Guo F, Ma XQ (2018) Soil nutrients dynamics in broadleaved forest and Chinese fir plantations in subtropical forests. J Trop For Sci 30:242–251

Seybold CA, Herrick JE, Brejda JJ (1999) Soil resilience: a fundamental component of soil quality. Soil Sci 164:224–234

Seybold CA, Hubbs MD, Tyler DD (2002) On-farm tests indicate effects of long-term tillage systems on soil quality. J Sustain Agric 19:61–73

Sharma KL, Mandal B (2009) Soil quality and its relevance for sustainable agriculture. J Indian Soc Soil Sci 57:572–586

Sharma KL, Mandal UK, Srinivas K et al (2005) Long-term soil management effects on crop yields and soil quality in a dryland alfisol. Soil Tillage Res 83:246–259

Sharma KL, Grace JK, Chandrika MS et al (2014) Effects of soil management practices on key soil quality indicators and indices in pearl millet (*Pennisetum americanum* (L) Leeke)–based system in hot semi-arid inceptisols. Commun Soil Sci Plant Anal 45:785–809

Sharma KL, Suma D, Chandrika J et al (2016) Soil quality assessment under restorative soil management practices in soybean (*Glycine max*) after six years in semi-arid tropical black lands of Central India. Commun Soil Sci Plant Anal 47:1465–1475

Shukla MK, Lal R, Ebinger M (2006) Determining soil quality indicators by factor analysis. Soil Tillage Res 87:194–204

Sinha NK, Chopra UK, Singh AK (2014) Cropping system effects on soil quality for three agro-ecosystems in India. Exp Agric 50:321–342

Sparling GP, Schipper LA, Bettjeman W, Hill R (2004) Soil quality monitoring in New Zealand: practical lessons from a 6-year trial. Agric Ecosyst Environ 104:523–534

Srinivasarao C, Lal R, Kundu S, Babu MBBP, Venkateswarlu B, Singh AK (2014) Soil carbon sequestration in rainfed production systems in the semiarid tropics of India. Sci Total Environ 487:587–603

SSSA (1987) Glossary of soil science terms. Soil Science Society of America, Madison

SSSA (1995) Statement on soil quality. Agronomy News June 1995. American Society of Agronomy/Crop Science Society of America/Soil Science Society of America, Madison

Stavi I, Lal R (2015) Achieving zero net land degradation: challenges and opportunities. J Arid Environ 112:44–51
Stenberg B, Pell M, Torstensson L (1998) Integrated evaluation of variation in biological chemical and physical soil properties. Ambio 27:9–15
Udawatta RP, Kremer RJ, Adamson BW, Anderson SH (2008) Variations in soil aggregate stability and enzyme activities in a temperate agroforestry practice. Appl Soil Ecol 39:153–160
Udawatta RP, Robert J, Kelly K, Nelson A, Jose S, Bardhan S (2014) Soil quality of a mature alley cropping agroforestry system in temperate north America. Commun Soil Sci Plant Anal 45:2539–2551
USDA-NRCS (1999) Soil quality test kit guide. Available at http://soils.usda.gov/sqi/assessment/files/test_kit_complete.pdf (verifi ed 30 Aug. 2010)
van Elsas JD, Jansson JK, Trevors JT (2007) Modern soil microbiology II. CRC Press, Boca Raton
Vasu D, Singh SK, Ray SK, Duraisami VP, Tiwary P, Chandran P, Nimkar AM, Anantwar SG (2016) Soil quality index as a tool to evaluate crop productivity in semi-arid Deccan plateau India. Geoderma 282:70–79
Vasu D, Tiwary P, Chandran P, Ray SK, Singh SK, Butte PS, Parhad V (2018) A conceptual model of natural land degradation based on regressive pedogenesis in semi-arid tropical environments. Land Degrad Dev 29(8):2254–2567
Vilček J, Koco S (2018) Integrated index of agricultural soil quality in Slovakia. J Maps 14:68–76
Warkentin BP, Fletcher HF (1977) Soil quality for intensive agriculture. In: Proceedings of international seminar on soil environ and fertilizer management in intensive agriculture, Society for Science of Soil and Manure – National Institute of Agricultural Science, Tokyo, pp 594–598
Waswa BS, Vlek PLG, Tamene LD, Okoth P, Mbakaya D, Aingore S (2013) Evaluating indicators of land degradation in smallholder farming systems of western Kenya. Geoderma 195–196:192–200
Wattel-Koekkoek EJW, van Vliet ME, Boumans LJM, Ferreira J, Spijker J, van Leeuwen TC (2012) Soil quality in the Netherlands for 2006–2010 and the change compared with 1993–1997: results of the National Soil Quality Monitoring Network RIVM report 680718003
Weerasekara CS, Udawatta RP, Gantzer CJ, Kremer RJ, Jose S, Veum KS (2017) Effects of cover crops on soil quality: selected chemical and biological parameters. Commun Soil Sci Plant Anal 48:2074–2082
Wienhold BJ, Pikul JL, Liebig MA, Mikha MM, Varvel GE, Doran JW, Andrews SS (2006) Cropping system effects on soil quality in the Great Plains: synthesis from a regional project. Renew Agric Food Syst 21:49–59
Yu P, Liu S, Zhang L, Li Q, Zhou D (2018) Selecting the minimum data set and quantitative soil quality indexing of alkaline soil under different land uses in north-eastern China. Sci Total Environ 616-617:564–571
Zalibekov ZG (2011) The arid regions of the world and their dynamics in conditions of modern climatic warming. Arid Ecosyst 1:1–7
Zhijun H, Selvalakhsmi S, Vasu D, Liu Q, Cheng H, Guo F, Ma X (2018) Identification of indicators for evaluating and monitoring the effects of Chinese fir monoculture plantations on soil quality. Ecol Indic 93:547–553
Zuber SM, Behnke GD, Nafziger ED, Villamil MB (2017) Multivariate assessment of soil quality indicators for crop rotation and tillage in Illinois. Soil Tillage Res 174:174–155

# Integrated Nutrient Management for Sustainable Crop Production and Improving Soil Health

3

Rajinder Singh Antil and Dev Raj

**Abstract**

Integrated nutrient management (INM) is a concept, which aims at the maintenance of soil health and plant nutrient availability in optimum amounts for sustaining soil health and crop productivity through optimization of the benefits from all possible sources of plant nutrients. INM could play an important role in increasing nutrient use efficiency (NUE), food grain production, and maintenance of soil health and increasing the farmer's income through integrated and balanced application of fertilizers. Cropping system is one of the important ingredients of sustainable agriculture system as it provides more efficient cycling of nutrients. Therefore, balanced fertilization must be based on the concept of INM for a cropping system rather than a crop, so that crop productivity of the system as a whole is sustained. Long-term studies conducted in different agroclimatic zones have established the benefits of INM. This chapter overviews the importance of different components of INM in improving NUE, crop productivity, and soil health.

**Keywords**

Balanced fertilization · Long-term experiments · Organic manures · Crop productivity · Soil health

R. S. Antil (✉)
Amity Food and Agricultural Foundation, Amity University, Noida, Uttar Pradesh, India

D. Raj
Department of Soil Science, CCS Haryana Agricultural University, Hisar, Haryana, India





R. S. Meena (ed.), *Nutrient Dynamics for Sustainable Crop Production*,
https://doi.org/10.1007/978-981-13-8660-2_3

## Abbreviations

| | |
|---|---|
| INM | Integrated nutrient management |
| Mt | Million tonne |
| NUE | Nutrient use efficiency |
| N | Nitrogen |
| P | Phosphorous |
| K | Potash |
| S | Sulfur |
| Zn | Zinc |
| B | Boron |
| Fe | Iron |
| Mn | Manganese |
| Mo | Molybdenum |
| HYV | High-yielding varieties |
| IPNS | Integrated Plant Nutrient Supply |
| INSAM | Integrated Nutrient Supply and Management |
| AICRP-IFS | All India Coordinated Research Project on Integrated Farming Systems |
| FYM | Farmyard manure |
| LTEs | Long-term experiments |
| BD | Bulk density |
| HC | Hydraulic conductivity |
| WSA | Water-stable aggregates |
| MWD | Mean weight diameter |
| IR | Infiltration rate |
| OC | Organic C |
| DOC | Dissolved organic carbon |
| MBC | Microbial biomass carbon |
| LFC | Light carbon fraction |
| HFC | Heavy Carbon Fraction |
| RDF | Recommended dose of fertilizer |

## 3.1 Introduction

Indian agriculture is no longer an unknown one. It has progressed rapidly in recent years and ranks now as the second largest food producer in the world, touching $367 billion in 2014. The country's agricultural production is more than that of the United States, which once supplied food grains to India to meet the domestic food shortage. Unknown to many, India's international trade in agricultural products fetches higher earnings for the country than trade in the services or manufacturing. Food grain production of India has increased from 50.8 (1950–1951) to 284.83 Mt (2017–2018). A fivefold increase in food grain production during the last 67 years combined with inadequate and imbalanced use of nutrients has led to extra mining of all the essential nutrients. Extra mining of nutrients will have to be checked in order to sustain the soil health. The maintenance of soil health is very important to ensure the food and nutritional security of the country. For efficient use of fertilizers, all nutrients must be applied in balanced proportions. The nitrogen (N)/phosphorus (P)/potash (K) consumption ratio (2016–2017) of India was 7.2:2.9:1 against the ideal ratio of 4:2:1. The distortion in NPK consumption ratio is more pronounced among the zones and states of India. The problem of imbalanced fertilizer use in case of secondary and micronutrients is even worse wherein the use is much less compared to the requirement of the crops. So, there is a need to narrow down the NPK consumption ratio to sustain the crop productivity and restore the soil health. *Continuous use of imbalanced fertilizers led to the deterioration in the soil health and stagnate the crop productivity* (Das et al. 2015; Buragohain et al. 2017).

Even though during the 1960s, India has become self-sufficient on the food front as against its large imports, but our soil has been extensively overexploited. If such a situation is continued for a longer time, then there are chances that our productive land may become unproductive. The green revolution technologies, viz., higher uses of chemical fertilizers and pesticides with the adoption of nutrient-responsive and high-yielding varieties (HYV) of crops, have increased the productivity of almost all the crops. However, during the last decades, the compounded growth rates for the production and productivity of major crops generally declined or stagnated compared to the 1980s (Table 3.1). The crop responses to fertilizers are also decreasing consistently (Table 3.2).

There are reports that farmers have to add higher quantities of fertilizers every year to obtain the same yield level as obtained in the previous year. It may be due to the decline in the soil organic matter content, imbalanced use of fertilizers, extra mining of nutrients, and deficiency of secondary and micronutrients. The use of organic manures along with chemical fertilizers may be an effective alternative approach for further improving crop yields and sustaining soil health (Walia et al. 2017; Meena and Yadav 2015).

**Table 3.1** Compound growth rates (% per annum) of production and productivity of crops

| Crop | Production | | | Productivity | | |
|---|---|---|---|---|---|---|
| | 1980–1981 to 1989–1990 | 1990–1991 to 1999–2000 | 2001 to 2009–2010 | 1980–1981 to 1989–1990 | 1990–1991 to 1999–2000 | 2001 to 2009–2010 |
| Rice (*Oryza sativa* L.) | 3.62 | 2.02 | 1.59 | 3.19 | 1.34 | 1.61 |
| Wheat (*Triticum aestivum* L.) | 3.57 | 3.57 | 1.89 | 3.00 | 1.83 | 0.68 |
| Pulses | 1.52 | 0.59 | 2.61 | 1.61 | 0.93 | 1.64 |
| Food grains | 2.85 | 2.02 | 1.96 | 2.74 | 1.52 | 2.94 |
| All major crops | 3.19 | 2.99 | 1.83 | 2.56 | 1.38 | 2.83 |

Source: Kumara et al. (2013)

**Table 3.2** Decline in crop response to fertilizer

| Period | kg food grain per kg nutrients (NPK) |
|---|---|
| 5th Plan (1974–1979) | 1:15 |
| 8th Plan (1992–1997) | 1:7.5 |
| 9th Plan (1997–2002) | 1:7 |
| 10th Plan (2002–2007) | 1:6.5 |
| 11th Plan (2007–2012) | 1:6 |

Source: FAI, Fertilizer Statistics (1974–1975, 1992–1993, 1997–1998, 2002–2003, 2007–2008); Kumara et al. (2013)

INM is not a new concept. It is an age-old practice when requirements of all the nutrients (primary, secondary, and micronutrients) were met through organic sources. In literature, a few terminologies, viz., Integrated Plant Nutrient Supply (IPNS) and Integrated Nutrient Supply and Management (INSAM), are also used to convey almost similar meaning as that of INM. The advantages of INM are:

- Sustain and improve crop productivity and soil health
- Prevent deficiencies of secondary and micronutrients
- Improve nutrient use efficiency
- Provide favorable effect on the physical, chemical, and biological properties of soils (Singh et al. 2012b)

The application of best nutrient management practices in diverse ecologies and production systems is thus critical to enhance food production and improve farm profitability and resource efficiency. Also, the dimensions of the challenges would require a holistic alliance of policy-makers, agricultural scientists, extension specialists, and the farmers to facilitate INM toward improving the soil health.

## 3.2 INM Definition/Concept

INM has been defined by different researchers as follows:

- INM is defined as the maintenance or adjustment of soil fertility and supply of plant nutrients to an optimum level for sustaining the desired crop productivity through optimization of benefit from all possible resources of plant nutrients in an integrated manner (Roy and Ange 1991).
- INM is used to maintain or adjust soil fertility and plant nutrient supply to achieve a given level of crop production. This is done by optimizing the benefits from all possible sources of plant nutrients (FAO 1998).
- INM is actually the technical and managerial component of achieving the objective of IPNS under farm situations. It takes into account all factors of soil and crop management including management of all other inputs such as water, agrochemicals, amendments, etc., besides nutrients (Goswami 1998; Meena and Meena 2017).

**Table 3.3** Extent of macronutrient deficiency in India

| | | % are of samples by category | | |
|---|---|---|---|---|
| Nutrient | No. of samples analyzed | Low | Medium | High |
| N | 3,650,004 | 63 | 26 | 11 |
| P | 3,650,004 | 42 | 38 | 20 |
| K | 3,650,004 | 13 | 37 | 50 |
| S | 27,000 | 40 | 35 | 25 |

Source: Motsara (2002)

## 3.3 Fertility Status of Soils

The inadequate and imbalanced fertilizer use has caused widespread nutrient (N, P, K, sulfur (S), zinc (Zn), and boron (B)) deficiencies and deterioration in soil health in many parts of India. It has been estimated that in India, 63, 42, 13, and 40% of soils were deficient in N, P, K, and S, respectively (Table 3.3). On an average, 49% of soils have been found deficient in Zn, 15% in iron (Fe), 3% in copper (Cu), 5% in manganese (Mn), 33% in B, and 13% in molybdenum (Mo) (Singh 2001).

## 3.4 Nutrient Removal and Balance in Soils

At the present level of production, the estimated NPK removal was about 28 Mt which results in a net negative balance of about 10 Mt. Organic manures and biofertilizers contribute to about 4 Mt, which means about 6 Mt negative balance has to be replenished by the soil (Antil and Narwal 2007; Yadav et al. 2017c). Recently, the Government of India has declared the target of doubling food grain production by 2025. It implies that for doubling the productivity, the nutrient removal would be more than double the present level to about 56 Mt. The gap between nutrient supply through all sources and removal would further escalate to more than 12 Mt from the present level of about 6 Mt, provided the contribution of organic and biofertilizer sources is also doubled. Thus, the soil health would further aggravate, which needs urgent attention. Although it is not possible to replenish 100% of nutrients removed by the crops every year, even then an attempt should be made to maximize the recycling of those nutrients which are likely to be deficient in the future. Thus, to meet this negative balance and to sustain the crop productivity and soil health on a long-term basis are possible only through the INM.

## 3.5 Nutrient Potentials of Organic Resources

India has a vast resource of organic input, and it is very difficult to assess its exact estimate, especially when production of residues, dung, etc. fluctuates every year. Further, the nutrient availability depends on the quality of the substrate technology used and value addition if any. The total available nutrient value of organic resources

**Table 3.4** Available organic nutrients in India

| Component | Potential availability (Mt) | Actual availability (Mt) | Nutrient value (Mt) |
|---|---|---|---|
| Crop residue | 603.46 | 201.11 | 4.865 |
| Animal dung | 791.66 | 287.45 | 3.474 |
| Green manure | 4.46 m ha | NA | 0.173 |
| Rural compost | 184.30 | 184.30 | 2.580 |
| City compost | 12.20 | 12.20 | 0.427 |
| Biofertilizer | 0.01 | 0.0094 | 0.370 |
| Others | 96.60 | NA | 0907 |
| Total | | | 12.796 |

Source: Bhattacharya (2007)

in India is 12.796 Mt, and the tappable amount is 8.952 Mt (after 30% deduction). The present utilization of organic nutrient resources has been estimated as 3.75 Mt (Bhattacharya 2007) of plant nutrient that can be made available for agricultural use (Table 3.4). Thus, about 25% of NPK requirement of Indian agriculture could be met by properly utilizing various organic resources (cattle dung, farmyard manure (FYM), crop residues, urban/rural wastes, and green manuring) which are readily available for agricultural use. Hence, there is an urgent need to refine the technologies available on the utilization of organic resources.

## 3.6 Components of INM

### 3.6.1 Balanced Fertilization

Balanced fertilization means rational use of fertilization and organic manures in such a manner that would ensure increased crop yields, improve quality of crops and cost/benefit ration, and have least adverse effect on the environment. Balanced fertilization must be based on the concept of INM for a cropping system (Goswami 1997) as this is the only viable strategy advocating accelerated and enhanced use of fertilizer with matching adoption of organic manures and fertilizers so that productivity is maintained for a sustainable agriculture. A balanced fertilization could be achieved through the application of multinutrients in balanced proportion from fertilizers, organic sources, biological sources, and more accurately and precisely through INM on a cropping system basis. Fertilizers continued to be the most important ingredient of INM. The dependence on fertilizers has been increasing constantly because of the need to supply large amounts of nutrients in intensive cropping with high productivity. Nonetheless, fertilizer consumption is not only inadequate but also imbalanced. At present, the consumption ratio of NPK (2016–2017) was about 7:3:1 against the ideal ratio of 4:2:1. The NPK use ratio is quite wide, whereas the application of K, S, and micronutrients is often ignored.

Utilization of fertilizer nutrients by the crops varies from 30 to 50% in the case of N, 15–20% in the case of P, and less than 5% in the case of micronutrients. Thus, a substantial amount of applied nutrients is lost through various pathways. Enhancing nutrient use efficiency should, therefore, be a prioritized area of research for the restoration and improvement of soil health and minimizing the cost of crop production. To increase the fertilizer use efficiency, crop yields and checks further mining of those nutrients, which are likely to be deficient in the future, the balanced amount of fertilizers based on soil testing should be applied. Long-term fertilizer experiment studies spread over a period of 30 years indicated that the application of balanced fertilizer and INM was the best tool for obtaining sustainability in crop yields of soybean and wheat (Tiwari 2008; Yadav et al. 2018a). Adoption of INM for a cropping system is the only viable strategy for accelerated and enhanced use of fertilizers with matching adoption of organic manures and biofertilizers, so that productivity is maintained for sustainable agriculture. Low or imbalanced fertilizer application is one of the important reasons for the low productivity (Singh et al. 2006; Meena and Lal 2018). Balanced fertilization along with organic manures and biofertilizers would be helpful to sustain crop yields and maintain soil fertility.

### 3.6.2 Organic Resources (Organic Manures, Crop Residues, Composts, Animal Dung, etc.)

Uses of chemical fertilizers alone deteriorate soil fertility and create unfavorable soil physical, chemical, and biological conditions in the intensive cropping system. It can be overcome by use of organics along with fertilizers for health and sustaining crop production. Organic manures have been the time-tested materials for improving the fertility and productivity of soils. Organic manures not only supply macro- and micronutrients but also help improving the physical, chemical, and biological conditions of the soils, and ultimately NUE would be improved. These manures, besides supplying nutrients to the first crop, also leave substantial residual effect on the succeeding crops in the system. Use of organic manures has been continuously declining in Indian agriculture.

#### 3.6.2.1 Organic Manures (Long-Term Experiment (LTE) Results)

The findings of LTEs carried out under AICRP-IFS (AICRP-CS Annual Reports 1992–1993 to 2001–2002) showed that a part of fertilizer N requirements of monsoon crop can be met by adding FYM with the annual production either at par with fertilizer application alone at recommended levels or slightly higher with the INM package (Table 3.5). It was further noticed at few locations that the fertilizer requirements of the winter wheat could be reduced to about 25% by substituting 25% N needs of the preceding monsoon crop through FYM. Eight years of study on INM in rice-wheat systems at Jabalpur (Vertisols) revealed that conjunctive use of 5 t FYM and 6 t green manure (*Parthenium*) with 90 kg N $ha^{-1}$ not only sustained the productivity but also saved nearly 90–100 kg fertilizer N $ha^{-1}$ $year^{-1}$. In addition to saving N, INM practices also improved the soil organic carbon and nutrient (available P and K) status of the soil (Table 3.6).

**Table 3.5** Effect of integrated nutrient supply through fertilizers and FYM on the productivity of crops under AICRP-CS

| Treatment | | Grain yield (t/ha) | | |
|---|---|---|---|---|
| Monsoon | Winter | Monsoon | Winter | Total |
| Parbhani (sorghum (*Sorghum bicolor*)-wheat) av. of 7 years | | | | |
| 100% NPK | 100% NPK | 2.97 | 2.64 | 5.62 |
| 50% NPK+ 50% N (FYM) | 100% NPK | 2.85 | 2.78 | 6.53 |
| Hanumangarh (pearl millet (*Pennisetum glaucum* L.)-wheat) av. of 5 years | | | | |
| 100% NPK | 100% NPK | 2.72 | 3.61 | 6.33 |
| 50% NPK+ 50% N (FYM) | 100% NPK | 2.56 | 2.40 | 6.38 |
| Ranchi (maize (*Zea mays* L.)-wheat) av. of 8 years | | | | |
| 100% NPK | 100% NPK | 2.92 | 2.71 | 5.63 |
| 75% NPK+ 25% N (FYM) | 100% NPK | 3.30 | 2.40 | 5.70 |
| Varansi (rice-wheat) av. of 5 years | | | | |
| 100% NPK | 100% NPK | 4.33 | 3.67 | 8.00 |
| 50% NPK+ 50% N (FYM) | 100% NPK | 4.71 | 4.02 | 8.72 |

Source: AICRP-IFS Reports (2005–2010); Das et al. (2014)

**Table 3.6** Average productivity (t ha$^{-1}$) of rice-wheat system and nutrient content under 7 years of INM in a vertisol (Jabalpur)

| | | | Soil nutrient content after 7 years | | |
|---|---|---|---|---|---|
| Treatment | Rice (t ha$^{-1}$) | Wheat (t ha$^{-1}$) | Org. C (%) | P (kg ha$^{-1}$) | K (kg ha$^{-1}$) |
| N90 | 4.42 | 4.19 | 0.58 | 21.1 | 138 |
| N180 | 5.08 | 4.70 | 0/71 | 18.7 | 125 |
| N90 + FYM | 4.95 | 4.49 | 0.74 | 40.1 | 230 |
| N90 + GM | 4.58 | 5.07 | 0.72 | 39.1 | 240 |
| Initial | | | 0.60 | 19.5 | 195 |

Source: Singh and Wanjari (2007); Singh et al. (2001); Singh et al. (2002)

Similar kind of advantageous effect of conjoint use of fertilizers and FYM was recorded in soybean/maize-wheat (Table 3.7) and other cropping systems (Singh et al. 2012a, b; Sudhir et al. 2004; Ashoka et al. 2017). Integration of NPK with FYM further increased the yield of both soybean and wheat. At Ranchi, use of lime as an ingredient of INM also significantly improved the productivity of the system.

The LTEs at Pantnagar (Mollisols), Barrackpore (Inceptisols), and Raipur (Vertisols) revealed that incorporation of FYM along with NPK gave the highest production of rice-wheat system (Singh et al. 2012b; Kumar et al. 2017b). Similar trends in the yield were also noted at the other locations. Combined use of fertilizers and organic manure (FYM) increased the productivity of the system with a significant residual effect on the subsequent wheat crop.

An LTE was initiated in winter 1967 to study the savings in fertilizer N at various doses of FYM and their modes of application in pearl millet-wheat cropping system. Results indicated the superiority of the combined use of FYM and N fertilizer in increasing the yield of both pearl millet and wheat crops compared to the application of fertilizer alone. Yield of both the crops responded linearly up to 120 kg N ha$^{-1}$

**Table 3.7** Effect of INM on the productivity (t/ha) of soybean/maize-wheat system under LTFE at different locations

| | Ranchi[a] | | Jabalpur[b] | | Palampur[c] | |
|---|---|---|---|---|---|---|
| Treatments | 1972 to 2009 | | 1972 to 2009 | | 1999 to 2009 | |
| Unfertilized | 0.61 | 0.69 | 0.81 | 1.24 | 0.29 | 0.38 |
| 100% N | 0.29 | 0.39 | 1.02 | 1.67 | 0.42 | 0.37[d] |
| 100% NP | 0.87 | 2.45 | 1.65 | 4.07 | 2.00 | 1.64 |
| 100% NPK | 1.50 | 2.80 | 1.82 | 4.42 | 3.24 | 2.29 |
| NPK + FYM | 1.87 | 3.33 | 2.00 | 4.85 | 4.66 | 3.10 |
| NPK + lime | 1.80 | 3.17 | – | – | 4.11 | 2.85 |
| CD (P = 0.05) | 0.21 | 0.39 | 0.26 | 0.44 | 0.71 | 0.50 |

Source: [a]Mahapatra et al. (2007); [b]Dwivedi et al. (2007); [c]Sharma et al. (2005)
[d]At present yields are zero

at all FYM doses ranging from 0 to 90 t $ha^{-1}$ $year^{-1}$ (Table 3.8). It may be due to the increased demand of N or by increased losses of N with the addition of FYM. Application of FYM to the winter (rabi) season crop has been found to be better as compared to the summer (kharif) season (Antil et al. 2011; Meena et al. 2016a). It might be due to higher losses of nutrients from the summer-applied manure owing to higher temperature.

To study the contribution of FYM on the grain yield of pearl millet and wheat crops, a linear regression was fitted between the intercept of the linear model and the soil organic C content. It has been observed that each unit (0.1%) increase in the soil C content increased the productivity of pearl millet by 272.6 kg $ha^{-1}$ and that of wheat by 1590.9 kg $ha^{-1}$ (Fig. 3.1). Thus, the demand for N by the crops has increased which affected the response. The unit response to fertilizer N (slope of the linear model) was also plotted against the soil C content. It was observed that in the case of pearl millet, the slope also increased linearly with increasing soil C content, but in the case of wheat, there was no specific trend (Fig. 3.1). Better $R^2$ values in wheat crop were due to the better season for its growth rather than pearl millet, which used to be influenced by rainfall. This experiment was started with a view to save fertilizer N as a consequence of FYM application. But the yield data indicated that by fixing the productivity of the cropping system, we can save fertilizer N, but due to the economic returns from the fertilizer N application, it is not worthwhile to reduce the N supply. However, we can save all other nutrients except N by applying FYM (Antil et al. 2011; Meena and Yadav 2014).

Keeping the results of the above LTE initiated in winter 1967, another LTE was initiated in 1995 to evaluate the impact of continuous application of fertilizers and organic manures (FYM, poultry manure, and press mud) in pearl millet-wheat cropping system. The lowest grain yield of pearl millet and wheat crops was recorded when either 15 t FYM or 7.5 t press mud or 5 t poultry manure $ha^{-1}$ was applied alone (Table 3.9). However, a significant increase in yield was obtained when organic manures were applied in combination with the recommended dose of N, which was comparable with the recommended dose of applied NP alone or NP applied in combination with organic manures, indicating that the amount of N

**Table 3.8** Grain yield (q $ha^{-1}$) of pearl millet and wheat as influenced by the modes and levels of FYM and fertilizer N (averaged from 1982–1983 to 2007–2008)

| | | Pearl millet | | | | Wheat | | | |
|---|---|---|---|---|---|---|---|---|---|
| | | Level of N (kg $ha^{-1}$) | | | | Level of N (kg $ha^{-1}$) | | | |
| FYM mode | FYM level (Mg $ha^{-1}$) | 0 | 60 | 120 | Mean | 0 | 60 | 120 | Mean |
| | 0 | 12.63 | 18.81 | 21.66 | 17.70 | 25.11 | 34.73 | 41.61 | 33.81 |
| Kharif | 15 | 15.54 | 22.23 | 24.10 | 20.62 | 33.53 | 40.25 | 45.46 | 39.75 |
| | 30 | 17.33 | 21.90 | 24.09 | 21.11 | 36.08 | 42.39 | 48.21 | 42.22 |
| | 45 | 17.88 | 22.73 | 26.50 | 22.37 | 38.47 | 43.78 | 50.19 | 44.15 |
| Rabi | 15 | 14.52 | 20.32 | 22.61 | 19.15 | 36.08 | 44.02 | 48.28 | 42.80 |
| | 30 | 16.12 | 22.69 | 23.95 | 20.92 | 39.94 | 44.70 | 51.06 | 45.23 |
| | 45 | 17.05 | 21.61 | 23.79 | 20.82 | 41.15 | 46.11 | 53.06 | 46.77 |
| Both | 15 | 17.35 | 21.65 | 24.77 | 21.26 | 40.26 | 46.08 | 52.11 | 46.15 |
| | 30 | 17.14 | 23.30 | 26.27 | 22.23 | 41.65 | 47.92 | 53.81 | 47.79 |
| | 45 | 19.55 | 24.24 | 27.18 | 23.66 | 43.91 | 48.71 | 55.12 | 49.25 |
| Mean | | 16.51 | 21.95 | 24.49 | | 37.62 | 43.87 | 49.89 | |
| LSD (0.05) | FYM mode: 3.12; N level: 2.53 | | | | | FYM mode: 2.98; N level: 2.43 | | | |

Source: Antil et al. (2011)

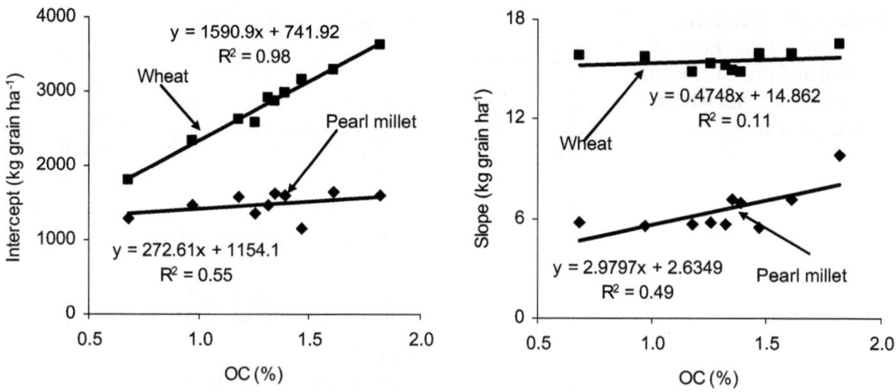


**Fig. 3.1** Influence of organic C on the intercept (left) and slope (right) of linear model in pearl millet and wheat after 30 cycles of pearl millet-wheat cropping system (Gupta et al. 2003)

**Table 3.9** Grain yield of pearl millet and wheat under different combinations of organic manures and fertilizers (average from 2000–01 to 2013–14)

| Organic manures | | Fertilizer (kg ha$^{-1}$) | | Yield (q ha$^{-1}$) | |
|---|---|---|---|---|---|
| Type of manure | Dose (t ha$^{-1}$) | N | $P_2O_5$ | Pearl millet | Wheat |
| No manure | 0 | 75 | 30 | 18.7 | 33.6 |
| | 0 | 150 | 60 | 22.8 | 47.2 |
| FYM | 15 | 0 | 0 | 15.6 | 23.9 |
| | 15 | 150 | 0 | 23.8 | 50.0 |
| | 15 | 150 | 30 | 25.6 | 54.2 |
| Poultry manure | 5 | 0 | 0 | 16.0 | 25.4 |
| | 5 | 150 | 30 | 24.0 | 48.0 |
| Press mud | 7.5 | 0 | 0 | 16.5 | 24.9 |
| | 7.5 | 75 | 30 | 23.2 | 42.2 |
| | 7.5 | 150 | 30 | 25.9 | 51.4 |
| LSD (P = 0.05) | | | | 2.12 | 2.61 |

Source: Kumara et al. (2013)

released by organic manures was not good enough to meet the remaining N requirement of the crop (Kumara et al. 2013; Datta et al. 2017b). Organic manures applied to the wheat crop also had a subsequent effect on the pearl millet crop.

The results of 3 years of field demonstrations at a farmer's field indicated that the application of 15 t FYM ha$^{-1}$ year$^{-1}$ in conjunction with RDF increased the productivity of different cropping systems (pearl millet-wheat, wheat-cotton, rice-wheat) compared to the application of RDF alone (Antil and Narwal 2007; Meena et al. 2015d). In addition to yield gains, integrated use of FYM and fertilizers improved the fertility status of the soil.

Singh et al. (2012a) evaluated the effect of balanced and imbalanced application of plant nutrients made in the rice-wheat system on crop productivity and soil

**Table 3.10** Treatment combinations and rates of nutrient and organic manures applied to rice and wheat crop each year

| Treatment no. | Treatment applied to Rice | Wheat |
|---|---|---|
| $T_1$ | Control | Control |
| $T_2$ | $N_{150}$ | $N_{150}$ |
| $T_3$ | $N_{150}$ $P_{75}$ | $N_{150}$ $P_{75}$ |
| $T_4$ | $N_{150}$ $P_{75}K_{75}$ | $N_{150}$ $P_{75}K_{75}$ |
| $T_5$ | $N*_{150}$ $P*_{75}K*_{75}Zn*_{25}$ | $N_{150}$ $P_{75}K_{75}$ |
| $T_6$ | $N_{150}$ $P_{75}K_{75}Zn_{25}$ + 15 t FYM $ha^{-1}$ | $N_{150}$ $P_{75}K_{75}$ |
| $T_7$ | $N_{150}$ + 7.5 t press mud $ha^{-1}$ | $N_{150}$ $P_{75}K_{75}$ |
| $T_8$ | $N_{75}$ $P_{37.5}K_{37.5}Zn_{25}$ + 20 t green manure $ha^{-1}$ | $N_{150}$ $P_{75}K_{75}$ |
| $T_9$ | $N_{150}$ $P_{75}K_{75}Zn_{25}$ + 7.5 t burnt rice husk $ha^{-1}$ | $N_{150}$ $P_{75}K_{75}$ |

*N, P, K, and Zn stand for N, $P_2O_5$, $K_2O$, and $ZnSO_4$, respectively, and applied in kg $ha^{-1}$ fertilizers. On the other hand, plots receiving combined application of inorganic fertilizers and organic manures received substantial amounts of micronutrients through manures
Source: Singh et al. (2012a)

**Table 3.11** Average yield and yield trends of rice in long-term experiment (1997–2009) as affected by different nutrient management practices

| Treatment | Average yield[a] t $ha^{-1}$ | Yield change, slope t $ha^{-1}$ | *t* statistics | *P* value |
|---|---|---|---|---|
| $T_1$ | 3.35 | −0.055 | −3.460 | 0.005 |
| $T_2$ | 6.54 | −0.114 | −8.357 | 0.004 |
| $T_3$ | 6.76 | −0.132 | −10.905 | 0.003 |
| $T_4$ | 6.82 | −0.134 | −11.413 | 0.002 |
| $T_5$ | 8.17 | 0.053 | 3.123 | 0.009 |
| $T_6$ | 8.92 | 0.054 | 1.439 | 0.045 |
| $T_7$ | 8.24 | 0.074 | 2.640 | 0.022 |
| $T_8$ | 8.22 | 0.048 | 3.312 | 0.002 |
| $T_9$ | 8.30 | 0.037 | 1.559 | 0.046 |

Source: Singh et al. (2012a)
[a]Average of 13 years of cropping

health. The treatments included various combinations of plant nutrients and are shown in Table 3.10. The trend in rice grain yield over a 13-year period varied markedly depending upon the nutrient management practices (Table 3.11). The rice grain yield decreased significantly in all the treatments where an imbalanced application of fertilizer nutrient was made. The rate of yield decline was lowest in control (0.055 t $ha^{-1}$ $year^{-1}$) and highest in $T_4$ (0.134 t $ha^{-1}$ $year^{-1}$) treatment. The grain yield of rice in $T_1$, $T_2$, $T_3$, and $T_4$ treatments decreased linearly with time, and this declining trend was significant (Table 3.11). The relationship between the initial yield and the yield decline in selected treatments was found significant with the $R^2$ values of 0.958 (Fig. 3.2). The decline in the rice grain yield in the present study can

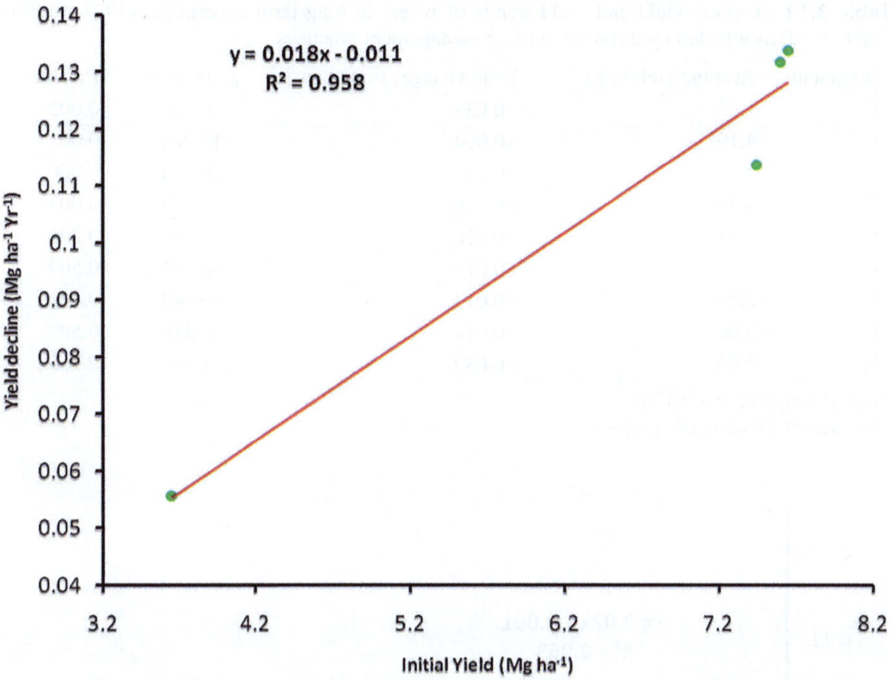


**Fig. 3.2** Relationship between first year (1997) rice yield and rice yield decline over 13-year period under selected treatments (Singh et al. 2012a)

be attributed to the gradual decline in soil organic carbon and decreased availability of micronutrients, particularly that of Zn. The rice grain yield tended to increase with time in all the treatments where a balanced dose of chemical fertilizers alone ($T_5$) or their combined use with organic manures was made ($T_6$, $T_7$, $T_8$, and $T_9$). However, the magnitude of increase was more in FYM and press mud-amended treatments.

Thirteen years of continuous cropping without the application of adequate quantity of nutrients or their imbalanced application ($T_1$, $T_2$, $T_3$, and $T_4$) resulted in a significant decrease in grain yield of wheat with time (Singh et al. 2012a; Meena et al. 2015e; Kumar et al. 2018b) and ranged from 0.038 t ha$^{-1}$ year$^{-1}$ in $T_1$ to 0.116 t ha$^{-1}$ year$^{-1}$ in $T_4$ treatment (Table 3.12). The decline in wheat yield was significantly correlated ($R^2 = 0.963$) with the initial yield (Fig. 3.3). The data on soil fertility parameters suggested that a gradual decline in soil organic matter and available Zn content of the soil were mainly responsible for the declining trend in rice and wheat yield. The wheat grain yield remained almost stable in treatments which received balanced application of nutrients ($T_5$) or their application with organic manure in preceding rice crop (FYM, press mud, and green manure) indicating a positive effect on the succeeding wheat crop.

**Table 3.12** Average yield and yield trends of wheat in long-term experiment (1997–1998 to 2009–2010) as affected by different nutrient management practices

| Treatment | Average yield[a] t $ha^{-1}$ | Yield change, slope t $ha^{-1}$ | *t* statistics | *P* value |
|---|---|---|---|---|
| $T_1$ | 1.65 | −0.038 | −6.995 | 0.002 |
| $T_2$ | 4.19 | −0.094 | −11.765 | 0.001 |
| $T_3$ | 4.26 | −0.105 | −10.124 | 0.006 |
| $T_4$ | 4.33 | −0.116 | −9.822 | 0.008 |
| $T_5$ | 5.33 | −0.026 | −1.108 | 0.291 |
| $T_6$ | 5.80 | −0.017 | −0.585 | 0.569 |
| $T_7$ | 5.56 | −0.013 | −0.587 | 0.469 |
| $T_8$ | 5.60 | −0.014 | −0.556 | 0.588 |
| $T_9$ | 5.53 | −0.032 | −1.169 | 0.266 |

Source: Singh et al. (2012a)
[a]Average of 13 years of cropping

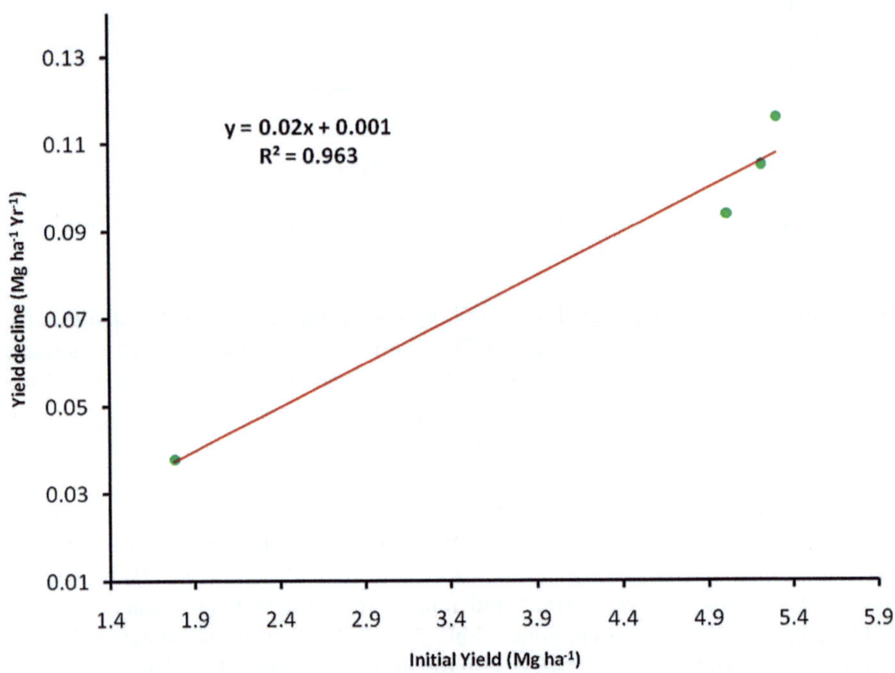


**Fig. 3.3** Relationship between first year (1997–1998) wheat yield and wheat yield decline over 13-year period under selected treatments (Singh et al. 2012a)

#### 3.6.2.2 Crop Residues

Crop residues are good sources of plant nutrients and are important components of INM. Crop residues, besides supplying nutrients to the current crop, leave sustainable residual effect on succeeding crop in the system. Recycling of crop residues is a viable strategy to meet at least a part of the nutrient requirement of different crops under various cropping systems. Total crop residue available in India is 603.46 Mt, out of which 201.11 Mt is available as nutrient for recycling with 4.865 Mt nutrient value (Bhattacharya 2007). About 30% N, 60–70% P, and 75% K-contained crop residues are available to the first crop and the rest to the subsequent crop. Crop residue addition improves the soil organic matter content, nutrient use efficiency, soil physical properties (structure and moisture retention), and microbial and enzymatic activity (Antil and Narwal 2007; Yadav et al. 2017b). Residue management under INM has considerable effect on soil microbial biomass C, which was enhanced if the residue is incorporated with the use of green manure (Jaipaul and Negi 2006; Meena et al. 2018a).

Disposal of rice straw in Trans- and Upper Gangetic Plains has emerged as a great problem. In these combined-harvested areas, farmers opt to burn the residues in situ, losing precious nutrients on one hand and polluting environment on the other. Recycling these residues back to fields helps to build stable organic matter in the soil and also to sustain crop yield levels. Stubbles left in the field even in traditional harvesting methods range from 0.5 to 1.5 t $ha^{-1}$ in case of different crops. When mechanical harvesting is done, this amount is much greater. Stubbles of coarse cereals such as sorghum, maize, pearl millet, etc., which are difficult to decompose, are normally collected and burnt during land preparation causing significant loss of plant nutrients. A 7-year study (Yadvinder-Singh et al. 2004) demonstrated that rice and wheat productivity was not adversely affected when rice residue was incorporated at least 10 days, preferably 20 days, prior to the establishment of the succeeding crop. This study showed that rice residue decomposition of about 25% during the pre-wheat fallow period was sufficient to avoid any detrimental effects on wheat yields.

### 3.6.2.3 Composts

The average quantity of rural and urban compost in India is 184.3 and 12.2 Mt having nutrient values of 2.56 and 0.427 Mt, respectively (Bhattacharya 2007). The effects of enriched (consists of cow dung, crop residue, rock phosphate, pyrite, urea) phosphocompost and ordinary compost were compared with FYM and biogas slurry with and without fertilizer in groundnut-wheat cropping system. It was observed that the pod yield of groundnut increased by 52% due to the application of enriched compost when supplemented with 50% NPK over ordinary compost (Antil and Narwal 2007; Meena et al. 2015c; Sofi et al. 2018). The grain and straw yields of wheat were significantly higher with the agro-industrial waste composts (sewage sludge, distillery effluent, press mud, and poultry waste composts) as compared to their raw materials. Compost-fertilized wheat grain yields were increased by 118% with poultry waste compost followed by press mud compost and recommended dose of NPK fertilizer when compared with unfertilized control (Table 3.13). Agro-industrial waste composts applied with NK (recommended dose) fertilizers except distillery effluent compost produced wheat grain yield comparable to that obtained with NPK (recommended dose) fertilizers, indicating a net saving of 100% of P fertilizer. Hence, instead of using fertilizer alone, the integrated use of compost and

**Table 3.13** Effect of different composts, organic amendments, and chemical fertilizer on grain and straw yield of wheat

| Treatment | Yield (g $pot^{-1}$) | |
|---|---|---|
| | Grain | Straw |
| Unfertilized control | 3.63b* | 4.86b |
| Sewage sludge | 4.26c | 6.48c |
| Sewage sludge compost | 7.05g | 9.28f |
| Distillery effluent | 3.45b | 4.53a |
| Distillery effluent compost | 5.70e | 6.81d |
| Press mud | 5.73e | 7.02d |
| Press mud compost | 7.21g | 10.05g |
| Poultry waste | 6.45f | 9.05f |
| Poultry waste compost | 7.92h | 10.86h |
| Chemical fertilizer (120 kg N + 60 kg $P_2O_5$ + 60 kg $K_2O$ $ha^{-1}$) | 7.14g | 10.24g |
| LSD (P = 0.05) | 0.32 | 0.30 |

Source: Antil et al. (2013)
*Different small letters within columns indicate significance at $P < 0.05$

fertilizer could be more effective and sustainable for wheat productivity (Antil et al. 2013; Dadhich and Meena 2014).

#### 3.6.2.4 Animal Dung

Total population of animal in India is 920.63 million with the dung production of 791.66 Mt having a nutrient (NPK) availability of 3.474 Mt. It has been estimated that about one third of the cattle dung produced is recycled in the fields, and the rest is burnt to meet the fuel demand which is a big loss. If this dung is properly managed, then the productivity of the soil can be increased. Therefore, farmers should be provided alternate sources of energy for cooking so that maximum dung could be used as manure. This problem can be solved if the dung is used in biogas plant. The major problem of direct application of biogas slurry is its transportation; however, the manurial value of biogas slurry is better than that of the compost. This problem can be managed by having two decomposing pits on both sides of the slurry outlet. The capacity of the pits should be sufficient to accommodate 6- month slurry, and all the wastes of the farm and household should be added it to be recycled. In this way, the farmers living in their farms can easily handle their dung properly.

### 3.6.3 Green Manuring

Green manures mobilize soil nutrient reserves, create conducive environment for soil microbes, and save on mineral nitrogen by fixing atmospheric N. Green manuring of dhaincha (*Sesbania aculeata*), mungbean (*Vigna radiata*), cowpea (*Vigna unguiculata*), and sun hemp (*Crotalaria juncea*) after harvesting of wheat in rice-growing areas of India saves 40–60 kg N $ha^{-1}$ and maintains soil fertility (Antil and

Narwal 2007; Verma et al. 2015c). Application of fertilizers ($N_{150}P_{75}K_{75}Zn_{25}$) along with 15 t FYM ha$^{-1}$ treatment produced higher yield than any other organic-amended treatment. Application of 50% of RDF ($N_{75}P_{37..5}K_{37..5}Zn_{25}$) with dhaincha green manuring produced rice grain yield comparable to that obtained with 100% of RDF ($N_{150}P_{75}K_{75}Zn_{25}$), indicating a saving of approximately 50% of fertilizers (Singh et al. 2012a; Dhakal et al. 2016; Kakraliya et al. 2018). The residual effect of FYM, press mud, green manuring, and burnt rice husk was also observed on the grain yield of succeeding wheat crop and resulted in an increase of 3.5, 2.8, 3.3, and 1.2 q wheat grain ha$^{-1}$, respectively.

### 3.6.4 Biofertilizers

Biofertilizers are cost-effective, eco-friendly, and renewable sources of plant nutrients to supplement chemical fertilizers in sustainable agricultural system. Biofertilizers have an important role in improving the nutrient supply and their availability for crop production. They help in increasing the biologically fixed atmospheric N and enhancing native P availability to crop. *Rhizobium* is the most well-known bacterial species that acts as the primary symbiotic fixer of N. *Rhizobium* is a potential biofertilizer for legumes, which saves about 25–50% of recommended dose of N and enriches soil with N for the succeeding crop. The free-living N-fixer, *Azotobacter*, imparts positive benefits to the crops through small increase in N input from BNF; development and branching of roots; production of plant growth hormones; enhancement in N, P, K, and Fe uptake; improved water status of the plants; increased nitrate-reductase activity; and production of antifungal compounds. Bacterial cultures of *Pseudomonas* and *Bacillus* species and fungal culture of *Aspergillus* species help to convert insoluble P into plant-usable forms and thus improve phosphate availability to the crops. Similarly, fungi like vesicular arbuscular mycorrhizae (VAM) increase nutrient uptake particularly that of P due to the increased contact of roots with a larger soil volume. Combined application of 20 t FYM ha$^{-1}$ + 100% RDF + *Azotobacter* spp. + *Pseudomonas striata* recorded significantly higher shelling (%), protein content, oil yield, and pod and haulm yield over control and application of only FYM (@ 10 t ha$^{-1}$) + *Azotobacter* spp. + *Pseudomonas striata* + 50% RDF (Ghosh et al. 2005; Meena et al. 2016b). The use of biofertilizers should be done along with fertilizers and organic manures in legume and nonlegume-based cropping systems in order to sustain the crop productivity.

### 3.6.5 Legumes in Rotation

Green manuring with leguminous crops is not only beneficial in enhancing the yield but also improving the fertility of soil. Incorporation and decomposition of legumes have a solubilizing effect of N, P, K, and micronutrients (Zn, Mn, Fe, and Cu) in the soil and mitigate the deficiency of different nutrient elements by way of recycling of nutrients, reducing the leaching and gaseous losses of N and increasing the efficiency of applied plant nutrients. Sun hemp (*Crotalaria juncea*) and dhaincha (*Sesbania aculeata*) are the most important common green manure crops. Legumes could prove an important ingredient of INM when grown for grain or fodder in a cropping system or when introduced for green manuring. Legumes grown as green manure, forage, or grain crops improved the productivity of the rice-wheat cropping system (RWCS) and rejuvenated soil fertility (Yadav et al. 2000; Verma et al. 2015b).

In the rice-wheat cropping sequence, incorporation of mungbean residues after picking the pods significantly increased the yield over fallow treatment. The legume should be introduced in cereal-based crop rotations; it would increase the yield and nutrient use efficiency in succeeding crops following legume and also reduce the mining of N from soils. Studies on INM in rice-groundnut system in acidic soils revealed that use of green manure along with blue green algae gave similar yield as was obtained with 60 kg $ha^{-1}$ fertilizer N and maintained higher available nutrient throughout the year (Ghosh et al. 2005; Meena et al. 2014).

Results from AICRP-IFS indicated that about 25 to 50% N can be saved under rice-wheat, rice-rice, rice-maize, maize-wheat, pearl millet-wheat, and sorghum-wheat cropping systems by growing mungbean as a catch crop. In spite of this, it is very difficult to accommodate a green manure crop within intensive cropping systems, and farmers are not interested to grow green manure crop as there is no direct cash benefit. Under such situations, growing a mungbean crop in summer and incorporation of the aboveground green biomass after picking of pod may serve as green manuring (Dwivedi et al. 2002; Datta et al. 2017a; Layek et al. 2018).

## 3.7 Effect of INM on Soil Health

### 3.7.1 Soil Physical Properties

Among the different physical properties of soil, bulk density (BD) has been considered as an important parameter for the assessment of soil health, mainly due to its relationships with the other soil state (strength and porosity) and rate (moisture retention and flow characteristics) variables. Soil aggregation, a physical property related to soil structure, is greatly influenced with the addition of organic resources. Hati et al. (2006) reported that addition of NPK along with FYM significantly improved soil aggregation, soil water retention, microporosity, and available water capacity and reduced the BD of the soil at 0–30 cm depth (Hati et al. 2006). The study suggests that addition of balanced fertilizers along with organic manures sustains a better soil physical environment and higher crop productivity. Das et al. (2014) reported that application of NPK fertilizers along with FYM or green gram residue + FYM or cereal residue improved the soil aggregation and structural stability and resulted in a higher C content in macroaggregates under the rice-wheat cropping system. The hydraulic conductivity (HC), water-stable aggregates (WSA) > 0.25 mm, and mean weight diameter (MWD) increased significantly with the addition of FYM. However, the addition of FYM reduced the BD. The values of

HC, WSA > 0.25 mm, and MWD were significantly higher when FYM was applied in both (rabi and kharif) the seasons compared to that applied either in rabi or kharif season (Antil et al. 2011; Kumar et al. 2018a). HC; moisture retention at 0, 0.1, 0.3, and 1.0 bar; and infiltration rate (IR) significantly increased with the increasing levels of FYM after 23 cycles of pearl millet-wheat cropping system (Table 3.14). On the other hand, dispersion percentage and BD decreased significantly with the application of FYM. Organic C (OC) was positively correlated with IR ($r = 0.97$), HC ($r = 0.89$), and moisture retention ($r = 0.94$) and negatively with BD ($r = -0.93$) and dispersion percentage ($r = -0.75$).

No significant differences in soil pH and EC were observed under different treatments after completion of the fourth cropping cycles; however, BD of soil almost remained close to the initial value under 100% organic and INM (50% organic +50% inorganic) treatments (Dubey et al. 2014; Meena et al. 2015b), while an increase in BD was observed under 100% inorganic applied treatment (Table 3.15).

The soil strength and IR increased significantly by incorporating green manure, wheat cut straw, and FYM in the fertilization schedule (Table 3.16). However, a reduction in BD was observed (Walia et al. 2010). Combined use of NPK and FYM in soybean (*Glycine max*)-wheat system resulted in 5.6% lower BD than NPK alone after the fourth cropping cycle (Bandyopadhyay et al. 2010; Yadav et al. 2018b). Reductions in BD due to the application of cattle manure (Nyamangara et al. 2001), poultry manure (Tejada and Gonzales 2008), and FYM (Bandyopadhyay et al. 2011) in LTEs have been observed. These reductions in BD could likely be attributed to the higher organic matter built up in soil (Hati et al. 2006; Varma et al. 2017a), better aggregation and consequent increase in total porosity, decrease in the degree of compaction (Leroy et al. 2008; Meena et al. 2018c), and increased root growth (Bandyopadhyay et al. 2010; Ram and Meena 2014).

### 3.7.2 Soil Chemical Properties

Several studies have shown that the application of organic manure in conjunction with fertilizers increased the soil OC and available N contents and their fractions more effectively than the application of fertilizers alone (Gong et al., 2009; Bijay-Singh 2018a; Meena et al. 2015a). Continuous application of FYM for a period of 33 years in pearl millet-wheat cropping system increased OC and available P and K content of the soil. Initial level of P and K of soil can be maintained with the application of 8.2 and 2.4 t FYM $ha^{-1}$ $year^{-1}$, respectively (Antil et al. 2011; Meena et al. 2017c). Due to the linear response of N application up to 120 kg N $ha^{-1}$ in the plots receiving 90 t FYM $ha^{-1}$, the savings in fertilizer N is not possible. These results indicate that application of 15 t FYM $ha^{-1}$ $year^{-1}$ is sufficient to maintain the nutrient (except N) status of P and K of soil to its initial level. Hence, the application of P and K fertilizers can be avoided with the application of organic manures.

Application of NP fertilizers for 16 years in pearl millet-wheat cropping system decreased the OC content of soil from its initial level. However, the application of organic manures alone or in combination with NP fertilizers increased the OC

**Table 3.14** Effect of FYM levels on some physical properties of soil after 23 cycles of pearl millet-wheat cropping system

| FYM level (Mg $ha^{-1}$) | OC (%) | BD (Mg $m^{-3}$) | Dispersion percentage | HC (cm $h^{-1}$) | IR (cm $h^{-1}$) after 30 min | 105 min | Moisture retention (at different bars) 0 | 0.1 | 0.3 | 1.0 |
|---|---|---|---|---|---|---|---|---|---|---|
| 0 | 0.94 | 1.48 | 56.94 | 0.28 | 0.6 | 0.3 | 42.7 | 24.8 | 16.9 | 10.6 |
| 15 | 1.22 | 1.44 | 53.24 | 0.41 | 0.9 | 0.6 | 46.5 | 27.4 | 18.0 | 12.5 |
| 30 | 1.30 | 1.32 | 50.16 | 0.41 | 1.8 | 1.2 | 48.0 | 28.4 | 18.3 | 13.4 |
| 45 | 1.36 | 1.31 | 48.47 | 0.45 | 2.1 | 1.4 | 49.6 | 28.8 | 19.4 | 13.8 |
| LSD (0.05) | 0.09 | 0.08 | 1.13 | 0.18 | ND | ND | 0.5 | 0.4 | 0.3 | 0.2 |

Source: Antil et al. (2011)

**Table 3.15** Effect of different nutrient management treatments on the properties of the soil on completion of fourth cropping cycle

| Treatment | pH | EC (dS $m^{-1}$) | OC (g $kg^{-1}$) | BD (Mg $m^{-3}$) | Available N (kg $ha^{-1}$) | Available P (kg $ha^{-1}$) | Available K (kg $ha^{-1}$) |
|---|---|---|---|---|---|---|---|
| Initial | 7.4 | 0.51 | 7.0 | 1.35 | 264 | 12.6 | 282 |
| 100% organic | 7.2 | 0.49 | 7.8 | 1.36 | 288 | 13.0 | 297 |
| 100% inorganic | 7.2 | 0.51 | 7.1 | 1.40 | 271 | 12.4 | 271 |
| 50% organic +50% inorganic | 7.2 | 0.50 | 7.4 | 1.37 | 278 | 12.7 | 291 |

Source: Dubey et al. (2014)

**Table 3.16** Effect of chemical fertilizers and organic manures on the physical properties of soil measured in 2007 after 23 years of cropping

| Fertilizer rate (% of recommended NPK) | | | | |
|---|---|---|---|---|
| Rice | Wheat | Bulk density (Mg $m^{-3}$) | Soil strength (MPa) | Infiltration rate (cm $h^{-1}$) |
| Control | Control | 1.45 | 1.30 | 2.72 |
| 100 | 100 | 1.46 | 1.43 | 6.06 |
| 50 + 50% N (FYM) | 100 | 1.39 | 0.97 | 4.48 |
| 50 + 50% N (WCS) | 100 | 1.42 | 1.07 | 3.24 |
| 50 + 50% N (GM) | 100 | 1.41 | 1.03 | 4.24 |
| LSD (0.05) | | 0.02 | 0.17 | 0.114 |

Source: Walia et al. (2010)

significantly (Table 3.17). Application of organic manures with or without NP fertilizers could not sustain the initial level of N. However, soil fertility with respect to P, K, and micronutrients can be maintained with the application of organic manures (15 t FYM or 7.5 t press mud or 5 t poultry manure $ha^{-1}$ $year^{-1}$) in conjunction with the recommended dose of N under pearl millet-wheat cropping systems. Hence, the application of P, K, and micronutrient fertilizers can be avoided with the application of organic manures (Kumara et al. 2013; Dadhich et al. 2015; Mitran et al. 2018). These results confirm the findings of the long-term experiment initiated in 1967 that we can save 100% of P, K, and micronutrient fertilizers, but we cannot save N through the combined application of organic manures with NP fertilizers.

The soil OC stock and C sequestration rate were calculated after 16 cycles of pearl millet-wheat cropping sequence. The results indicated that the soil OC stock decreased with greater magnitude (26, 127.4 kg $ha^{-1}$) in 75 kg N + 30 kg $P_2O_5$ $ha^{-1}$ for 16 years (Table 3.18), whereas it was 71, 129.9 kg $ha^{-1}$ with 15 t FYM $ha^{-1}$.

**Table 3.17** Effect of long-term application of organic manures and fertilizers on organic C (%), available N, P, K, Zn, Mn, Fe, and Cu (mg kg$^{-1}$) status of soil after 16 cycles of pearl millet-wheat cropping sequence

| Type of manure | Dose (t ha$^{-1}$) | Fertilizer (kg ha$^{-1}$) N | Fertilizer (kg ha$^{-1}$) $P_2O_5$ | OC | N | P | K | Zn | Mn | Fe | Cu |
|---|---|---|---|---|---|---|---|---|---|---|---|
| No manure | | 75 | 30 | 0.36 | 64.5 | 8.9 | 173.4 | 0.72 | 3.30 | 5.02 | 0.48 |
| | | 150 | 60 | 0.43 | 67.5 | 9.6 | 178.4 | 0.76 | 3.55 | 5.03 | 0.50 |
| FYM | 15 | 0 | 0 | 1.01 | 82.6 | 26.0 | 294.4 | 1.84 | 4.29 | 7.29 | 0.78 |
| | 15 | 150 | 0 | 0.98 | 82.7 | 26.9 | 298.4 | 1.80 | 4.57 | 7.27 | 0.81 |
| | 15 | 150 | 30 | 1.12 | 84.6 | 28.0 | 302.4 | 1.79 | 4.59 | 7.27 | 0.76 |
| Poultry manure | 5 | 0 | 0 | 0.59 | 78.4 | 28.6 | 247.4 | 1.65 | 3.81 | 6.82 | 0.72 |
| | 5 | 75 | 30 | 0.58 | 80.9 | 29.0 | 253.4 | 1.72 | 3.92 | 6.80 | 0.69 |
| Press mud | 7.5 | 0 | 0 | 0.73 | 95.7 | 27.1 | 253.6 | 1.73 | 4.00 | 6.99 | 0.74 |
| | 7.5 | 75 | 30 | 0.76 | 95.7 | 27.2 | 261.4 | 1.79 | 4.01 | 7.10 | 0.77 |
| | 7.5 | 150 | 30 | 0.75 | 98.3 | 27.3 | 267.4 | 1.81 | 4.14 | 8.09 | 0.77 |
| Initial in 1995 | | | | 0.39 | 98 | 12.6 | 217 | – | – | – | – |
| LSD (P = 0.05) | | | | 0.05 | 2.2 | 0.2 | 2.6 | 0.01 | 0.064 | 0.043 | 0.020 |

Source: Kumara et al. (2013)

**Table 3.18** Long-term effects of organic manures and fertilizers on soil organic carbon stock (kg ha$^{-1}$) and carbon sequestration rate (kg ha$^{-1}$ year$^{-1}$) in surface soil (0–15 cm) after 16 cycles of pearl millet and wheat cropping sequence

| Organic manures | | Fertilizer (kg ha$^{-1}$) | | | |
|---|---|---|---|---|---|
| Type of manure | Dose (t ha$^{-1}$) | N | $P_2O_5$ | Soil OC stock | C sequestration rate |
| No manure | 0 | 75 | 30 | 668 | −18.8 |
| | 0 | 150 | 60 | 774 | 25.0 |
| FYM | 15 | 0 | 0 | 1818 | 387.5 |
| | 15 | 150 | 0 | 1764 | 368.8 |
| | 15 | 150 | 30 | 2016 | 456.3 |
| Poultry | 5 | 0 | 0 | 972 | 93.8 |
| Manure | 5 | 150 | 30 | 1044 | 118.8 |
| Press mud | 7.5 | 0 | 0 | 1314 | 212.5 |
| | 7.5 | 75 | 30 | 1368 | 225.0 |
| | 7.5 | 150 | 30 | 1350 | 231.3 |
| LSD (P = 0.05) | | | | 112 | 17.3 |

Source: Kumara et al. (2014)

Taking into consideration the amount of FYM, poultry manure and press mud were added during the period of 16 years of soil OC stock of soil such as 71, 129.9, 36, 288.0, and 49, 448.4 kg ha$^{-1}$, respectively (Kumara et al. 2014; Meena et al. 2017a). The major portion of soil OC is retained through clay-organic matter interactions, indicating the importance of the inorganic part of the soil as substrate to bind the organic carbon. However, recommended doses of NP fertilizers tended to have more soil OC stock in soil compared to half recommended doses of NP fertilizers and further increase of soil OC stock in combined application of fertilizers with FYM, poultry manure, and press mud.

The C sequestration rate was shown in negative trends with greater magnitude (−18.8 kg ha$^{-1}$ year$^{-1}$) in 75 kg N + 30 kg $P_2O_5$ ha$^{-1}$ for 16 years (Table 3.18), whereas it was 387.5 kg ha$^{-1}$ year$^{-1}$ with 15 t FYM ha$^{-1}$. Taking into consideration the combined application of fertilizers with FYM, poultry manure and press mud were added during the period of 16 years of C sequestration rate of soil, such as 456.3, 118.8, and 225.0 kg ha$^{-1}$ year$^{-1}$, respectively (Kumara et al. 2014). The increase in C sequestration rate of soil with the addition of organic manures plus NP fertilizers might be due to better crop growth with accompanying higher root biomass generation and higher return of plant residues on the surface. The study established that the regular application of recommended doses of NP fertilizer under pearl millet-wheat cropping sequence reduced the negative soil OC trends.

Application of FYM as a component of INM increased the OC content of the soil, and the increase ranged from 0.03 to 0.06% year$^{-1}$ under different cropping systems (pearl millet-wheat, wheat-cotton, rice-wheat). The cumulative increase in OC content of the soil under different cropping systems (pearl millet-wheat, wheat-cotton, rice-wheat) has been presented in Fig. 3.4a. The available P content of the soil also increased in all the demonstrations at the farmer's field in all the cropping

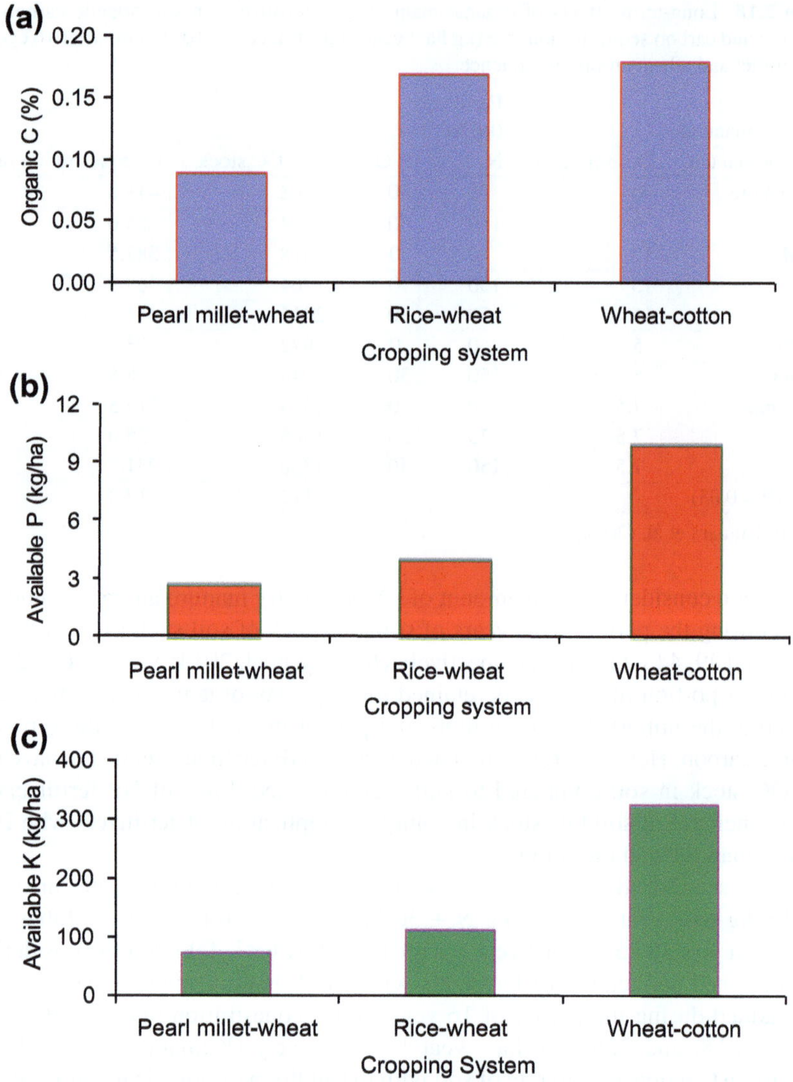


**Fig. 3.4** Influence of FYM on the buildup of organic C (**a**), available P (**b**), and K (**c**) in soil as the components of INM under various wheat-based cropping systems

systems, and the rate of increase was ranging from 0.9 to 3.3 kg ha$^{-1}$ year$^{-1}$. The cumulative increase in the available P content of the soil has been presented in Fig. 3.4b. Continuous application of FYM can increase the available P content of the soil to a level, which is sufficient to meet the requirement of the wheat crop. The maintenance of the available K content of the soil with fertilizer application is very difficult because it leads to luxury consumption. It has been observed in the long-term field experiments as well as in the demonstrations at farmer's fields that there is sufficient buildup of the available K content of the soil due to the application of

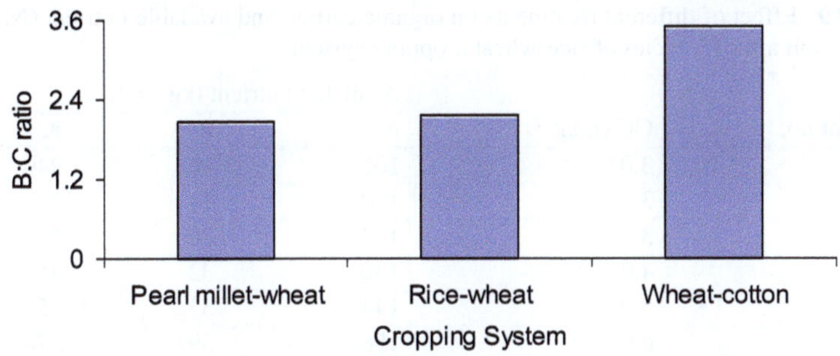


**Fig. 3.5** Influence of INM on B:C ratio of various wheat-based cropping systems

FYM. The rate of increase ranged from 25 to 109 kg $ha^{-1}$ $year^{-1}$. The cumulative increase in the available K content of the soil has been presented in Fig. 3.4c. It has been observed that the plots receiving FYM did not show the deficiency of Zn and Fe in any crop.

It exhibits not only the higher productivity and maintenance of soil fertility but is also economical, and the benefit/cost ratio of FYM use ranged from 2.07 to 3.5 (Antil and Narwal 2007; Sihag et al. 2015) in different cropping systems (pearl millet-wheat, wheat-cotton, rice-wheat) (Fig. 3.5). Thus, this technology can be easily followed by the farming community for increasing the productivity of different cropping systems.

The application of N alone or in combination with P or NPK or NPKZn maintained the initial levels of available N after 13 cycles of rice-wheat cropping (Singh et al. 2012a; Yadav et al. 2017a). The combined application of inorganic fertilizers and FYM or press mud or green manure ($T_6$ to $T_8$) increased the available N significantly over the initial level, and this increase varied from 14 to 24 kg $ha^{-1}$. The available P content of the soil decreased significantly in control ($T_1$) and in $T_2$ treatment where no P application was made. However, application of P through mineral fertilizers ($T_3$, to $T_5$) or its combined use with organic manures ($T_6$ to $T_9$) showed an increase of 6–18 kg $ha^{-1}$ in available P. Application of K through inorganic fertilizers alone ($T_4$ and $T_5$) or their combined use with organic sources showed an increase of 54–76 kg $ha^{-1}$ in the available K status of the soil, while in the absence of K application ($T_1$ to $T_3$), the available K content decreased between 81 and 101 kg $ha^{-1}$ over its initial value in 13 years of continuous rice-wheat cropping (Singh et al. 2012a; Meena et al. 2017b).

Seven years of rice-wheat cropping system without any fertilizers or organic amendments decreased (4.4 to 3.5 g $kg^{-1}$) the total soil C significantly over its initial level (Sekhon et al. 2009; Verma et al. 2015a). It indicates that C added through plant residues in the rice-wheat cropping system was not sufficient to maintain the soil C content to its initial levels.

Thirteen years of rice-wheat cropping system without any fertilization also decreased the OC content of soil from its initial status of 4.2 g $kg^{-1}$ soil to 3.0 g $kg^{-1}$ soil (Table 3.19). Balanced application of inorganic fertilizers ($T_5$) resulted in

**Table 3.19** Effect of different treatments on organic carbon and available nutrient (N, P, and K) content of soil after 13 cycles of rice-wheat cropping system

| | | Available nutrient (kg ha$^{-1}$) | | |
|---|---|---|---|---|
| Treatment no. | OC (g kg$^{-1}$) | N | P | K |
| $T_1$ | 3.0 | 105 | 16 | 240 |
| $T_2$ | 3.7 | 140 | 14 | 228 |
| $T_3$ | 3.8 | 142 | 30 | 224 |
| $T_4$ | 4.0 | 140 | 32 | 356 |
| $T_5$ | 5.0 | 144 | 33 | 359 |
| $T_6$ | 6.9 | 156 | 39 | 378 |
| $T_7$ | 7.4 | 160 | 42 | 365 |
| $T_8$ | 5.7 | 150 | 36 | 381 |
| $T_9$ | 4.9 | 146 | 34 | 362 |
| Initial values | 4.2 | 136 | 24 | 305 |
| LSD (0.05) | 0.9 | 11 | 8 | 16 |

Source: Singh et al. (2012a)

significantly higher accumulation of soil OC as compared $N_{150}$ ($T_2$), $N_{150}P_{75}$ ($T_3$), and $N_{150}P_{75}K_{75}$ ($T_4$) treatments.

Continuous application of FYM ($T_6$), press mud ($T_7$), and green manure ($T_8$) for 13 years increased soil OC by 1.9, 2.4, and 0.7 g kg$^{-1}$, respectively, in surface soil over inorganic fertilizer-only treatment ($T_5$). This increase in soil carbon accounted for 14.5, 15.8, and 5.3% of C added through FYM, press mud, and green manure, respectively. The available N decreased significantly with time in control over their initial status in soil (Table 3.13). Application of organic manures in conjunction with fertilizer under rice-wheat cropping system improved the OC, available P, K, and the micronutrient content of the soil.

Thirty-one years of maize-wheat-cowpea (fodder) cropping system influenced the soil OC (0–30 cm soil depth) significantly. The highest (6.5 g kg$^{-1}$) soil OC (0–15 cm) was observed in the NPK + FYM treatment. Significantly higher soil OC was found in the NPK treatment as compared to the control and N treatment. The soil OC was similar to the initial level (4.4 g kg$^{-1}$) in the control and N-fertilized plots after 31 years, whereas in FYM and NPK + FYM treatments, the soil OC increased over the initial level by 38.6 and 63.6%, respectively (Hati et al. 2006; Kumar et al. 2017a).

Bijay-Singh (2018b) did a meta-analysis of the data published on crop yields and soil parameters in maize-wheat, rice-rice, and rice-wheat cropping systems from LTEs. The buildup of soil OC in the upland system was less as compared to the lowland systems. The application of the recommended doses of NPK fertilizers resulted in the buildup of soil OC over no-fertilizer application in all the three cropping systems (Table 3.20). Decrease in soil OC content in the no-fertilizer application (control) from the initial values might be due to cultivation of the soil in the maize-wheat cropping system.

**Table 3.20** Average soil OC content at the start (initial) of long-term experiments on maize-wheat, rice-wheat, and rice-rice cropping systems and in no-fertilizer (N, P, and K) control and optimum N, P, and K fertilizer level treatments

| Cropping system | Number of experiments | Duration (years) | Soil OC (g $kg^{-1}$) | | |
|---|---|---|---|---|---|
| | | | Initial | No-fertilizer control | Optimum N, P, and K fertilizer levels |
| Maize-wheat | 12 | 6–25 | 6.4 | 5.8 | 6.8 |
| Rice-wheat | 10 | 9–27 | 14.3 | 14.9 | 16.3 |
| Rice-rice | 23 | 6–26 | 16.7 | 18.1 | 19.6 |

Source: Bijay-Singh (2018a, b)

### 3.7.3 Soil Biological Properties

Generally, enzyme activities in the soil are closely related to the organic matter content. Application of balanced amounts of nutrients and manures improved the organic matter and MBC content of soils, which corresponded with higher enzyme activity (Mandal et al. 2007). Soil microbial biomass constitutes a transformation matrix for organic matter in soil and acts as an active reservoir for plant-available nutrients. It is established that soil microbial carbon and soil enzymes respond more quickly to changes in environment sand agronomic practices than does the soil organic carbon. The dissolved organic carbon (DOC), microbial biomass carbon (MBC), light carbon fraction (LFC), and heavy carbon fraction (HFC) content of soil increased significantly, increasing the levels of FYM. However, the increase in DOC, MBC, LFC, and HFC was highest in 15 Mg $ha^{-1}$ followed by 10 and 5 Mg $ha^{-1}$ (Kumara 2013; Varma et al. 2017b; Gogoi et al. 2018). Application of NPK fertilizers alone or in combination with FYM and green manuring improved soil microbial biomass C, urease activity, and alkaline phosphatase activity but had little effect on the dehydrogenase activity (Goyal et al. 1999).

Application of organic manures alone or in combination with N or NP fertilizers significantly increased dehydrogenase, alkaline phosphatase, and urease activity as compared to the application of fertilizers alone (Sheoran 2015; Dhakal et al. 2015). However, combined application of organic manures and fertilizers significantly decreased the dehydrogenase and urease activity compared with the organic manures applied alone, while alkaline phosphatase activity showed the reverse trend (Table 3.21).

## 3.8 Effect of INM on NUE

Use efficiency of different nutrients continues to be extremely low, and its enhancement has remained a prime concern at all times. Long-term studies indicated that INM helped in enhancing the use efficiency of N at different locations (Dwivedi et al. 2016; Meena et al. 2018b). In Inceptisols (Ludhiana), the N use efficiency in maize recorded under 100% N (alone) was 16.7%, which increased to 23.5, 36.4,

**Table 3.21** Effect of long-term application of organic manures and fertilizers on soil enzyme activity

| Type of manure | Dose (Mg ha$^{-1}$) | Fertilizer (kg ha$^{-1}$) N | $P_2O_5$ | Dehydrogenase activity (μg TPF g$^{-1}$ 24 h$^{-1}$) | Alkaline phosphatase activity (μg PNP g$^{-1}$ h$^{-1}$) | Urease activity (μg$NH_4^+$-N g$^{-1}$ h$^{-1}$) |
|---|---|---|---|---|---|---|
| No manure | 0 | 75 | 30 | 36.53 | 572.25 | 58.81 |
| | 0 | 150 | 60 | 32.32 | 580.75 | 65.02 |
| FYM | 15 | 0 | 0 | 63.71 | 684.00 | 87.72 |
| | 15 | 150 | 0 | 59.75 | 742.75 | 76.27 |
| | 15 | 150 | 30 | 50.48 | 733.50 | 71.25 |
| Poultry manure | 5 | 0 | 0 | 48.62 | 704.45 | 85.20 |
| | 5 | 150 | 30 | 40.13 | 756.00 | 75.75 |
| Press mud | 7.5 | 0 | 0 | 58.14 | 664.63 | 97.60 |
| | 7.5 | 75 | 30 | 44.86 | 675.65 | 83.49 |
| | 7.5 | 150 | 30 | 39.44 | 673.20 | 71.28 |
| C.D. (P = 0.05) | | | | 5.69 | 17.91 | 7.88 |

Source: Sheoran (2015)

**Table 3.22** N use efficiency in different crops as affected by long-term nutrient supply with and without FYM

| Soil type | Location | Crop | N use efficiency (%) N | NP | NPK | NPK + FYM |
|---|---|---|---|---|---|---|
| Inceptisol | Ludhiana | Maize | 16.7 | 23.5 | 36.4 | 40.2 |
| | | Wheat | 32.0 | 50.6 | 63.1 | 67.8 |
| Alfisol | Palampur | Maize | 6.4 | 34.7 | 52.6 | 63.7 |
| | | Wheat | 1.9 | 35.6 | 50.6 | 72.6 |
| Mollisol | Pantnagar | Rice | 37.5 | 40.7 | 44.4 | 61.7 |
| | | Wheat | 42.4 | 46.1 | 48.4 | 47.9 |

Source: Singh et al. (2012b)

and 40.2% on integration with P, PK, and FYM, respectively (Table 3.22). Similar trend was noted in Mollisols (Pantnagar) under, and agroecological regions need to be worked out, and nutrient supply packages with optimum application rates of organic manures and fertilizers should be developed.

## 3.9 Constraints in Adoption of INM Technology

The constraints in the adoption of INM technology are as follows:

- Cost and nonavailability of FYM
- Difficulties in growing green manure crops
- Nonavailability of biofertilizers
- Nonavailability of soil-testing facilities

- High cost of chemical fertilizers
- Nonavailability of water
- Lack of knowledge and poor advisory services
- Nonavailability of quality seeds

## 3.10 Suggestions for Improving Adoption of INM Technology

The following are the suggestions for improving adoption and implementation of the INM technology:

- Create mass awareness among the farmers for recycling and use of organic resources and for preparation of quality compost and FYM.
- Educate the farmers through media, mass contact and awareness program, and on-farm demonstration about the benefits of INM.
- Educate the farmers about the nature of soil and its importance for mankind and need to preserve it for posterity.
- Loan facilities should be provided to farmers by financial agencies to encourage adoption of INM.
- Develop promotional literatures in local languages and distribute to the farmers.
- Advise the farmers to incorporate crop residues into soil to the extent possible.
- Information on balanced use of fertilizers based on soil test, biofertilizers, green manuring, and information on compost-making techniques should be transferred to farmers.
- Motivate KVKs, fertilizer industry, and NGOs to propagate the usefulness of INM.

## 3.11 Priorities for Future Research

- Contribution of organic resources during the crop growing season and residual nutrients from their applications to soil nutrient budget should be assessed.
- Efficient management of crop residues in different cropping systems should be investigated under different agroecosystems.
- Role of industrial wastes in improving nutrient use efficiency should be evaluated under different agroecosystems.
- The major component of the system that needs attention is recycling of solid waste and vermicomposting.
- Beneficial effect of inclusion of legume on soil health needs to be quantified under cereal-cereal cropping systems.
- Organics, wastes, residues, and biofertilizers should be identified for various location-specific cropping systems.
- Location-specific INM technology should be developed and effectively transferred to the farmers for sustaining crop productivity of various cropping systems.

- Causes of decline in the rate of growth of total production under various cropping systems associated with the deterioration of soil health and reduction of soil OC stocks should be studied, and possible measures should be found out.
- Dynamics of soil organic matter and C sequestration should be studied regularly under LTEs in different cropping and agroecosystems.

## 3.12 Conclusions

Based on the results of short and LTEs conducted in different soil-crop environments, it is concluded that INM has a great potential to meet the growing demands of nutrients, to achieve maximum yields, and to sustain the crop productivity on a long-term basis without any adverse effect on the environment. The soil health can be maintained with the use of balanced fertilization along with organic resources, biofertilizers, and green manures besides getting higher yield. The success of INM depends on the different components of INM and how precisely they are used. The INM could play a major role in improving the nutrient use efficiency, crop yields, soil health, and socioeconomic status of small and marginal farmers. About 25% nutrient requirement of Indian agriculture could be met if various sources of organic resources (dung, FYM, crop residues, urban/rural waste, and green manuring) are properly utilized.

## References

AICRP-CS Annual Reports (1992–93 to 2001–02) AICRP on Cropping Systems, PDCSR, Modipuram, India

AICRP-IFS Reports (2005–2010) AICRP on Integrated Farming Systems, PDFSR, Modipuram, India

Antil RS, Narwal RP (2007) Integrated nutrient management for sustainable soil health and crop productivity. Indian J Fert 3(9):111–112

Antil RS, Narwal RP, Singh B, Singh JP (2011) Long-term effects of FYM and N application on soil health and productivity under pearl millet – wheat cropping system. Indian J Fert 7(7):14–32

Antil RS, DevRaj, Narwal RP, Singh JP (2013) Evaluation of maturity and stability parameters of composts prepared from a wide range of organic wastes and their response to wheat. Waste Biomass Valoriz 4:95–104

Ashoka P, Meena RS, Kumar S, Yadav GS, Layek J (2017) Green nanotechnology is a key for eco-friendly agriculture. J Clean Prod 142:4440–4441

Bandyopadhyay KK, Misra AK, Ghosh PK, Hati KM (2010) Effect of integrated use of farmyard manure and chemical fertilizers on soil physical properties and productivity of soybean. Soil Tillage Res 110:115–125

Bandyopadhyay PK, Saha S, Mallick S (2011) Comparison of soil physical properties between a permanent fallow and a long-term rice–wheat cropping with inorganic and organic inputs in the humid subtropics of eastern India. Commun Soil Sci Plant Anal 42:435–449

Bhattacharya P (2007) Prospect of organic nutrient resources utilization in India. Indian J Fert 3:93–107

Bijay-Singh (2018a) Optimizing fertilizer use to maintain soil health. In: Reicosky D (ed) Managing soil health for sustainable agriculture Vol. 2: monitoring and management. Burleigh Dodds Science Publishing, Cambridge. (ISBN:9781786761927. www.bdspublishing.com)

Bijay-Singh (2018b) Are nitrogen fertilizers deleterious to soil health? Agronomy 8:48. https://doi.org/10.3390/agronomy8040048

Buragohain S, Sharma B, Nath JD, Gogaoi N, Meena RS, Lal R (2017) Impact of ten years of biofertilizer use on soil quality and rice yield on an inceptisol in Assam, India. Soil Res. https://doi.org/10.1071/SR17001

Dadhich RK, Meena RS (2014) Performance of Indian mustard (*Brassica juncea* L.) in response to foliar spray of thiourea and thioglycollic acid under different irrigation levels. Indian J Ecol 41(2):376–378

Dadhich RK, Meena RS, Reager ML, Kansotia BC (2015) Response of bio-regulators to yield and quality of Indian mustard (*Brassica juncea* L. Czernj. and Cosson) under different irrigation environments. J App Nat Sci 7(1):52–57

Das B, Chakraborty D, Singh VK, Aggarwal P, Singh R, Dwivedi BS (2014) Effect of integrated nutrient management practice on soil aggregate properties, its stability and aggregate-associated carbon content in an intensive rice-wheat system. Soil Tillage Res 136:9–18

Das D, Dwivedi BS, Meena MC, Singh VK (2015) Integrated nutrient management for improving soil health and crop productivity. Indian J Fert 11(4):64–83

Datta R, Baraniya D, Wang YF, Kelkar A, Moulick A, Meena RS, Yadav GS, Ceccherini MT, Formanek P (2017a) Multi-function role as nutrient and scavenger off reeradical in soil. Sustain MDPI 9:402. https://doi.org/10.3390/su9081402

Datta R, Kelkar A, Baraniya D, Molaei A, Moulick A, Meena RS, Formanek P (2017b) Enzymatic degradation of lignin in soil: a review. Sustain MDPI 9:1163. https://doi.org/10.3390/su9071163, 1–18

Dhakal Y, Meena RS, De N, Verma SK, Singh A (2015) Growth, yield and nutrient content of mungbean (*Vigna radiata* L.) in response to INM in eastern Uttar Pradesh, India. Bangladesh J Bot 44(3):479–482

Dhakal Y, Meena RS, Kumar S (2016) Effect of INM on nodulation, yield, quality and available nutrient status in soil after harvest of green gram. Legum Res 39(4):590–594

Dubey R, Sharma RS, Dubey DP (2014) Effect of organic, inorganic and integrated nutrient management on crop productivity, water productivity and soil properties under various rice-based cropping systems in Madhya Pradesh, India. Int J Curr Microbiol App Sci 3:381–389

Dwivedi BS, Shukla AK, Singh VK (2002) PDCSR Annual Report (2001–02), PDCSR, Modipuram

Dwivedi AK, Singh M, Kauraw DL, Wanjari RH, Chauhan SS (2007) Impact of fertilizer and manure use for three decades on crop productivity and sustainability, and soil quality under soybean-wheat system on vertisols in Central India. IISS, Bhopal, p 51

Dwivedi BS, Singh VK, Meena MC, Dey A, Datta SP (2016) Integrated nutrient management for enhancing nutrient use efficiency. Indian J Fert 12(44):62–71

FAI, Fertilizer Statistics (1974–75, 1992–93, 19997–98, 2002–03, 2007–08) The Fertilizer Association of India, New Delhi

FAO (1998) Guide to efficient plant nutrition management. Land and Water Development Division, FAO, Rome, pp 1–19

Ghosh PK, Ghosh RK, Bandyopadhyay KK, Mishra AK, Manna MC (2005) Role of integrated plant nutrient supply for sustainable production in groundnut based cropping system. Fert News 50:45–53

Gogoi N, Baruah KK, Meena RS (2018) Grain legumes: impact on soil health and agroecosystem. In: Meena et al (eds) Legumes for soil health and sustainable management. Springer, Singapore. https://doi.org/10.1007/978-981-13-0253-4_16

Gong W, Yan X, Wang J, Hu T, Gong Y (2009) Long-term manure and fertilizer effects on soil organic matter fractions and microbes under a wheat–maize cropping system in northern China. Geoderma 149:318–324

Goswami NN (1997) Balanced use of plant nutrients. Fert News 42(4):15–19

Goswami NN (1998) Some thoughts on concept, relevance and feasibility of IPNS under Indian condition. In: Integrated plant nutrient supply system for sustainable productivity bulletin no. 2. IISS, Bhopal, pp 3–9

Goyal S, Chander K, Mundra MC, Kapoor KK (1999) Influence of inorganic fertilizers and organic amendments on soil organic matter and soil microbial properties under tropical conditions. Biol Fertil Soils 29:196–200

Gupta AP, Narwal RP, Antil RS (2003) Influence of soil organic matter on the productivity of pearl millet – wheat cropping sequence. Arch Agron Soil Sci 49:325–332

Hati KM, Swarup A, Singh D, Misra AK, Ghosh PK (2006) Long-term continuous cropping, fertilization and manuring effects on physical properties and organic carbon content of a sandy loam soil. Aus J Soil Res 44:487–495

Jaipaul, Negi MS (2006) In: Sharma et al (eds) Balanced fertilization for sustainable crop production. CSKHPKV/ICAR, Palampur/New Delhi, pp 96–97

Kakraliya SK, Singh U, Bohra A, Choudhary KK, Kumar S, Meena RS, Jat ML (2018) Nitrogen and legumes: a meta-analysis. In: Meena RS et al (eds) Legumes for soil health and sustainable management. Springer, Singapore. https://doi.org/10.1007/978-981-13-0253-4_9

Kumar S, Meena RS, Pandey A, Seema (2017a) Soil acidity management and an economics response of lime and sulfur on sesame in an alley cropping system. Int J Curr Microb Appl Sci 6(3):2566–2573

Kumar S, Meena RS, Yadav GS, Pandey A (2017b) Response of sesame (*Sesamum indicum* L.) to sulphur and lime application under soil acidity. Int J Plant Soil Sci 14(4):1–9

Kumar S, Meena RS, Bohra JS (2018a) Interactive effect of sowing dates and nutrient sources on dry matter accumulation of Indian mustard (*Brassica juncea* L.). J Oilseed Brassica 9(1):72–76

Kumar S, Meena RS, Lal R, Yadav GS, Mitran T, Meena BL, Dotaniya ML, EL-Sabagh A (2018b) Role of legumes in soil carbon sequestration. In: Meena RS et al (eds) Legumes for soil health and sustainable management. Springer, Singapore. https://doi.org/10.1007/978-981-13-0253-4_4

Kumara BH (2013) Long-term effects of organic manures and fertilizers on soil organic carbon pools and nitrogen fractions. Ph.D. thesis, Department of Soil Science, CCS Haryana Agricultural University Hisar, India

Kumara BH, Antil RS, DevRaj (2013) Long-term effects of nutrient management on soil health and crop productivity under pearl millet – wheat cropping system. Indian J Fert 9(12):86–97

Kumara BH, Antil RS, DevRaj (2014) Impact of long-term manures and fertilizer application on carbon sequestration and its efficiency under pearl millet – wheat cropping sequence. Int J Farm Sci 4(1):21–26

Layek J, Das A, Mitran T, Nath C, Meena RS, Singh GS, Shivakumar BG, Kumar S, Lal R (2018) Cereal+Legume intercropping: an option for improving productivity. In: Meena RS et al (eds) Legumes for soil health and sustainable management. Springer, Singapore. https://doi.org/10.1007/978-981-13-0253-4_11

Leroy BLM, Herath HMSK, Sleutel S, De Neve S, Gabriels D, Reheul D, Moens M (2008) The quality of exogenous organic matter: short term effects on soil physical properties and soil organic matter fractions. Soil Use Manag 24:139–147

Mahapatra P, Singh RP, Singh BP, Mishra B, Sarkar AK (2007) SSAC (BAU) technical bulletin no. 4, BAU, pp 1–75

Mandal A, Patra AK, Singh D, Swarup A, Masto RE (2007) Effect of long-term application of manure and fertilizer on biological and biochemical activities in soil during crop development stages. Bioresour Technol 98:3585–3592

Meena RS, Lal R (2018) Legumes and sustainable use of soils. In: Meena RS et al (eds) Legumes for soil health and sustainable management. Springer, Singapore. https://doi.org/10.1007/978-981-13-0253-4_1

Meena H, Meena RS (2017) Assessment of sowing environments and bio-regulators as adaptation choice for clusterbean productivity in response to current climatic scenario. Bangladesh J Bot 46(1):241–244

Meena RS, Yadav RS (2014) Phonological performance of groundnut varieties under sowing environments in hyper arid zone of Rajasthan, India. J App Nat Sci 6(2):344–348

Meena RS, Yadav RS (2015) Yield and profitability of groundnut (*Arachis hypogaea* L) as influenced by sowing dates and nutrient levels with different varieties. Legum Res 38(6):791–797
Meena RS, Yadav RS, Meena VS (2014) Response of groundnut (*Arachis hypogaea* L.) varieties to sowing dates and NP fertilizers under Western Dry Zone of India. Bangladesh J Bot 43(2):169–173
Meena RS, Dhakal Y, Bohra JS, Singh SP, Singh MK, Sanodiya P (2015a) Influence of bioinorganic combinations on yield, quality and economics of mungbean. Am J Exp Agric 8(3):159–166
Meena RS, Meena VS, Meena SK, Verma JP (2015b) The needs of healthy soils for a healthy world. J Clean Prod 102:560–561
Meena RS, Meena VS, Meena SK, Verma JP (2015c) Towards the plant stress mitigate the agricultural productivity: a book review. J Clean Prod 102:552–553
Meena RS, Yadav RS, Meena H, Kumar S, Meena YK, Singh A (2015d) Towards the current need to enhance legume productivity and soil sustainability worldwide: a book review. J Clean Prod 104:513–515
Meena RS, Yadav RS, Reager ML, De N, Meena VS, Verma JP, Verma SK, Kansotia BC (2015e) Temperature use efficiency and yield of groundnut varieties in response to sowing dates and fertility levels in Western Dry Zone of India. Am J Exp Agric 7(3):170–177
Meena H, Meena RS, Singh B, Kumar S (2016a) Response of bio-regulators to morphology and yield of clusterbean [*Cyamopsis tetragonoloba* (L.) Taub.] under different sowing environments. J App Nat Sci 8(2):715–718
Meena RS, Bohra JS, Singh SP, Meena VS, Verma JP, Verma SK, Shiiag SK (2016b) Towards the prime response of manure to enhance nutrient use efficiency and soil sustainability a current need: a book review. J Clean Prod 112:1258–1260
Meena RS, Gogaoi N, Kumar S (2017a) Alarming issues on agricultural crop production and environmental stresses. J Clean Prod 142:3357–3359
Meena RS, Kumar S, Pandey A (2017b) Response of sulfur and lime levels on productivity, nutrient content and uptake of sesame under guava (*Psidium guajava* L.) based agri-horti system in an acidic soil of eastern Uttar Pradesh, India. J Crop Weed 13(2):222–227
Meena RS, Meena PD, Yadav GS, Yadav SS (2017c) Phosphate solubilizing microorganisms, principles and application of microphos technology. J Clean Prod 145:157–158
Meena H, Meena RS, Lal R, Singh GS, Mitran T, Layek J, Patil SB, Kumar S, Verma T (2018a) Response of sowing dates and bio regulators on yield of clusterbean under current climate in alley cropping system in eastern UP. Indian Leg Res 41(4):563–571
Meena RS, Kumar V, Yadav GS, Mitran T (2018b) Response and interaction of *Bradyrhizobium japonicum* and *Arbuscular mycorrhizal* fungi in the soybean rhizosphere: a review. Plant Growth Regul 84:207–223
Meena BL, Fagodiya RK, Prajapat K, Dotaniya ML, Kaledhonkar MJ, Sharma PC, Meena RS, Mitran T, Kumar S (2018c) Legume green manuring: an option for soil sustainability. In: Meena RS et al (eds) Legumes for soil health and sustainable management. Springer, Singapore. https://doi.org/10.1007/978-981-13-0253-4_12
Mitran T, Meena RS, Lal R, Layek J, Kumar S, Datta R (2018) Role of soil phosphorus on legume production. In: Meena RS et al (eds) Legumes for soil health and sustainable management. Springer, Singapore. https://doi.org/10.1007/978-981-13-0253-4_15
Motsara MR (2002) Available N, P and K status of Indian soils as depicted by soil fertility maps. Fert News 47:15–21
Nyamangara J, Gotosa J, Mpofu SE (2001) Cattle manure effects on structural stability and water retention capacity of a granite soil in Zimbabwe. Soil Tillage Res 62:157–162
Ram K, Meena RS (2014) Evaluation of pearl millet and mungbean intercropping systems in arid region of Rajasthan (India). Bangladesh J Bot 43(3):367–370
Roy RN, Ange AL (1991) Integrated Plant Nutrition System (IPNS) and sustainable agriculture. FAI, New Delhi, pp SV/1-1–SV/1-12
Sekhon KS, Singh JP, Mehla DS (2009) Soil organic C pools after seven years of manures and mineral fertilizers application in a rice-wheat rotation. Arch Agron Soil Sci 55:197–206

Sharma SP, Singh MV, Subehia SK, Jain PK, Kaushal V, Verma TS (2005) Research bulletin no. 2. AICRP-LTFE, IISS/Department of Soils, CSK HPKV, Bhopal/Palampur, pp 1–88
Sheoran S (2015) Effects of organic manures and fertilizers on soil properties under pearl millet-wheat cropping system: a long-term study. M.Sc. thesis, Department of Soil Science, CCS Haryana Agricultural University Hisar, India
Sihag SK, Singh MK, Meena RS, Naga S, Bahadur SR, Gaurav YRS (2015) Influences of spacing on growth and yield potential of dry direct seeded rice (*Oryza sativa* L.) cultivars. Ecoscan 9(1–2):517–519
Singh MV (2001) Evaluation of current micronutrient stocks in different agro-ecological zones of India for sustainable crop production. Fert News 46:25–42
Singh M, Wanjari RH (2007) AICRP on Long-term Fertilizer Experiments (LTFE) to study changes in soil quality, crop productivity and sustainability research bulletin. IISS, Bhopal
Singh M, Singh VP, Reddy KS (2001) Effect of integrated use of fertilizer N and FYM or green manure on transformation of NK and S and productivity of rice-wheat system on a vertisols. J Indian Soc Soil Sci 49:430–435
Singh M, Tripathi AK, Reddy DD (2002) Potassium balance and release of kinetics of nonexchangeable K in Typic Haplustert as influenced by cattle manure application under soybean-wheat system. Aus J Soil Res 40:533–541
Singh RK, Ghosh PK, Bandyopadhyay KK, Mishra AK, Mandal KG, Hati KM (2006) Integrated plant nutrient supply for sustainable production in soybean-based cropping system. Indian J Fert 1:25–32
Singh JP, Kaur J, Mehla DS, Narwal RP (2012a) Long-term effects of nutrient management on soil health and crop productivity under rice-wheat cropping system. Indian J Fert 8(8):28–48
Singh M, Dwivedi BS, Datta SP (2012b) Integrated nutrient management for enhancing productivity, nutrient use efficiency and environmental quality. In: Soil science in the service of nation. ISSS, New Delhi, pp 55–67
Sofi PA, Baba ZA, Hamid B, Meena RS (2018) Harnessing soil Rhizobacteria for improving drought resilience in legumes. In: Meena RS et al (eds) Legumes for soil health and sustainable management. Springer, Singapore. https://doi.org/10.1007/978-981-13-0253-4_8
Sudhir K, Singh MV, Jayaprakash SM (2004) Soil quality, crop production and sustainability experiences under long-term finger millet-maize cropping in alfisol. IISS, Bhopal, pp 1–130
Tejada M, Gonzales JL (2008) Influence of two organic amendments on the soil physical properties, soil losses, sediments and runoff water quality. Geoderma 145:325–334
Tiwari KN (2008) Future plant nutrition research in India. J Indian Soc Soil Sci 56:327–336
Varma D, Meena RS, Kumar S (2017a) Response of mungbean to fertility and lime levels under soil acidity in an alley cropping system in Vindhyan Region, India. Int J Chem Stud 5(2):384–389
Varma D, Meena RS, Kumar S, Kumar E (2017b) Response of mungbean to NPK and lime under the conditions of Vindhyan Region of Uttar Pradesh. Legum Res 40(3):542–545
Verma JP, Jaiswal DK, Meena VS, Meena RS (2015a) Current need of organic farming for enhancing sustainable agriculture. J Clean Prod 102:545–547
Verma JP, Meena VS, Kumar A, Meena RS (2015b) Issues and challenges about sustainable agriculture production for management of natural resources to sustain soil fertility and health: a book review. J Clean Prod 107:793–794
Verma SK, Singh SB, Prasad SK, Meena RN, Meena RS (2015c) Influence of irrigation regimes and weed management practices on water use and nutrient uptake in wheat (*Triticum aestivum* L. Emend. Fiori and Paol.). Bangladesh J Bot 44(3):437–442
Walia MK, Walia SS, Dhaliwal SS (2010) Long-term effect of integrated nutrient management on properties of Typic Ustochrept after 23 cycles of an irrigated rice-wheat system. J Sustain Agric 34:724–743
Walia MK, Dhaliwal SS, Walia SS, Shukla AK (2017) Impact of integrated nutrient management on soil health and rice yields in India – a review. Indian J Fert 13(7):28–37
Yadav RL, Singh SR, Prasad K, Dwivedi BS, Batta RK, Singh AK, Patil NG, Chaudhary SK (2000) In: Yadav JSP, Singh GB (eds) Natural resource management for agricultural production in India. ICMNR, New Delhi, pp 775–870

Yadav GS, Babu S, Meena RS, Debnath C, Saha P, Debbaram C, Datta M (2017a) Effects of godawariphosgold and single supper phosphate on groundnut (*Arachis hypogaea*) productivity, phosphorus uptake, phosphorus use efficiency and economics. Indian J Agric Sci 87(9):1165–1169

Yadav GS, Lal R, Meena RS, Babu S, Das A, Bhomik SN, Datta M, Layak J, Saha P (2017b) Conservation tillage and nutrient management effects on productivity and soil carbon sequestration under double cropping of rice in North Eastern Region of India. Ecol Indic. http://www.sciencedirect.com/science/article/pii/S1470160X17305617

Yadav GS, Lal R, Meena RS, Datta M, Babu S, Das LJ, Saha P (2017c) Energy budgeting for designing sustainable and environmentally clean/safer cropping systems for rainfed rice fallow lands in India. J Clean Prod 158:29–37

Yadav GS, Das A, Lal R, Babu S, Meena RS, Saha P, Singh R, Datta M (2018a) Energy budget and carbon footprint in a no-till and mulch based rice–mustard cropping system. J Clean Prod 191:144–157

Yadav GS, Das A, Lal R, Babu S, Meena RS, Patil SB, Saha P, Datta M (2018b) Conservation tillage and mulching effects on the adaptive capacity of direct-seeded upland rice (*Oryza sativa* L.) to alleviate weed and moisture stresses in the North Eastern Himalayan Region of India. Arch Agron Soil Sci. https://doi.org/10.1080/03650340.2018.1423555

Yadvinder-Singh, Bijay S, Ladha JK, Khind CS, Gupta RK, Meelu OP, Pasuquin E (2004) Long-term effects of organic inputs on yield and soil fertility in the rice-wheat rotation. Soil Sci Soc Am J 68:845–853

# Management of Micronutrients in Soil for the Nutritional Security

4

Dileep Kumar, K. P. Patel, V. P. Ramani, A. K. Shukla, and Ram Swaroop Meena

**Abstract**

Availability of soil micronutrient is a major limiting factor in crop productivity and its quality. The micronutrient deficiencies of zinc (Zn) 40%, iron (Fe) 12.6%, copper (Cu) 4.5%, manganese (Mn) 6.0%, and boron (B) 22.8% in soils have been reported across the country. The manganese deficiency is emerging extremely fast, particularly in wheat crops grown after rice in Haryana (12%) and Punjab (18%) due to leaching of Mn from the upper surface of the coarse-textured soils. In acid soils of India, the majority of the soil samples indicated a sufficient supply of Cu, Fe, and Mn, low deficiencies of Zn (30%), and higher deficiencies of B (46%) and Mo (50%). Application of soil or foliar spray of Zn, Mo, and B and foliar spray of Fe and Mn has been recommended as the most suitable method for the management of micronutrients for the better nutrition of the crops. The average response of Zn application to cereals, oilseeds, and pulses was around 20, 18, and 24%, respectively. The average yield increase due to iron (ferrous sulfate) has been recorded as 450 kg $ha^{-1}$ in chickpea, 780 kg $ha^{-1}$ in wheat, and up to 1500 kg $ha^{-1}$ in paddy. The average yield increase recorded in paddy and wheat is 360 kg $ha^{-1}$ and 560 kg $ha^{-1}$, respectively, due to Mn supplementation in Punjab. The average yield increase in cereals and pulses crops was recorded up to 400 kg $ha^{-1}$ due to boron application in the northeast region.

D. Kumar (✉) · V. P. Ramani
Micronutrient Research Project, Anand Agricultural University, Anand, Gujarat, India

K. P. Patel
BACA, Anand Agricultural University, Anand, Gujarat, India

A. K. Shukla
All India Coordinated Research Project on Micronutrients ICAR-Indian Institute of Soil Science, Bhopal, MP, India

R. S. Meena
Department of Agronomy, Institute of Agricultural Sciences (BHU), Varanasi, Uttar Pradesh, India



R. S. Meena (ed.), *Nutrient Dynamics for Sustainable Crop Production*,
https://doi.org/10.1007/978-981-13-8660-2_4

Cereals, mainly rice and wheat, are inherently very low in concentration of Zn and Fe in grain, particularly when grown under Zn- and Fe-deficient soils. Deficiency of these nutrients in soil is affecting crop productivity, quality of food, and human nutrition. In rice out of the total micronutrients absorbed by the crop, only 31% Zn, 33% B, 18% Fe, 9% Mn, and 67% Cu remain in grains, and they are removed from the field. In cereals, Fe uptake varies from 150 to 1200 g $ha^{-1}$ $year^{-1}$. Micronutrient malnutrition now afflicts over two billion peoples across the world which causes health problems especially in women and children in developing countries. Besides agronomic strategy, microbial and physiological interventions help to mobilize micronutrients from source to sink and resulted in micronutrient-dense grain production with an increase in crop yields which helps to combat malnutrition in animals and humans. Hence, there is need to improve micronutrient quality through fortifying the grains with micronutrients.

**Keywords**
Animal · Fortification · Human health · Micronutrients · Malnutrition

## Abbreviations

| | |
|---|---|
| B | Boron |
| $CaCO_3$ | Calcium carbonate |
| Cl | Chlorine |
| Cu | Copper |
| EDTA | Ethylenediaminetetraacetic acid |
| Fe | Iron |
| $FeSO_4$ | Iron sulfate |
| K | Potassium |
| Mn | Manganese |
| Mo | Molybdenum |
| N | Nitrogen |
| Ni | Nickel |
| NPK | Nitrogen phosphorus potassium |
| P | Phosphorus |
| S | Sulfur |
| Zn | Zinc |
| $ZnSO_4$ | Zinc sulfate |

## 4.1 Introduction

Green Revolution has made the country self-sufficient in the food grain production. The increased production and productivity have greatly enhanced the demand on soil for nutrients (Thiyagarajan 2002). The greater crop production per unit area has resulted in larger depletion of soil-available micronutrients because traditional fertilizer practices were designed to meet these needs for only major nutrients like nitrogen, phosphorous (P), and potassium (K) (Sharma and Kumar 2016; Ashoka et al. 2017). Ultimately, micronutrient deficiency especially Zn, Fe, and B has become a limiting factor for crop production in different states of the country (Shukla et al. 2014). There are eight essential plant nutrient elements for better crop defined as micronutrients like B, Zn, Mn, Fe, Cu, Mo, Cl, and Ni. They constitute in total less than 1% of the dry weight of most plants. In order to release the genetic potential yields of crops, there is need to correcting micronutrient deficiencies in soil.

It is estimated that worldwide about 2 billion peoples suffer from Fe, Zn, and other micronutrient deficiencies (Black 2003; WHO 2016). The problem is most severe in developing countries. Diets in these countries depend on staple crops such as cereals and coarse grain crops which are poor in mineral micronutrients and vitamin content, and consequently micronutrient deficiencies are widespread among these populations. The chronic lack of micronutrients can cause severe but often invisible health problems, especially among women and young children (Kennedy et al. 2003; Black et al. 2013; FAO 2015; Jangir et al. 2017). On the other hand, increasing crop yields in different agricultural systems as a result of chemical fertilizer application has been accomplished with reduced micronutrient concentration in the grain of the different crops.

Literature is abounding on crop response to micronutrients in India, especially to Zn and B, and has been thoroughly reviewed (Rattan et al. 2008; Singh et al. 2011; Murthy 2011; Shukla and Behera 2012; Yadav et al. 2018b). A good response of crops to Zn is obtained throughout the country, response to B is more in eastern states of India, and response to Mn has strengthen up for wheat in Punjab (Singh et al. 2011). Soil application and foliar sprays are the most commonly used methods of Zn application. Results from field experiments revealed the superiority of soil application of Zn over foliar application (Rathore et al. 1995). The lower efficiency of the foliar mode is primarily due to delayed cure of the deficiency and the low concentration of Zn in spray solution.

Thus, correcting micronutrient deficiency is essential for both maintaining soil fertility and harnessing the genetic yield potential of crops with enriched micronutrient content in edible part. It will also help in producing quality food and reducing malnutrition in animal and human beings. According to an estimate, the current micronutrient application to crops may need to be doubled by 2050 to meet the food demand of increasing population of the country through intensive cropping on marginal lands.

## 4.2 Global Scenario of Micronutrient Deficiency

It is predictable that world human population will hike to 9.7 billion by the year 2050, and India's population is projected to overtake that of China and will rise to 1.6 billion from its existing level of 1.2 billion (Sreeja 2014). The major cause of micronutrient deficiency is related with food and nutritional security (Meenakshi et al. 2010; Ghaly and Alkoaik 2010; Zadeh and Begum 2011; Kumar et al. 2017b). Micronutrients are major limitation across the world and control crop productivity as well as produce quality (especially micronutrient concentration). Zinc and iron deficiency in soil is a serious problem that affects many cultivated soils. It is estimated that about half of the cultivated soils in the world contains reduced amounts of bioavailable zinc (Fig. 4.1).

Due to lower soil organic matter and soil moisture as well as high levels of pH and $CaCO_3$, the problem of micronutrient deficiency is aggravated significantly in arid and semiarid regions (Cakmak 2008; Gonçalves et al. 2010; Yadav et al. 2018a). The low availability of these metals in the soil limits Zn uptake by plants, resulting in significant decreases in crop productivity as well as nutritional quality of food. Micronutrient deficiency (Zn, 40%; Fe, 12.6%; Cu, 4.5%; Mn, 6.0%; and B, 22.8%) in soils is noted across the country. In states like Gujarat, Bihar, and Madhya Pradesh, Zn deficiency is almost consistent during the last four decades despite making efforts for popularization of Zn application in various crops. Interestingly, over the last 15–20 years, when Zn fertilizer in soils was used more commonly, deficiencies of Mn and Fe emerged very fast in the intensive rice–wheat cropping system. Manganese deficiency is emerging very fast, particularly in wheat crops

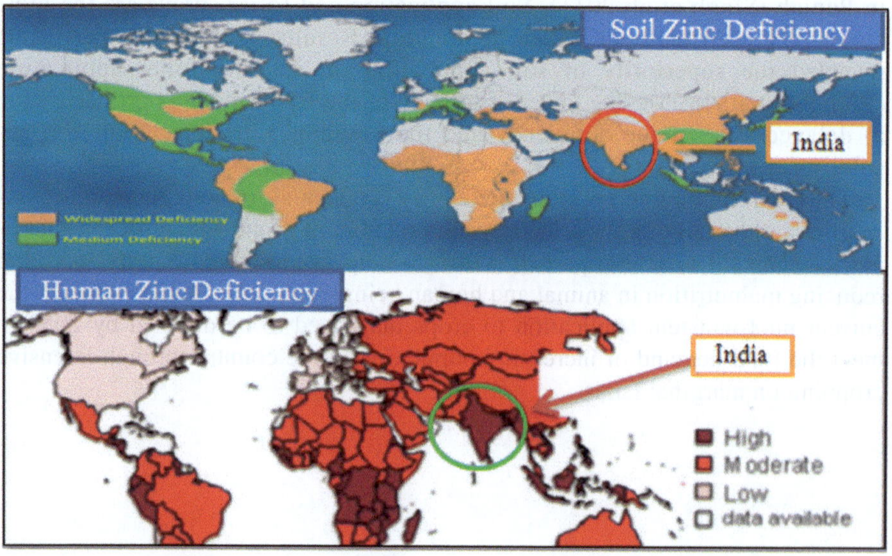


**Fig. 4.1** Soil and human and zinc deficiency. (Modified from Cakmak 2011)

grown after rice in Haryana (12%) and Punjab (18%) due to continuous leaching of Mn from the surface of the coarse-textured soils.

## 4.3 Micronutrient Deficiency in Soils of India

Deficiency of micronutrient has become a major constraint to the crop productivity, stability, and sustainability of crops in many Indian soils and may further deteriorate due to global warming (Kumar et al. 2011; Datta et al. 2017a). The problem is in the areas like arid and semiarid regions where soil pH value and $CaCO_3$ content are high and soil organic matter content is low (Katyal and Vlek 1985; Takkar 1996; Singh 2008). In order to assess the micronutrient status of soil, a large number of soil and plant samples were collected and analyzed for micronutrients, viz., Zn, Mn, Fe, Cu, Mo, and B, under the aegis of Indian Council of Agricultural Research (ICAR), All India Coordinated Research Project of Micro and Secondary Nutrients and Pollutant Elements in Soils and Plants [AICRP (MSPE)], and other state agencies. The Zn-deficient areas have been delineated, and soil fertility maps have been developed for the country (Fig. 4.2). Surface (0–15 cm depth) and some profile soil samples were also collected to assess the extent of micronutrient deficiencies in different soil types (Katyal and Sharma 1991). Though soils of India are sufficient in total micronutrient contents, but their availability in soil is very low. The concentration of the micronutrients of soil varied widely with respect to the types of soil, cropping system, and management conditions.

The analysis of more than 2.0 lakhs soil samples by AICRP-MSPE during 2011–2017 revealed widespread Zn and Fe deficiency in soils. On average, 36.5 and 12.2% soils are deficient in Zn and Fe, respectively. The spatial distribution of Zn deficiency varied from state to state. More than 50% soils of the states like Tamil Nadu (63.3%), Rajasthan (56.5%), Madhya Pradesh (57.1%), and Goa (55.3%) exhibited Zn deficiency, while the states like Arunachal Pradesh, Himachal Pradesh, Meghalaya, Mizoram, Nagaland, Tripura, and Uttarakhand had Zn deficiency in less than 10% of soils. More than 20% soils had Fe deficiency in states like Rajasthan (34.4%), Gujarat (25.9%), Haryana (21.7%), and Maharashtra (23.1%), while the states like Uttar Pradesh, Telangana, Andhra Pradesh, Bihar, Goa, and Tamil Nadu had deficiency in 10–20% of soils. Owing to the wide variations in physicochemical properties of soil, variations in the proportion of Zn-deficient soils have been observed in different districts within the states (Shukla and Tiwari 2016).

Iron is present in soil in different forms like the pool of immediately available Fe, the available Fe, Fe available on decomposition, and potential medium- to long-term sources of available Fe (Katyal and Deb 1982). Indian soils are comparatively rich in plant-available Fe and its availability in some states like Gujarat, Haryana, Maharashtra, Telangana, and Andhra Pradesh is posing threat to the crop production

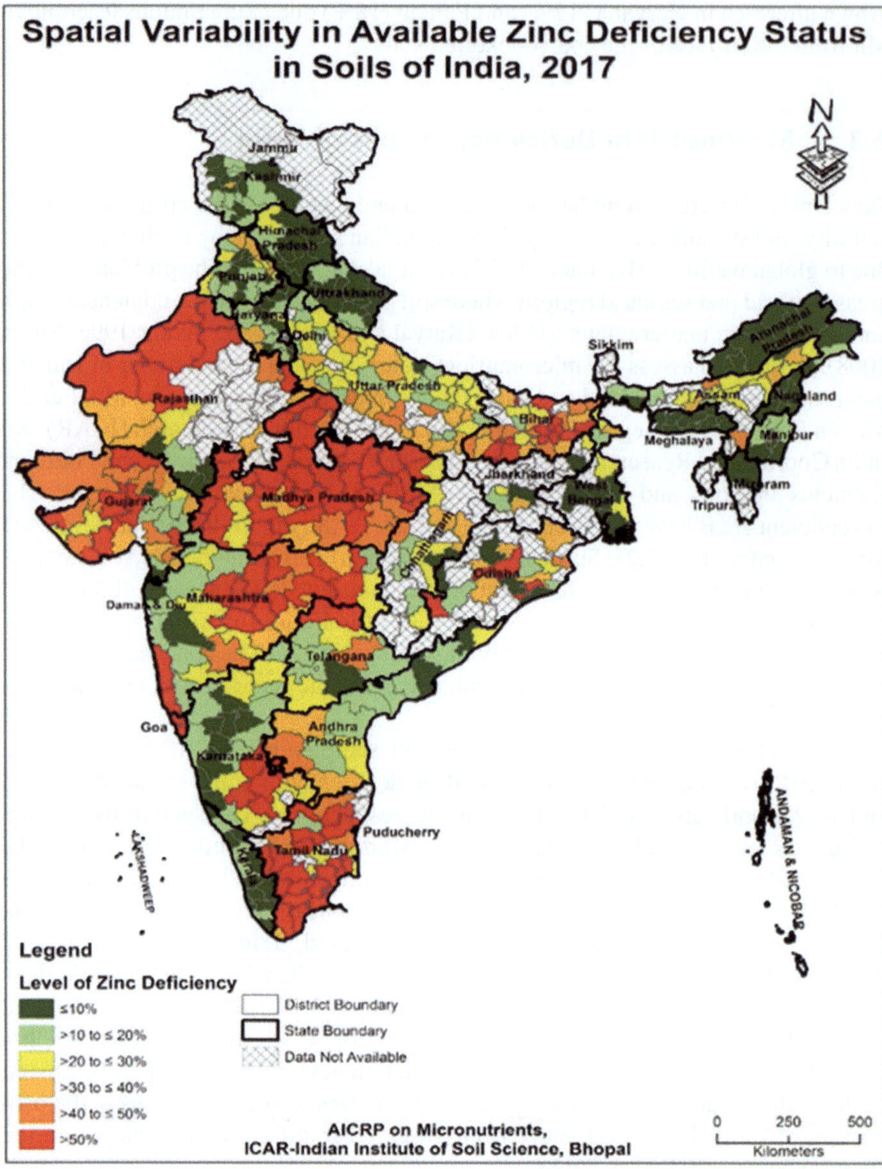


**Fig. 4.2** Zinc deficiency status in soils of different states of India

(Fig. 4.3), in soils of the different states of the country, Fe varies from 0.00 to 34.4% soil with minimum at Assam where not a single sample was deficient in Fe and maximum at Rajasthan (34.3%). Overall Fe deficiency in India stayed close to 12%, however, in some of the states like Gujarat, Haryana, Maharashtra, Telangana and

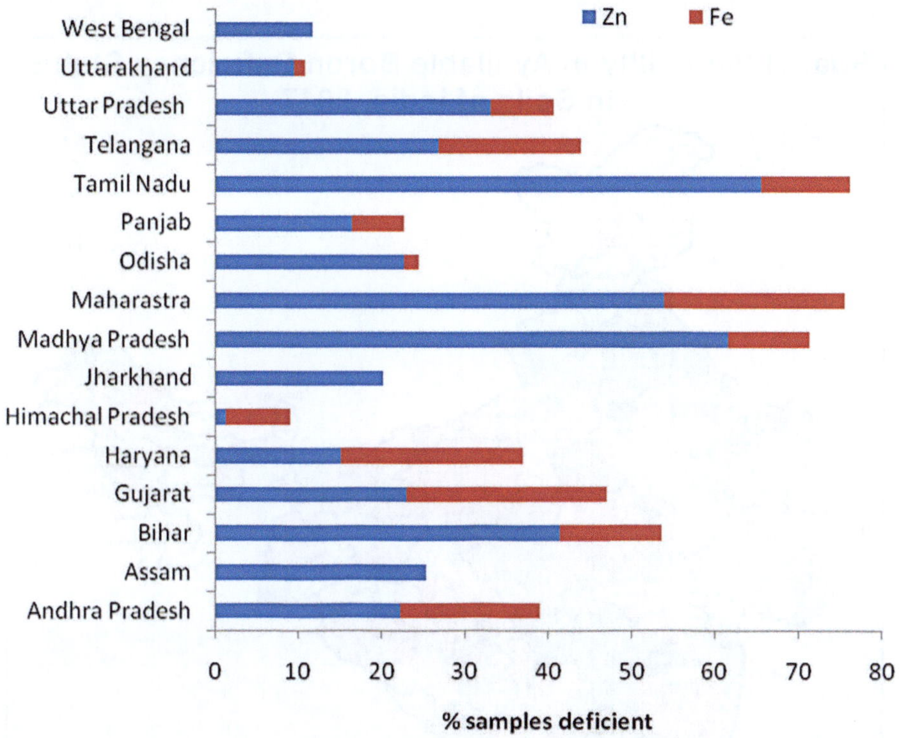


**Fig. 4.3** Zn and Fe deficiency status in soils of different states of India

Andhra Pradesh the deficiency exhibited 25.9%, 21.7%, 23.1%, 17.6% and 17.2% respectively (Shukla and Tiwari 2016).

Boron deficiency is becoming a serious constraint to sustainable agricultural productivity. In Indian soils, B deficiency ranged from 2.9 to 60.0%. Deficiency of B is more common in highly acid soils of Jharkhand (60.0%), Nagaland (54.3%), Odisha (48.9%), Jammu and Kashmir (48.9%), and Meghalaya (47.9%), whereas it was negligible in the soils of Andhra Pradesh (2.9%), Rajasthan (3.0%), and Madhya Pradesh (4.3%) (Shukla and Tiwari 2016; Dhakal et al. 2015). The spatial distribution of B deficiency from state to state is presented in Figs. 4.4a and 4.4b.

The availability of micronutrients under floodplain alluvial agroclimatic zone of West Bengal, India, may be influenced by the different agroclimatic features (ACF) and also variation in soil texture, pH, and organic matter content, which may ultimately affect the optimum crop yield (Ray and Banik 2016; Gogoi et al. 2018). It is obvious that the soil micronutrient availability varied with variations in soil physicochemical properties in different agroclimatic features in the alluvial zone of West Bengal. It was observed that soils were mostly deficient in Zn (20–33%) and B (60–92%), which underlines the need of Zn and B application in order to enhance the crop productivity.

a

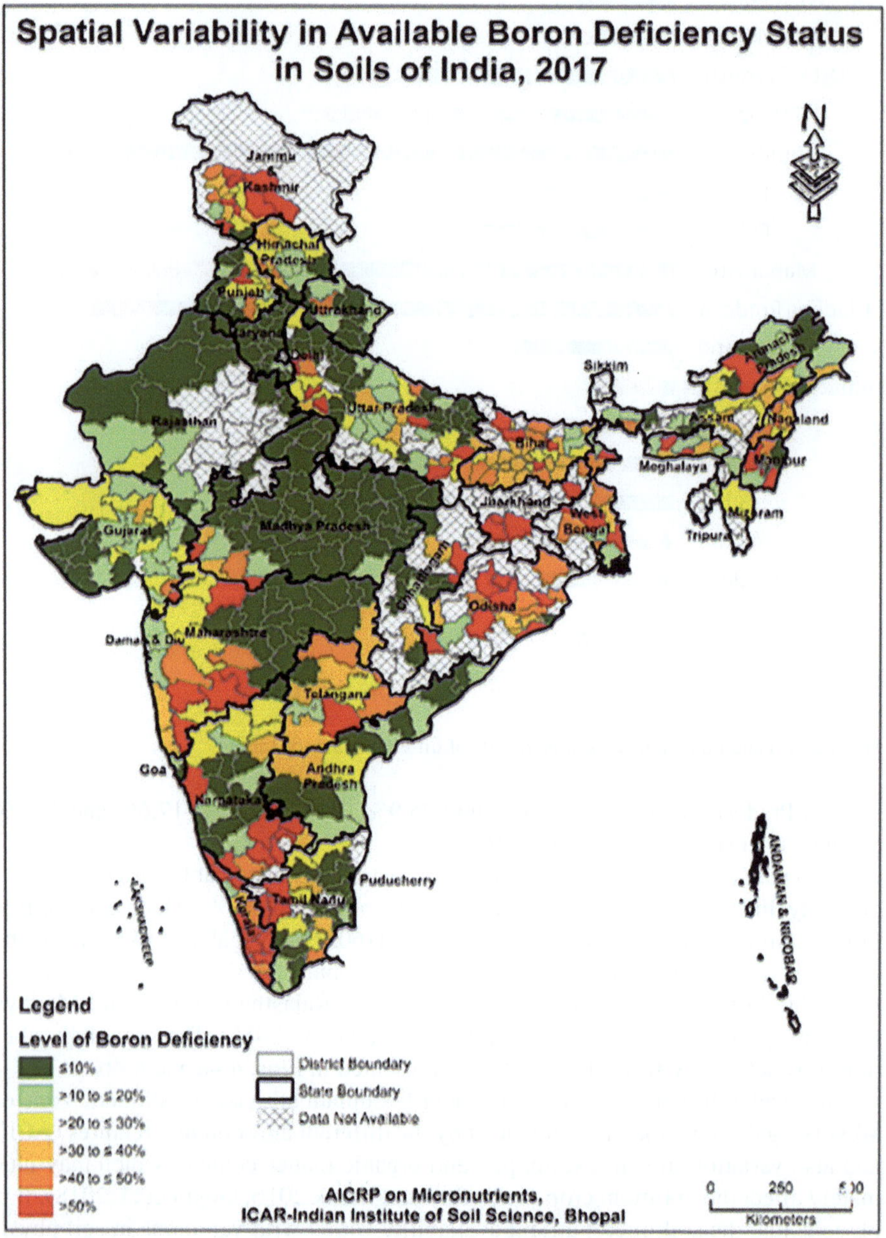


**Fig. 4.4a** Status of B deficiency in soils of different states of India

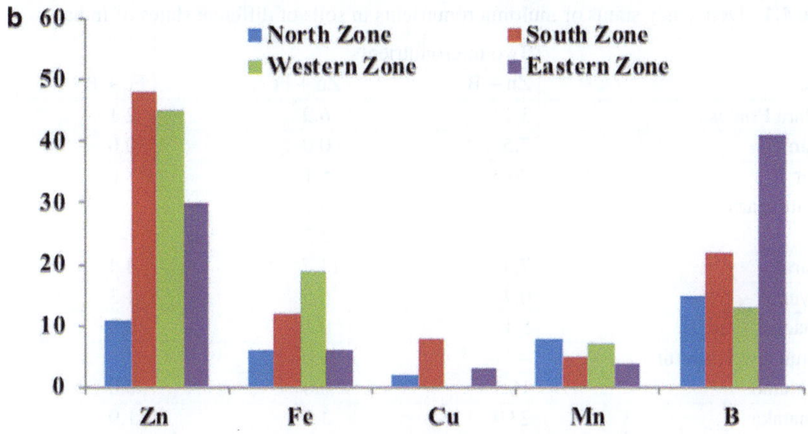


**Fig. 4.4b** Status of micronutrient deficiency in different zones of India (2011–2016)

### 4.3.1 Multimicronutrient Deficiency

Deficiencies of multiple micronutrients in crops in Indian soils due to depletion in fertility are an emerging issue in agriculture. From the last two decades, multimicronutrient deficiencies for Zn + Fe (5.85%), Fe + B (3.0%), and Zn + B (9.8%) have been noticed (Table 4.1). Among the micronutrient combinations, the deficiency of Zn + B was much higher in states like Karnataka (23.4%), Bihar (20.3%), Tamil Nadu (13.3%), and Odisha (14.0%). Deficiency of Zn + Fe was prevalent in the area of Rajasthan (23.3%), Gujarat (11.7%), and Maharashtra (10.1%). Another combination of Fe + B deficiency was highest in Maharashtra (9.8%) (Shukla and Tiwari 2016; Yadav et al. 2017b).

## 4.4 Micronutrient Deficiencies in Crops

Steady growth of crop yields during the past few decades compounded the problem by progressively depleting micronutrient pools. Some common farming practices in agriculture (such as liming acid soils) contribute to widespread occurrence of micronutrient deficiencies in crops by decreasing the availability of the micronutrients present in the soil. The problems in alleviating micronutrient deficiencies include difficulties to identify field crop symptoms, variation in soil micronutrient status, soil pH, intensity, seasonal fluctuations in the levels, temperature regimes in the region, and inadequate facilities and field tests to validate soil and plant micronutrients in the region (IRRI). Assessment of micronutrient deficiency can be made through visual leaf symptoms and soil and plant analyses. Crop response to the application of micronutrients through soil/foliar not only confirms the deficiencies but also helps in determining nutrient needs of crops. Visible symptoms of Zn and Fe deficiency in some crops are depicted in Plates 4.1 and 4.2.

**Table 4.1** Deficiency status of multimicronutrients in soils of different states of India

| | Two micronutrients | | |
|---|---|---|---|
| State | Zn + B | Zn + Fe | Fe + B |
| Andhra Pradesh | 3.1 | 6.9 | 2.1 |
| Assam | 7.5 | 0.0 | 0.0 |
| Bihar | 20.3 | 5.1 | 8.0 |
| Chhattisgarh | – | 3.1 | – |
| Goa | – | 2.2 | – |
| Gujarat | 7.3 | 11.7 | 4.4 |
| Haryana | 0.7 | 6.5 | 1.3 |
| Himachal Pradesh | 5.1 | 0.1 | 0.3 |
| Jammu and Kashmir | – | 0.0 | – |
| Jharkhand | 10.1 | 0.0 | 0.0 |
| Karnataka | 23.4 | 3.9 | 1.9 |
| Kerala | 4.9 | 0.9 | 0.4 |
| Madhya Pradesh | 0.9 | 5.8 | 0.2 |
| Maharashtra | 7.9 | 10.1 | 9.8 |
| Manipur | 3.1 | 0.0 | 0.0 |
| Odisha | 14.0 | 3.8 | 3.4 |
| Panjab | 4.2 | 3.7 | 2.2 |
| Rajasthan | 3.3 | 23.3 | 0.4 |
| Tamil Nadu | 13.3 | 8.2 | 2.0 |
| Telangana | 8.9 | 4.7 | 4.1 |
| Uttar Pradesh | 8.0 | 4.7 | 1.4 |
| Uttarakhand | 0.9 | 0.2 | 0.0 |
| West Bengal | 4.9 | 0.0 | 0.0 |
| **All India** | **9.8** | **5.8** | **3.0** |

**Plate 4.1** Symptoms of Zn deficiency in maize and wheat crop (interveinal chlorosis of young leaves with some greenness to veins, short internodes and small leaves, resetting or whirling of leaves)

**Plate 4.2** Symptoms of Fe deficiency in groundnut and potato crop (plants may include interveinal chlorosis in younger leaves; leaves may turn white as it is the constituent of chlorophyll)

## 4.5 Micronutrient Management Strategies

Micronutrient management varies with crops, type of soil, severity of deficiency, and method, time, and frequency of application. The large variation in rate of micronutrient application has emanated from the sensitivity of crops to soil type and deficiency status, soil environment, sources and their residual effects, and method of application. Consistent monitoring of the soils under different crops and cropping systems acts as a useful guide in determining the frequency, rates, sources, and time of application of the micronutrients. Thus, many aspects need to be considered when interpreting the results of the studies. While planning replenishment of the micronutrients removed by the crop and/or depleted from soil, it has to be taken into consideration. For the management of the micronutrient, there is need to maintain effective balance between demand set by the plants and supply from the soil (Shukla et al. 2016; Dadhich and Meena 2014) of these nutrients which is both crop and soil specific (Fig. 4.5).

### 4.5.1 Zinc (Zn)

A huge number of crops and cropping system based on rice, wheat, maize, sorghum, cotton, soybean, sugarcane, potato, mustard, groundnut, green gram, black gram, and chickpea occupy most of the areas across the country and have shown high responses to Zn application. Zinc fertility status continued to decline, and as a result, a number of responsive trials reached 72% during 2001–2010. Therefore, use of Zn fertilizer proved a highly profitable option in 58% of cultivated soils in India during the year 1967–1984, which has increased to 63% during the year 1985–2001 and 72% during the year 2002–2011 (Shukla and Behera 2011; Buragohain et al. 2017; Kumar et al. 2018b).

In intensively cultivated rice–wheat cropping system, Zn application is needed once each year. Generally, in Zn-deficient lowland rice areas, Zn is applied to rice at 5 kg Zn or 25 kg $ZnSO_4.7H_2O$ $ha^{-1}$, and wheat is grown on residual Zn fertility (Shukla and Behera 2011). Seed priming with Zn can improve crop emergence, stand establishment, subsequent growth, and yield. For example, priming *Echinacea*

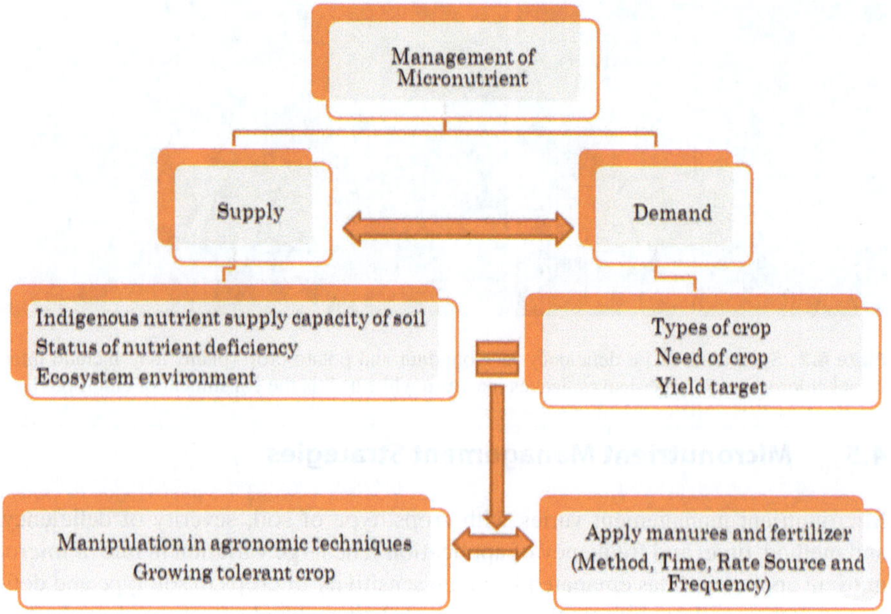


**Fig. 4.5** Steps in micronutrient management in soil–plant system. (Modified from Shukla et al. 2016)

*purpurea* (L.) seed with 0.05% $ZnSO_4$ solution increased germination and field emergence by 38 and 41%, respectively (Babaeva et al. 1999). Similarly, in common bean (*Phaseolus vulgaris* L.), seed priming with Zn significantly improved yield and related traits (Kaya et al. 2007). In barley (*Hordeum vulgare* L.), seed priming with Zn improved germination and seedling development (Ajouri et al. 2004). Recently Zn-coated urea has been reported as a noble source of Zn for rice and wheat (Shivay et al. 2008a, b).

The optimal rate of Zn application to rice was higher (22 kg $ha^{-1}$) in highly sodic (pH >10.0) and in floodplain soils (Takkar and Nayyar 1981) compared to 11 kg $ha^{-1}$ in moderately alkaline soils (pH 9.4–9.7; Takkar and Singh 1978) and 2.5 kg $ha^{-1}$ in sandy alkaline soil (Takkar et al. 2004; Kumar et al. 2018a).

#### 4.5.1.1 Soil Management

Soil application of Zn fertilizer increased grain Zn content in several cereal crops by a factor of two to three, depending on species and crop genotype. The type of soil also influences the extent of increase in Zn content in grain as a consequence of soil Zn fertilization. In contrast to Zn, soil application of inorganic Fe fertilizers to Fe-deficient soils is usually ineffective because of quick conversion of Fe (II) into unavailable Fe (III) forms. In contrast, an application of synthetic Fe chelates for correction of Fe deficiency is effective but expensive and is likely to be even more uneconomic if the aim is to increase Fe concentration in the grain rather than to increase yield. Rengel et al. (1999) noticed that the most effective fertilization could

**Table 4.2** Zn concentration in wheat grain as influenced by soil Zn fertilization

| Zn fertilization mg Zn $kg^{-1}$ soil | Yield (g $plant^{-1}$) | Grain Zn content ($mg\ kg^{-1}$) | Grain Zn content (ng Zn per seed) | Grain Zn content (μg Zn per plant) |
|---|---|---|---|---|
| 0 | 1.00 ± 0.17 | 9.1 ± 0.4 | 233 ± 35 | 9 ± 2 |
| 0.05 | 2.20 ± 0.13 | 9.9 ± 0.6 | 307 ± 25 | 22 ± 2 |
| 0.2 | 2.24 ± 0.16 | 14 ± 0.7 | 404 ± 11 | 31 ± 2 |
| 0.8 | 2.51 ± 0.30 | 83 ± 4 | 2540 ± 95 | 205 ± 16 |
| 3.2 | 1.70 ± 0.03 | 145 ± 5 | 3750 ± 330 | 245 ± 7 |

Zn content of the seed sown = 600 ng $seed^{-1}$ (Cultivar Excalibur)

**Table 4.3** Effect of foliar application of Zn and Fe on their content in wheat grain

| *Variety* | *Zn content (μg $g^{-1}$)* | | | | *Fe content (μg $g^{-1}$)* | | | |
|---|---|---|---|---|---|---|---|---|
| | *−Zn* | *+Zn* | *Mean* | *% increase* | *−Fe* | *+Fe* | *Mean* | *% increase* |
| PDW 274 | 31.7 | 40.3 | 36.0 | 27.1 | 36.6 | 47.7 | 42.2 | 30.3 |
| PDW 291 | 31.2 | 43.3 | 37.3 | 38.8 | 38.1 | 45.6 | 41.8 | 19.7 |
| PBW 343 | 32.9 | 38.6 | 35.7 | 17.3 | 40.8 | 50.9 | 45.8 | 24.8 |
| PBW 502 | 31.8 | 42.8 | 37.3 | 37.7 | 38.9 | 47.9 | 43.4 | 23.1 |
| PBW 550 | 30.4 | 40.0 | 35.2 | 31.6 | 42.1 | 47.6 | 44.8 | 13.1 |
| Mean | 31.6 | 41.0 | 36.3 | 21.9 | 39.3 | 47.9 | 43.6 | 21.9 |
| | *Zn* | | | | *Fe* | | | |
| | *Variety* | | *NS* | | *Variety* | | *NS* | |
| CD (5%) | Foliar spray | | 2.3 | | Foliar spray | | 2.3 | |
| | Interactions | | NS | | Interactions | | NS | |

Four foliar applications of 0.5% each of Zn and Fe separately at different growth stages

be via soil (for Zn) and foliar (for Fe) to increase micronutrient density in grain of field crops (Table 4.2).

#### 4.5.1.2 Foliar Fertilizer Applications

Absorption of Zn from nutrient solution was more efficient than from soil; foliar application of Zn was even more effective than application to the rhizosphere in providing Zn for transport to soybean grains, indicating that foliar spraying with Zn in field-grown crops can be effective in increasing Zn content in grains. Dhaliwal et al. (2009) reported that increase in Zn content in different varieties of wheat varied from 17.3 to 38.8%, whereas increase in Fe content of different varieties of wheat varied from 13.1 to 30.1% when Zn and Fe were applied through foliar for four times at 0.5% of $ZnSO_4$/$FeSO_4$ each at different growth stages (Table 4.3). Foliar supplementation with $ZnSO_4$ or $FeSO_4$ generally increases crop yields considerably more than Zn or Fe concentration in grain (Gupta 2005).

#### 4.5.1.3 Seed Treatment

Seeds of cereal crops can be very well treated with different sources of micronutrients before sowing in order to increase their content in harvested grains. Seed treatment with micronutrients is gaining momentum in recent times in research programs.

**Table 4.4** Response of priming seeds with solution of zinc sulfate on grain yield, content, and uptake of zinc in wheat

| | Seed treatment | | |
|---|---|---|---|
| Grain | Nonprimed | Primed | LSD |
| Yield (t $ha^{-1}$) | 2.52 | 2.86 | 0.123 |
| Zn content (mg $kg^{-1}$) | 27.8 | 31.1 | 1.88 |
| Zn uptake (g $ha^{-1}$) | 70.0 | 88.7 | 7.44 |

Yilmaj et al. (1997) observed that soil + foliar application of zinc registered maximum Zn content in wheat grain, while similar increase in Zn content of whole shoot of wheat was found under seed + foliar application. Harris et al. (2008) also observed significant enhancement in yield, Zn content, and its uptake by wheat due to seed treatment with 0.3% Zn solution over no seed treatment (Table 4.4).

### 4.5.2 Iron (Fe)

Interrelated strategies for soil and crop management are attractive not only for improving growing conditions for different crops but also for exploiting potential of plant for Fe mobilization as well as utilization by crop. Research development in soil and crop management strategies has provided the means to resolve complex plant Fe nutritional problems through rhizosphere fertilization and water regulation, managing cropping systems and screening for Fe-efficient genotypes, etc. Some simple and effective soil management practices such as root feeding and bag fertilization have been popularized and widely used by local farmers in India to improve the Fe content in food grains.

#### 4.5.2.1 Soil Management

In the past several years, farmers have applied mineral fertilizers to soil in order to improve the health of their plants; however, it is very difficult to apply Fe, because most inorganic Fe in soil is inaccessible to plants. Numerous studies have showed that application of inorganic Fe fertilizers to Fe-deficient soils has been found beneficial for many crops. When $FeSO_4$ applied to calcareous soils, it quickly reacts with $CaCO_3$ and form Fe oxides that are less available for plant uptake (Vempati and Loeppert 1988). Although application of synthetic Fe chelates in soil, such as Fe-EDTA or Fe-EDDHA, has shown better results than inorganic Fe salts, such as $FeSO_4$, these compounds are expensive, and the outcome is rarely fruitful (Shenker and Chen 2005; Varma et al. 2017).

#### 4.5.2.2 Crop Management

Uptake of micronutrient and its transportation to the edible parts of plants can be increased by foliar fertilizer applications. Leaf-applied substances can enter leaves either by penetration of the leaf cuticle or via the stomatal pathway. Foliar application is increasingly used to alleviate micronutrient deficiencies; many studies have been focused on uptake and distribution of a single micronutrient in fruit tree, corn, and wheat (Haslett et al. 2001; Godsey et al. 2003; Dadhich et al. 2015). It was remarkable that Fe concentrations in the polished rice from optimal foliar fertilization in the confirmation experiment were markedly increased by 2.6 times comparing with no foliar application (Fang et al. 2008; Verma et al. 2015b; Mitran et al. 2018). In another case study, spraying of 2500 mg $l^{-1}$ $FeSO_4$ solution two to three times during the flowering period increased the yield of soybean by 17.5–22.9% (Han et al. 1994). The most widely used Fe sources for foliar spraying are inorganic Fe forms and chelates mixed with inorganic Fe forms. Chelated forms of Fe are usually more effective in reducing Fe chlorosis than are inorganic forms (Vempati and Loeppert 1988). There are a few studies that aimed to increase yields and Fe concentrations in grains by different Fe fertilizers. Although foliar fertilization enhanced crop yields to a greater extent than it increased the Fe concentration in grains, it might be the only available fertilization practice that can slightly increase Fe concentration in grain (Rengel et al. 1999; Frossard et al. 2000). Some simple and cost-effective practices to improve Fe nutrition of fruit crops have been put forward and widely used by local farmers in China, such as trunk $Fe^{2+}$ injection (Liu et al. 2002). These technical approaches could play a significant role in improving Fe nutrition of crops in the short term, resulting in good financial returns for farmers.

### 4.5.3 Boron (B)

Boron availability in soil to plant is mainly related to the total B content and other soil properties such as pH, $CaCO_3$, organic matter contents, nutrient interactions, varieties, and environmental factors, which influence the emergence of B deficiency or toxicity in plants (Sakal et al. 1996; Saha and Singh 1997). It has been noticed that an increase in organic carbon (OC) content from 0.50 to 0.75% enhances the fixation of B in soils by 48–60%. Therefore, association of B with OC prevents its leaching and thereby ensures its higher availability to crop plants (Katyal and Vlek 1985). High concentrations of B were recorded in the saline soils of the Indo-Gangetic plain and moderate levels in Vertic Ustochrepts of Rajasthan and Madhya Pradesh (Mathur et al. 1964; Saha et al. 1998; Ram and Meena 2014).

Borax ($Na_2B_4O_7.10H_2O$) as soil application, boric acid ($H_3BO_3$) properly as foliar application, disodium octaborate tetrahydrate ($Na_2B_8O_{13}.4H_2O$), and boronated single superphosphate, these are the B carriers included in the FCO (2003). Solubor ($Na_2B_4O_7.5H_2O + Na_2B_{10}O_{18}.10H_2O$) can be used for both soil and foliar application because of its higher solubility.

A large number of crops have responded to B fertilization. Precise B fertilization is important for the normal growth, yield, and quality of produce due to a very narrow range of B deficiency and toxicity in soil and plants (Singh and Goswami 2013; Datta et al. 2017b). Boron application at 0.75 kg $ha^{-1}$ and 1.5 kg $ha^{-1}$ in spring sunflower was effective, and the crop responded well up to the second dose, and the higher (1.5 kg $ha^{-1}$) level gave the highest seed yield (2.01 t $ha^{-1}$), which was 13.5

and 6.3% more than that of the control and 0.75 kg B $ha^{-1}$, respectively (Shekhawat and Shivay 2008; Sofi et al. 2018).

Soil application of 20 kg sodium tetraborate (14% B) to supply 2.8 kg B $ha^{-1}$ or two foliar applications with 0.2% solution of this salt was found equally effective for increasing grain yield of soybean (Dwivedi et al. 1990). Deficiency of boron is also invariably corrected by its soil application depending upon the soil type (Arora et al. 1985; Sakal et al. 1988; Ali and Monoranjan 1989). The rate varying between 1.0 and 2.5 kg B $ha^{-1}$ has been found to be optimum for different crops in calcareous soils of Bihar (Sakal et al. 1988; Sinha et al. 1991).

Foliar application of 0.2% boron at flowering along with three irrigations at branching, pre-flowering, and pod development stages is optimum for realizing optimum yield of summer mungbean on a sandy loam alluvial soil in West Bengal (Mondal et al. 2012). Similarly, soil- and foliar-applied boron have been found to have significant effect on growth and yield of groundnut (Ansari et al. 2013; Yadav et al. 2017c). They clearly demonstrated that solubor as soil application at 10 kg $ha^{-1}$ can be applied to achieve better land utilization, high yield, as well as productivity and profitability than other treatments under rainfed sandy loam soils.

### 4.5.4 Manganese (Mn)

Manganese deficiency has emerged in states of Punjab, Haryana, and western Uttar Pradesh where rice–wheat system is practiced on highly permeable coarse-textured soils. During rice cultivation, solubility of Mn increases due to reduction under submerged condition and leaches down to lower soil layers in 4–7 years of cultivation. The subsequent wheat and berseem crops suffer due to Mn deficiency. Studies conducted in Punjab revealed that adoption of deep tillage practices can ameliorate Mn deficiency in coarse-textured soils and enhance wheat yield and uptake of Mn by wheat. Crop responses to application of Mn either through soil application or foliar spray have been reported by Takkar et al. (1989) in different parts of the country. Soil application of manganese sulfate ($MnSO_4$) at 50 kg $ha^{-1}$ is recommended for getting the optimum crop response. However, foliar spray of 0.5% $MnSO_4$ solution to wheat, oat, and berseem was more economical to soil application. Generally, 3–4 sprays of $MnSO_4$ solution are required to correct Mn deficiency in wheat and oat.

Soaking of potato tubers in 0.05% $MnSO_4$ solution for 3 h proved 2.7 times more effective than soil application of 20 kg $ha^{-1}$ of $MnSO_4$ and 11% more effective than two foliar sprays of 0.2% $MnSO_4$ solution in increasing the tuber yield (Sharma and Grewal 1988). Manganese-efficient cultivars found to have higher harvest index as well as grain yield as compared to Mn-inefficient cultivars (Jhanji et al. 2013). The magnitude of response of 53 cultivars to Mn application on coarse-textured Mn-deficient field decreased successively as the rating of the tolerance increased and there were no significant responses in the most tolerant categories (Takkar and Walker 1993).

## 4.6 Micronutrient Fertilizers in Indian Agriculture

Last five-decade intensive research has been conducted on micronutrient fertilizer application schedule in crops and cropping system. But still, no systematic information is available on the use of micronutrient fertilizers in India. About 125 small-scale units are manufacturing $ZnSO_4$ in the country with a capacity to produce around 0.175 million tons per annum. It is estimated that, in order to ameliorate micronutrient deficiencies by the year 2025, India will need to apply a 324,000 t Zn, 130,000 t Fe, 11,000 t Cu, 3900 t B, and 22,000 t Mn as fertilizers annually (Takkar et al. 1997). The efficiency of fertilizer particularly zinc management through 5 R principles (*Right* source, *Right* rate, *Right* place, *Right* time, and *Right* method) could be improved through fertilization practices that include an application of micronutrients according to crop requirements (Rattan 2017; Verma et al. 2015c). An adequate supply of credit for farmers and distributors is necessary to ensure the availability of fertilizers when and where they are required.

In India, among micronutrient fertilizers, maximum $ZnSO_4$ is produced. During the year 1991–92, production of $ZnSO_4$ was 58,440 t against installed capacity of 157,050 t (PDIL 1996). Consumption of $ZnSO_4$ increased from few hundred tons (t) in 1970s to attain peak at 194,406 t in 2013–2014, and recently in 2016–2017, it has been 179,824 t (Fig. 4.6). Similarly the consumption of FeSO4, $MnSO_4$, and $CuSO_4$ in 2016–2017 was 21,658, 4287, and 1609 t, respectively (FAI 2017). Numerous studies have attempted to examine the role of price and non-price factors in the growth of fertilizer use in India (Fig. 4.6; Raju 1989; Kundu and Vishist 1991; Rabobank 2005; Sihag et al. 2015).

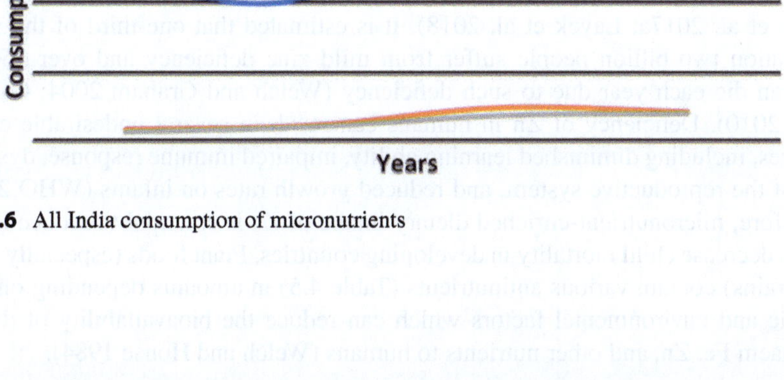


**Fig. 4.6** All India consumption of micronutrients

## 4.7 Improving Micronutrient Nutrition in Staple Food Crops

Micronutrient fertilizers can have significant effects on the accumulation of nutrients in edible plant products (Grunes and Allaway 1985; Allaway 1986). It is expected that almost half of the Indian soils are deficient in plant-available Zn (Singh 2009; Shukla and Behera 2011), leading to reductions in crop production and also nutritional quality of the harvested grains (Shukla and Behera 2012; Verma et al. 2015a). As cereal grains contain inherently very low amount of Zn, growing them on potentially Zn-deficient soils further decreases grain Zn concentrations. As cereal-based foods, rice and wheat are the major source of daily calorie intake in developing countries like India; hence widespread occurrence of Zn deficiency is reported in human. Studies conducted under All India Micronutrient Project in Nalgonda and Ranga Reddy districts in Andhra Pradesh indicated that soils having low Zn status produced plant grains with lower Zn content. People feeding on such grains and other vegetation showed lower Zn content in their blood plasma compared to areas which had high available Zn status and lower Zn deficiency in soil (Singh 2009). Severe Fe anemia was found in 34% in adolescent girls of Bikaner, Rajasthan, and Gujarat (Seshadari 1998). The concentration of Zn, Cu, Fe, and Mn in drinking water and soil is correlated with dental caries in 1516 children age between 7 and 17 years in 10 rural areas in the district of Ludhiana (Guaba 1983). Crops require minerals and organic materials to transform nutrients into forms that plants can use for growth. Without minerals and soil organic matter, it is impossible to sustain a healthy crop which is the basis for the nutrition values of animals and human. The bioavailability of minerals for plant growth has been significantly decreased as a result of accelerated withdrawal of minerals from the soil without corresponding additions and has severely impacted on human health.

## 4.8 Role of Micronutrient for Soil–Plant–Animal–Human Continuum

It has been estimated that approximately 3.7 billion populations are Fe-deficient, with 2 billion of these highly deficient in Fe that they can be described as being anemic (WHO). In addition, 35% of all children in the world between 0 and 5 years old suffer from Zn/Fe deficiencies, 250 million suffer from vitamin A deficiency, and 260 million suffer from iodine (I) or selenium (Se) deficiencies (Cababallero 2002; Yadav et al. 2017a; Layek et al. 2018). It is estimated that one-third of the world population two billion people suffer from mild zinc deficiency and over 450,000 children die each year due to such deficiency (Welch and Graham 2004; Cakmak et al. 2010). Deficiency of Zn in humans can result in several undesirable consequences, including diminished learning ability, impaired immune response, dysfunction of the reproductive system, and reduced growth rates on infants (WHO 2003). Therefore, micronutrient-enriched dietary intake of Zn is essential to reduce illness and to decrease child mortality in developing countries. Plant foods (especially seeds and grains) contain various antinutrients (Table 4.5) in amounts depending on both genetic and environmental factors which can reduce the bioavailability of dietary non-haem Fe, Zn, and other nutrients to humans (Welch and House 1984).

**Table 4.5** Antinutrients in plant foods that reduce Fe and Zn bioavailability and examples of major dietary sources

| Antinutrients | Essential micronutrients metal-inhibited | Major dietary sources |
|---|---|---|
| Phytic acid or Phytin | Fe, Zn, Cu, Ni | Whole legume seeds and cereal grains |
| Fibers (e.g., cellulose, hemicellulose, lignin, cutin, suberin, etc.) | Fe, Zn, Cu | Whole cereal grain products (e.g., wheat, rice, maize, oat, barley, and rye) |
| Certain tannins and other polyphenolics | Fe | Tea, coffee, beans, sorghum |
| Hemagglutinins (e.g., lectins) | Fe | Most legumes and wheat |
| Goitrogens | I | *Brassicas* and *Alliums* |
| Heavy metals (e.g., Cd, Hg, Pb, etc.) | Fe, Zn | Contaminated leafy vegetables and roots |

Graham et al. (2001)

**Table 4.6** Examples of substances in foods that promote Fe, Zn, and vitamin A bioavailability and major dietary sources

| Substance | Nutrient | Major dietary sources |
|---|---|---|
| Certain organic acids (e.g., ascorbic acid, fumarate, malate, citrate) | Fe and/or Zn | Fresh fruits and vegetables |
| Hemoglobin | Fe | Animal meats |
| Certain amino acids (e.g., methionine, cysteine, histidine) | Fe and/or Zn | Animal meats |
| Long-chain fatty acids (e.g., palmitate) | Zn | Human breast milk |
| Selenium | I | Sea foods, tropical nuts |
| Β-carotene | Fe | Green and orange vegetables |
| Inulin and other nondigestible carbohydrates (prebiotics) | Fe, Zn | Chicory, garlic, onion, wheat, Jerusalem artichoke |

Graham et al. (2001)

Some dietary substances that promote the bioavailability of Fe and Zn in the presence of antinutrients are also known (Table 4.6). Their levels are influenced by both genetic and environmental factors.

### 4.8.1 Agronomic Strategies

Agronomic manipulation is an inexpensive and simple approach which can be utilized to enrich genetically inefficient cultivars by application of micronutrient fertilizers at different rates, methods, and crop growth stages (Shukla and Tiwari 2014; Dhakal et al. 2016). Fertilizer studies focusing specifically on increasing Zn concentration of grain (or other edible parts) are, however, very rare, although a large number of studies are available on the role of soil- and foliar-applied Zn fertilizers

**Table 4.7** Effect of micronutrient application on grain yield and grain Zn/Fe/Mn concentration of different groups of cultivars at different locations in India

| Crops | | Efficient cultivars | | | | Inefficient cultivars | | | |
|---|---|---|---|---|---|---|---|---|---|
| **Zinc (Zn)** | | | | | | | | | |
| | | **Grain yield ($t\ ha^{-1}$)** | | **Grain Zn ($mg\ kg^{-1}$)** | | **Grain yield ($t\ ha^{-1}$)** | | **Grain Zn ($mg\ kg^{-1}$)** | |
| | | **−Zn** | **+Zn** | **−Zn** | **+Zn** | **−Zn** | **+Zn** | **−Zn** | **+Zn** |
| **1.** | **Bhopal** | | | | | | | | |
| A. | Pigeonpea | 1.41 | 1.54 | 32.6 | 43.8 | 1.06 | 1.41 | 35.1 | 48.2 |
| B. | Wheat | 3.72 | 3.87 | 41.0 | 47.8 | 2.85 | 3.37 | 43.0 | 56.3 |
| **2.** | **Hyderabad** | | | | | | | | |
| A. | Rice (dehusked) | 5.98 | 6.18 | 11.0 | 16.7 | 5.36 | 7.92 | 9.5 | 16.9 |
| B. | Maize | 5.04 | 6.13 | 24.2 | 27.4 | 4.39 | 6.59 | 23.7 | 29.5 |
| **3.** | **Pantnagar** | | | | | | | | |
| A. | Wheat | 3.71 | 3.95 | 20.3 | 43.1 | 3.26 | 4.23 | 15.1 | 43.8 |
| **Iron (Fe)** | | | | | | | | | |
| | | **Grain yield ($t\ ha^{-1}$)** | | **Grain Fe ($mg\ kg^{-1}$)** | | **Grain yield ($t\ ha^{-1}$)** | | **Grain Fe ($mg\ kg^{-1}$)** | |
| | | **−Fe** | **+Fe** | **−Fe** | **+Fe** | **−Fe** | **+Fe** | **−Fe** | **+Fe** |
| **4.** | **Anand** | | | | | | | | |
| A. | Pigeonpea | 2.50 | 2.42 | 34.1 | 36.0 | 2.27 | 2.55 | 33.7 | 38.5 |
| B. | Chickpea | 3.15 | 3.27 | 59.0 | 62.8 | 2.36 | 2.91 | 56.0 | 67.5 |
| **5.** | **Pusa** | | | | | | | | |
| A. | Maize | 5.19 | 5.55 | 46.8 | 66.2 | 5.22 | 6.22 | 41.3 | 63.2 |
| **Manganese (Mn)** | | | | | | | | | |
| | | **Grain yield ($t\ ha^{-1}$)** | | **Grain Mn ($mg\ kg^{-1}$)** | | **Grain yield ($t\ ha^{-1}$)** | | **Grain Mn ($mg\ kg^{-1}$)** | |
| | | **−Mn** | **+Mn** | **−Mn** | **+Mn** | **−Mn** | **+Mn** | **−Mn** | **+Mn** |
| **6.** | **Ludhiana** | | | | | | | | |
| A. | Rice | 6.71 | 6.85 | 41.4 | 53.0 | 4.83 | 5.51 | 31.6 | 44.2 |
| B. | Wheat | 5.02 | 5.45 | 25.0 | 33.1 | 4.24 | 5.20 | 19.9 | 30.2 |

in correction of Zn deficiency and increasing plant growth and yield (Rengel et al. 1999; Kakraliya et al. 2018).

In India, through NAIP-funded project on micronutrient enrichments, efforts have been made to identify genetically efficient cultivars of cereals and pulses for Zn and Fe to develop options for micronutrient biofortification. Genetically efficient and inefficient cultivars were identified based on Yield Efficiency and Uptake Efficiency Index. Interestingly, the genetically inefficient cultivars were agronomically highly efficient. Thus, the efficient cultivars could be utilized by breeders for QTL identification and developing high-yielding micronutrient enriched cultivars (genetic biofortification), while the inefficient cultivars were for agronomic biofortification to dense the grains of highly responsive cultivars with micronutrients. Application of micronutrient either through soil, foliar, or both could increase yield as well as concentration (Table 4.7). However, the variation in yield and concentration was driven by the type of crop and genetic makeup of the cultivars (Shukla et al. 2012). In case of efficient cultivars, the application of micronutrient had little effect on yield, but increase in Zn and Fe concentration was registered in all the crops. At Bhopal, the Zn concentration in efficient cultivars of pigeonpea increased by 34%, while it was 16.6% in case of wheat.

Experiment conducted for Zn enrichment in rice and maize at Hyderabad showed that Zn application could increase 51.8% Zn concentration in rice grain and 13% in maize. In case of inefficient cultivars, both grain yield and micronutrient (Zn and Fe) concentration increased with the application of micronutrient. This happens because the genetically inefficient cultivars are agronomically highly efficient and thus responded to external application of Zn. At Bhopal, application of Zn in inefficient cultivars enhanced the grain yield of pigeonpea by 33%, while that of wheat by 18%. The increment in Zn concentration was more in case of pigeonpea (37.2%) than that of wheat (30.9%). Inefficient cultivars of rice grown at Hyderabad showed 47.8% increase in yield and 77.8% increase in Zn concentration. In case of maize, the increase in Zn concentration was less than the rice, while grain yield enhanced by one and a half-fold. At Pantnagar, wheat yield increased up to 29.8%, but concentration increased approximately three times.

Effect of Fe application on grain yield and Fe concentration in grain was studied in pigeonpea and gram at Anand, Gujarat and in maize at PUSA, Bihar. Similar to Zn application, Fe could hardly influence the yield of efficient cultivars, but it had significant effect on Fe concentration in grain. The efficient cultivars of pigeonpea and chickpea grown at Anand exhibited 10 and 6% increase in Fe concentration, respectively, while in case of Fe-inefficient cultivars, pigeonpea and chickpea yield had increased by 14 and 20%, respectively, and the density of Fe concentration in both crops enhanced by 20%. At Pusa, seed loading with Fe enhanced by 46% in efficient cultivars of maize. In case of inefficient cultivars, increase in grain yield of maize was recorded 19% and concentration by 53%. The efficiency of micronutrients depends on the right method, right rate, and right time of application. There are different approaches to improve the micronutrient content of the edible part. One is to increase the efficiency of uptake and transport into edible tissue and second is to increase the amount of bioavailable micronutrient accumulation in the plant. Manganese concentration enhanced by 28 and 39.8% in efficient and inefficient rice cultivars, respectively, while in wheat this increase was recorded by 32 and 52%.

Strategies for micronutrient (Zn, Fe, and Mn) enrichment in different crops were developed using several permutation and combinations of nutrient management options. The cultivars of different crops identified as efficient may be grown in soils low in specific micronutrients. Of the several strategies used, soil plus three foliar feeding has been identified as best option for grain enrichment with Zn, in soils having low Zn status. In adequate Zn soils, two to three foliar sprays are sufficient to increase grain Zn concentration in rice, wheat, and pigeonpea. Foliar spray of K along with Zn was also an effective strategy for enhancing grain Zn concentration in pigeonpea. Among the Zn management strategies, soil plus foliar feeding was superior over foliar or soil application alone. Zinc applied to previous crop also contributed significantly to grain yield, and it was at par with treatment receiving soil Zn application in previous crop. In efficient group of cultivars, the grain yield remains unaffected due to either of Zn management strategies.

Grain Zn accumulation mechanism varied with efficiency of cultivars, its ability to translocate Zn from soil as well as from shoot to grain. The mechanism for increased root uptake in wheat may be related to proliferation of crown root growth, exudation of organic acids or phytosiderophores, and increased tolerance to Zn deficiency. In low Zn soil, Zn application to soil is inevitable in order to mitigate Zn deficiency at early growth stage, while in excess Zn soil, foliar feeding is an effective option in enhancing grain Zn concentration in wheat.

### 4.8.2 Physiological Mechanism for Absorption and Translocation of Fe in Gram

The experiment was conducted in micro-plot having Fe-deficient soil with the Fe treatments, viz., control, Fe soil (20 kg Fe $ha^{-1}$), and Fe spray (0.5% ferrous sulfate at flower initiation, pod development, and grain-filling stage), as well as nipping (at 45 DAS) and defoliation (at 45 DAS, 20% leaves) as physiological interventions to study the effect of physiological interventions on yield and Fe content in grain of gram. Average Fe concentration in gram grain increased by 2.9 (control 48.2 and highest 49.6 mg $kg^{-1}$) and 13.2% (control 50.2 and highest 56.8 mg $kg^{-1}$), respectively, over control in case of Fe-efficient and Fe-inefficient group of varieties. Soil application and foliar feeding showed beneficial effect to improve Fe content in grain. Among physiological interventions, defoliation was found superior over nipping to increase Fe content in gram grain (Fig. 4.7).

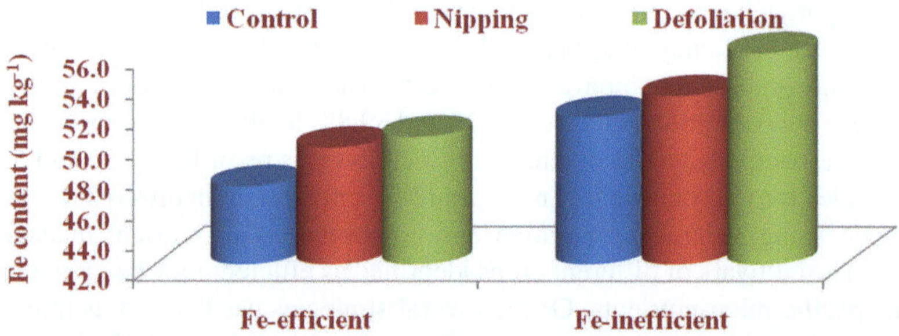


**Fig. 4.7** Effect of physiological treatments on Fe content in grain of Fe-efficient and Fe-inefficient groups of gram varieties

### 4.8.3 Bioavailability of Enriched Cereals Using Rat Model

Soils improve human health through the nutrients taken up by plants and animals that eat those plants, nutrients that are needed for adequate nutrition as humans consume the plants or animals. Enhanced micronutrient content in grain does not imply that it will be available to human and animal (Shukla et al. 2016; Kumar et al. 2017a). Therefore, Fe bioavailability was assessed under AICRP on micronutrient (NAIP) using rat models at Anand. The consumption of the experimental diets by rats was based on the AIN-93G (Reeves et al. 1993). The results revealed that Fe intake was more from pigeonpea-based diet and from seeds of inefficient cultivars that contain high Fe. The excretion of Fe was also higher in rats fed with Fe-enriched grain due to excess intake, but Fe supplied through enriched grain was bioavailable to animals as good as Fe supplied through standard sources.

The absorption of Fe was the highest, i.e., 416.15 ± 26.24 μg $day^{-1}$ in rats fed with standard diet followed by pigeonpea-efficient group (347.09 ± 29.21 μg $day^{-1}$) which did not differ from it. However, $D_1$ and $D_2$ groups differed ($p < 0.05$) either from $D_0$ or $D_3$ group. The iron concentration in the liver due to standard and pigeonpea-efficient diets (Table 4.8) was found statistically similar, but they differed from $D_1$ or $D_2$ groups. The iron content in the kidney of experimental animals fed with standard diet (34.52 ± 1.29 μg $g^{-1}$) was statistically ($p < 0.05$) higher than those fed with pigeonpea-based diets, i.e., $D_1$, $D_2$, and $D_3$; however, they did not differ among themselves. The Fe content of femur in the rats on standard and pigeonpea-based diets varied from 87.94 ± 2.03 to 99.95 ± 1.82 μg $g^{-1}$ which was significantly ($p < 0.05$) higher in rats fed with standard diet compared to control pigeonpea, pigeonpea-inefficient, and pigeonpea-efficient diets. The data on iron absorption and its deposition in the liver in pigeonpea-efficient ($D_3$) diet was comparable with standard diet. Thus, bioavailability of Fe from pigeonpea-efficient variety was comparable with standard diet comprising of ferrous sulfate as iron source. The iron intake (Table 4.9) of the rats from standard diet and pigeonpea-based diets varied from 702.90 ± 32.72 to 795.09 ± 46.61 μg $day^{-1}$ which was statistically similar.

## 4.9 Conclusion

The agricultural production scenario in the country is facing several challenges so far as enhancement in production and quality of the produce is concerned. The ICAR and SAUs in the country have very carefully taken up the program of micronutrient research after realizing the importance of its management long back in 1970s when Zn deficiency as *khaira* disease was noticed in *tarai* region of Uttar Pradesh and reported by YL Nene (1966). Almost all the parts have been covered by AICRP on MNS and pollutant elements to find out the deficient areas in different micronutrients. The whole picture has emerged indicating widespread deficiency of

**Table 4.8** Liver, kidney, and femur weight and Fe content in control and experimental animals fed with pigeonpea-based experimental diets

| Diet | Liver | | Kidney | | Femur | | |
|---|---|---|---|---|---|---|---|
| | Wet weight (g) | Fe content (μg $g^{-1}$) | Wet weight (g) | Fe content (μg $g^{-1}$) | Wet weight (g) | Fe content (μg) | Fe content (μg $g^{-1}$) |
| $D_0$ | 9.22 ± 0.67 | $57.60^b$ ± 1.14 | 1.39 ± 0.08 | $34.52^b$ ± 1.29 | 1.40 ± 0.07 | 139.44 ± 6.22 | $99.95^b$ ± 1.82 |
| $D_1$ | 8.92 ± 0.89 | $49.05^a$ ± 1.04 | 1.27 ± 0.11 | $28.07^a$ ± 0.22 | 1.29 ± 0.08 | 114.86 ± 8.38 | $88.76^a$ ± 1.27 |
| $D_2$ | 7.68 ± 0.44 | $52.45^a$ ± 0.88 | 1.28 ± 0.06 | $30.58^a$ ± 0.54 | 1.21 ± 0.08 | 105.95 ± 6.08 | $87.94^a$ ± 2.03 |
| $D_3$ | 9.08 ± 0.63 | $57.59^b$ ± 0.85 | 1.34 ± 0.05 | $29.77^a$ ± 1.05 | 1.30 ± 0.06 | 120.41 ± 5.15 | $92.56^a$ ± 0.94 |

*$D_0$* $FeSO_4$ (dried/anhydrous), *$D_1$* pigeonpea control, *$D_2$* pigeonpea-inefficient, and *$D_3$* pigeonpea-efficient
$^{ab}$Means with different superscripts in columns for a parameter differ significantly ($p < 0.05$)

**Table 4.9** Apparent absorption (μg $day^{-1}$) of iron in rats fed with pigeonpea-based experimental diets

| Diet | Fe intake | Fe excretion | Absorption |
|---|---|---|---|
| $D_0$ | 752.99 ± 39.89 | 336.85[a] ± 20.36 | 416.15[b] ± 26.24 |
| $D_1$ | 795.09 ± 46.61 | 494.51[b] ± 19.43 | 300.58[a] ± 31.03 |
| $D_2$ | 702.90 ± 32.72 | 433.43[ab]±25.11 | 269.48[a] ± 13.92 |
| $D_3$ | 748.14 ± 30.20 | 401.04[ab]±36.97 | 347.09[ab]±29.21 |

[ab]Means with different superscripts in columns for a parameter differ significantly ($p < 0.05$)

Zn followed by B, Fe, Mn, and Cu in specific regions of the country. Accordingly, strategies to combat with the deficiencies in soils and plants have been considered by identifying suitable agrotechnologies of micronutrient supplementation in soils and plants. The cheapest and efficient sources; their rate, frequency, and method of application; and possible remedial measures have been worked out for different crops and cropping systems.

It has been noticed that the straight inorganic fertilizers containing micronutrient elements are cheaper sources although the chelated sources are more efficient but still costlier venture for farmers. It has also been noticed that the micronutrient fertilizer efficiency is very less, i.e., 3–5% in many cases, but still they carry the residual effect and therefore need not apply to all the crops, but their recommendations are more meaningful when considered for the cropping systems. Also, native sources of micronutrients need to be exploited as they are unavailable to the plant roots due to one or other soil-related constraints like salinity/sodicity, high $CaCO_3$ content, low OC, excess acidity, etc. Therefore, the role of amendments like gypsum and lime and use of organics are greater in mitigating the micronutrient deficiencies in different regions having various soil types.

There is an emerging deficiency of multimicronutrients in some pockets of the country, but it could be managed with site-specific nutrient management or using customized micronutrients containing fertilizers. It is a need of the time to meet the requirement of crops for their micronutrient demand for dense micronutrient seeds/grains especially of Zn, Fe, Mn, etc., which are nutritionally important trace elements. There are several approaches to tackle this issue in which genetic breeding approach is one to solve the problem once for all in a long way and agronomical approach is a short-term approach which can give relief on temporary basis but needs attention every year with regard to its supplementation.

Although the researches on micronutrient management have been carried out so exhaustively from basic and application sides, the correct usage still needs attention of extension agencies and government policy planners. The fertilizer production and consumption of major nutrients like NPK monitoring and data recording systems are nicely arranged, while in case of micronutrients, no systematic information is available, and therefore, it becomes difficult to target the areas for priority attention. The quality products in markets are also not available for one or other reasons, and therefore furious products are sold in the market, and farmers are not only cheated, but they lose their confidence from the importance of micronutrient usage in their agriculture. The refined information based on GPS mapping could be utilized to pinpoint the areas deficient in micronutrients for their supplementation program. There are areas of developing science like nanotechnology which may be utilized to provide benefit to the farmers by developing efficient Nano product with its technology of use. However, the steps toward nutrient-based subsidy and other policies of organic farming and FCO role to tackle certain issues will help to solve these problems in days to come favorably so that micronutrient management will have its real impact on agricultural production and quality produce which will further help in mitigating the problems of malnutrition in human and animals and sustain soil health.

## 4.10 Future Research Need

More information on the transformation and availability of micronutrients for different soils and the effect of manipulating the soil physical environment and its moisture regimes on plant-available micronutrients need to be generated. The residual availability of various sources of micronutrients for a cropping system needs to be worked out. In order to enhance micronutrient use efficiencies, the use of nanotechnology options along with testing of new micronutrient products may be taken up. The physiological and microbiological interventions will be quite useful further for enhancing the micronutrient availability so as to produce micronutrient-dense seeds/grains. There is also need to study on mobilization of native micronutrients using indigenous microorganism. A nationwide program of micronutrient fertilization could be highly successful in both increasing their concentration in food crops and improving daily intake of trace elements in the population to targeted levels besides increase in crop yield.

**Acknowledgment** The authors are thankful to the AICRP-MSPE, ICAR, and SAUs for the data utilization from the researches carried out at different centers/stations for the publication.

## References

Ajouri A, Asgedom H, Becker M (2004) Seed priming enhances germination and seedling growth of barley under conditions of P and Zn deficiency. J Plant Nutr Soil Sci 167:630–636

Ali SJ, Monoranjan R (1989) Effect of NPK and micronutrient in controlling sterility in wheat. Fertil News 34:35–36

Allaway WH (1986) Soil–plant–animal and human interrelationships in trace element nutrition. In: Mertz W (ed) Trace elements in human and animal nutrition. Academic Press, Orlando, pp 465–488

Ansari MA, Prakash N, Singh IM, Sharma PK, Punitha P (2013) Efficacy of boron sources on productivity, profitability and energy use efficiency of groundnut (*Arachis hypogaea*) under north east hill regions. Indian J Agric Sci 83(9):959–963

Arora CL, Singh B, Takkar PN (1985) Secondary and micronutrient deficiency in crops. Progr Farm 21(8):139

Ashoka P, Meena RS, Kumar S, Yadav GS, Layek J (2017) Green nanotechnology is a key for eco-friendly agriculture. J Clean Prod 142:4440–4441

Babaeva EY, Volobueva VF, Yagodin BA, Klimakhin GI (1999) Sowing quality and productivity of *Echinacea purpurea* in relation to soaking the seed in manganese and zinc solutions. Iz vestiya Timiryazevskoi Sel'skokhozyaistvennoi Akademii 4:73–80

Black R (2003) Micronutrient deficiency – an underlying cause of morbidity and mortality. Bull World Health Organ 81(03808):79

Black RE, Victora CG, Walker SP, Bhutta ZA, Christian P, de Onis M, Ezzati M, Grantham-McGregor S, Katz J, Martorell R, Uauy R, the Maternal and Child Nutrition Study Group (2013) Maternal and child undernutrition and overweight in low-income and middle-income countries. Lancet 382(9890):427–451

Buragohain S, Sharma B, Nath JD, Gogaoi N, Meena RS, Lal R (2017) Impact of ten years of bio-fertilizer use on soil quality and rice yield on an inceptisol in Assam, India. Soil Res. https://doi.org/10.1071/SR17001

Cababallero B (2002) Impact of micronutrient deficiencies on growth: the stunting syndrome. Ann Nutr Metab 46:8–17

Cakmak I (2008) Enrichment of cereal grains with zinc: agronomic or genetic biofortification. Plant Soil 302:1–17

Cakmak I (2011) Zinc plays critical role in plant growth. Available http://www.zinc.org/crops/resourceserve/zincplayscriticalroleinplant growth. Accessed 17 Apr 2011

Cakmak I, Pfeiffer WH, McClafferty B (2010) Biofortification of durum wheat with zinc and iron. Cereal Chem 87(1):10–20

Dadhich RK, Meena RS (2014) Performance of Indian mustard (*Brassica juncea* L.) in Response to foliar spray of thiourea and thioglycollic acid under different irrigation levels. Indian J Ecol 41(2):376–378

Dadhich RK, Meena RS, Reager ML, Kansotia BC (2015) Response of bio-regulators to yield and quality of Indian mustard (*Brassica juncea* L. Czernj. and Cosson) under different irrigation environments. J Appl Nat Sci 7(1):52–57

Datta R, Baraniya D, Wang YF, Kelkar A, Moulick A, Meena RS, Yadav GS, Ceccherini MT, Formanek P (2017a) Multi-function role as nutrient and scavenger off reeradical in soil. Sustain MDPI 9:402. https://doi.org/10.3390/su9081402

Datta R, Kelkar A, Baraniya D, Molaei A, Moulick A, Meena RS, Formanek P (2017b) Enzymatic degradation of lignin in soil: a review. Sustain MDPI 9:1163. https://doi.org/10.3390/su9071163, 1–18

Dhakal Y, Meena RS, De N, Verma SK, Singh A (2015) Growth, yield and nutrient content of mungbean (*Vigna radiata* L.) in response to INM in eastern Uttar Pradesh, India. Bangladesh J Bot 44(3):479–482

Dhakal Y, Meena RS, Kumar S (2016) Effect of INM on nodulation, yield, quality and available nutrient status in soil after harvest of green gram. Legum Res 39(4):590–594

Dhaliwal SS, Sadana US, Manchanda JS, Dhadli HS (2009) Biofortification of wheat grains with zinc and iron in *Typic Ustochrept* soils of Punjab. Indian J Fertil 5(11):13-16 & 19-20

Dwivedi GK, Dwivedi M, Pal SS (1990) Mode of application of micronutrients in soybean – wheat crop sequence. J Indian Soc Soil Sci 38(4):58–63

FAI (2017) Specialty fertilizers and micronutrients statistics 2016–17, 6th edn. The Fertilizers Association of India, New Delhi

Fang Y, Wang L, Xin Z, Zhao L, An X, Hu Q (2008) Effect of foliar application of zinc, selenium, and iron fertilizers on nutrients concentration and yield of rice grain in China. J Agric Food Chem 56:2079–2084

FAO (2015) The state of food insecurity in the world – meeting the 2015 international hunger targets: taking stock of uneven progress. Food and Agriculture Organization of the United Nations, Rome

FCO (2003) Fertilizer Control Order, (1985) and the Essential Commodities Act, 1955, Revised up to 2003. FAI, New Delhi

Frossard E, Bucher M, Mächler F, Mozafar A, Hurrell R (2000) Potential for increasing the content and bioavailability of Fe, Zn and Ca in plants for human nutrition. J Sci Food Agric 80:861–879

Ghaly AE, Alkoaik FN (2010) Extraction of protein from common plant leaves for use as human food. Am J Appl Sci 7:331–342

Godsey CB, Schmidt JP, Schlegel AJ, Taylor RK, Thompson CR, Gehl RJ (2003) Correcting iron deficiency in corn with seed row-applied iron sulfate. Agron J 95:160–166

Gogoi N, Baruah KK, Meena RS (2018) Grain legumes: impact on soil health and agroecosystem. In: Meena RS et al (eds) Legumes for soil health and sustainable management. Springer. https://doi.org/10.1007/978-981-13-0253-4_16

Gonçalves AC Jr, Nacke H, Marengoni NG, Carvalho EA, Coelho GF (2010) Produtividade e componentes de produção da soja adubada com diferentes doses de fósforo, potássio e zinco. Cienc Agrotec 34:660–666

Graham RD, Welch RM, Bouis HE (2001) Addressing micronutrient malnutrition through enhancing the nutritional quality of staple foods: principles

Grunes DL, Allaway WH (1985) Nutritional quality of plants in relation to fertilizer use. In: Engelstad OP (ed) Fertilizer technology and use. SSSA, Madison, pp 589–619

Guaba K (1983) Master's thesis. Post Graduate Institute of Medicine, Education and Research, Chandigarh

Gupta AP (2005) Micronutrient status and fertilizer use scenario in India. J Trace Elem Med Biol 18:325–331

Han C, Li C, Li R, Wang Q, Wu M, Xiao C (1994) Technical approach to application of iron fertilization on soybean. Oil Crop China 16:43–45. in Chinese

Harris D, Rashid A, Miraj G, Arif M, Yunas M (2008) On-farm seed priming with zinc in chickpea and wheat in Pakistan. Plant Soil 306:3–10

Haslett BS, Reid RJ, Rengel Z (2001) Zinc mobility in wheat: uptake and distribution of zinc applied to leaves or roots. Annu Bot 87:379–386

Jangir CK, Kumar S, Lakhran H, Meena RS (2017) Towards mitigating malnutrition in pulses through biofortification. Trends Biosci 10(17):2999–3002

Jhanji S, Sadana US, Sekhon NK, Khurana MPS, Sharma A, Shukla AK (2013) Screening diverse wheat genotypes for manganese efficiency based on high yield and uptake efficiency. Field Crop Res 154:127–132

Kakraliya SK, Singh U, Bohra A, Choudhary KK, Kumar S, Meena RS, Jat ML (2018) Nitrogen and legumes: a meta-analysis. In: Meena RS et al (eds) Legumes for soil health and sustainable management. Springer. https://doi.org/10.1007/978-981-13-0253-4_9

Katyal JC, Deb DL (1982) Review of soil research in India, part I. Indian Society of Soil Science, New Delhi, pp 146–159

Katyal JC, Sharma BD (1991) DTPA-extractable and total Zn, Cu, Mn and Fe in Indian soils and their association with some soil properties. Geoderma 49:165–179

Katyal JC, Vlek PLG (1985) Micronutrient problems in tropical Asia. Fertil Res 7:131–150

Kaya M, Atak M, Khawar KM, Çiftçi CY, Özcan S (2007) Effect of pre-sowing seed treatment with zinc and foliar spray of humic acids on yield of common bean (*Phaseolus vulgaris* L.). Int J Agric Biol Eng 7:875–878

Kennedy G, Nantel G, Shetty P (2003) The source of "hidden hunger": global dimensions of micronutrient deficiencies. Food Nutr Agric 32:8–16

Kumar M, Swarup A, Patra AK, Chandrakala JU (2011) Micronutrient fertilization under rising atmospheric $CO_2$ for micronutrient security in India. Indian Journal of Fertilizers 7(7):52–60

Kumar S, Meena RS, Pandey A, Seema (2017a) Soil acidity management and an economics response of lime and sulfur on sesame in an alley cropping system. Int J Curr Microb App Sci 6(3):2566–2573

Kumar S, Meena RS, Yadav GS, Pandey A (2017b) Response of sesame (*Sesamum indicum* L.) to sulphur and lime application under soil acidity. Int J Plant Soil Sci 14(4):1–9

Kumar S, Meena RS, Bohra JS (2018a) Interactive effect of sowing dates and nutrient sources on dry matter accumulation of Indian mustard (*Brassica juncea* L.). J Oilseed Bras 9(1):72–76

Kumar S, Meena RS, Lal R, Yadav GS, Mitran T, Meena BL, Dotaniya ML, EL-Sabagh A (2018b) Role of legumes in soil carbon sequestration. In: Meena RS et al (eds) Legumes for soil health and sustainable management. Springer. https://doi.org/10.1007/978-981-13-0253-4_4

Kundu TR, Vishist DC (1991) Demand for intermediate inputs in Indian agriculture. Indian J Agric Econ 46(2):152–158

Layek J, Das A, Mitran T, Nath C, Meena RS, Singh GS, Shivakumar BG, Kumar S, Lal R (2018) Cereal+legume intercropping: an option for improving productivity. In: Meena RS et al (eds) Legumes for soil health and sustainable management. Springer. https://doi.org/10.1007/978-981-13-0253-4_11

Liu W, Du L, Liu D (2002) A review on diagnosis for iron nutrition status of fruit trees and fertilization correction for Fe-deficiency chlorosis. Chin Agric Sci Bull 18:67–69. in Chinese

Mathur CM, Moghe VB, Talati MR (1964) Distribution of boron in soils of western Rajasthan irrigated with high boron waters. J Indian Soc Soil Sci 12:319–324

Meena RS, Lal R (2018) Legumes and sustainable use of soils. In: Meena RS et al (eds) Legumes for soil health and sustainable management. Springer. https://doi.org/10.1007/978-981-13-0253-4_1

Meenakshi JV, Johnson NL, Manyong VM, DeGroote H, Javelosa J et al (2010) How cost effective is biofortification in combating micronutrient malnutrition? An ex ante assessment. World Dev 38:64–75

Mitran T, Meena RS, Lal R, Layek J, Kumar S, Datta R (2018) Role of soil phosphorus on legume production. In: Meena RS et al (eds) Legumes for soil health and sustainable management. Springer. https://doi.org/10.1007/978-981-13-0253-4_15

Mondal C, Bandopadhyay P, Alipatra A, Banerjee H (2012) Performance of summer mungbean [*Vigna radiata* (L.) Wilczek] under different irrigation regimes and boron levels. J Food Legum 25(1):37–40

Murthy IYLN (2011) Zinc response to oilseed crop. Indian J Fertil 7(10):104–116

Nene YL (1966) Symptoms, causes and control of *khaira* disease in paddy. Bull Indian Phytopathol Soc 3:97–101

PDIL (1996) Market survey of micronutrients in India. Job No. 3886, Prepared by projects and Development India Limited. NOIDA for Ministry of Agriculture, GOI

Rabobank (2005) Indian fertilizer industries: a snapshot of urea and DAP business. Rabobank International, F&A Research and Advisory

Raju S (1989) Fertilizers use in Andhra Pradesh: an analysis of factors affecting consumption. Artha Vijnana 31(4):11–19

Ram K, Meena RS (2014) Evaluation of pearl millet and mungbean intercropping systems in arid region of Rajasthan (India). Bangladesh J Bot 43(3):367–370

Rathore GS, Khamparia RS, Gupta GP, Dubey SB, Sharma BL, Tomar US (1995) Twenty-five years of micronutrient research in soils and crops of Madhya Pradesh. Research Bulletin. Department of Soil Science and Agricultural Chemistry, JNKVV, Jabalpur, 101 pp

Rattan (2017) Quo Vadis zinc research: golden march with green revolution. 44th RV Tamhane Memorial Lecture delivered on 12 December during 83rd Annual Convection of the Indian Society of Soil Science held at Amity University, Kolkata

Rattan RK, Datta SP, Katyal JC (2008) Micronutrient management–research achievements and future challenges. Indian J Fertil 4(12):93–118

Ray SK, Banik GC (2016) Available micronutrient status in relation to soil properties in some villages under four agro-climatic features of West Bengal. J Indian Soc Soil Sci 64(2):169–175

Reeves PG, Nielsen FH, Fahey GC (1993) AIN-93 purified diets for laboratory rodents: Final report of the American Institute of Nutrition *ad-hoc* writing committee on the reformulation of the AIN-76Arodent diet. J Nutr 123:1939–1951

Rengel Z, Batten GD, Crowley DE (1999) Agronomic approaches for improving the micronutrient density in edible portions of field crops. Field Crop Res 60:27–40

Saha JK, Singh MV (1997) Effect of $CaCO_3$ on the adsorption and desorption of boron in swell-shrink soils in Madhya Pradesh. J Indian Soc Soil Sci 45:57–61

Saha JK, Singh MV, Sharma BL (1998) Available boron status in major soils of Madhya Pradesh. J Indian Soc Soil Sci 46:478–479

Sakal R, Singh AP, Sinha RB, Bhogal NS (1988) Annual progress reports. ICAR All India Coordinated Scheme of Micro-and Secondary Nutrients in Soils and Crops of Bihar, Research Bulletin, Department of Soil Science, RAU, Samastipur, Bihar, India

Sakal R, Singh AP, Sinha RB, Bhogal NS (1996) Twenty-five years of research on micro and secondary nutrients in soils of Bihar. Rajendra Agricultural University, Pusa, 208pp

Seshadari S (1998) Community trials of effectiveness of daily vs weekly supplementation. MS University, Vadodara

Sharma UC, Grewal JS (1988) Relative effectiveness of methods of micro-nutrient application to potato. J Indian Soc Soil Sci 36(1):128–132

Sharma U, Kumar P (2016) Micronutrient research in India: extent of deficiency, crop responses and future challenges. Int J Adv Res 4(4):1402–1406

Shekhawat K, Shivay YS (2008) Effect of nitrogen sources, sulphur and boron levels on productivity, nutrient uptake and quality of sunflower (*Helianthus annuus*). Indian J Agron 53(2):129–134

Shenker M, Chen Y (2005) Increasing iron availability to crops: fertilizers, organo-fertilizers and biological approaches. Soil Sci Plant Nutr 51:1–17

Shivay YS, Kumar D, Prasad R, Ahlawat IPS (2008a) Relative yield and zinc uptake by rice from zinc sulphate and zinc oxide coatings onto urea. Nutr Cycl Agroecosyst 80:181–188

Shivay YS, Prasad R, Rahal A (2008b) Relative efficiency of zinc oxide and zinc sulphate enriched urea for spring wheat. Nutr Cycl Agroecosyst 82:259–264

Shukla AK, Behera SK (2011) Zinc management in Indian agriculture – past, present and future. Indian J Fertil 7(10):14–33

Shukla AK, Behera SK (2012) Micronutrient fertilizers and higher productivity. Indian J Fertil 8(4):100–117

Shukla AK, Tiwari PK (2014) Progress report of AICRP on micro and secondary nutrients and pollutant elements in soils and plants. IISS, Bhopal

Shukla AK, Tiwari PK (2016) Progress report of AICRP on micro and secondary nutrients and pollutant elements in soils and plants. IISS, Bhopal

Shukla AK, Behera SK, Shivay YS, Singh P, Singh AK (2012) Indian J Agron 57:124–131

Shukla AK, Tiwari PK, Prakash C (2014) Micronutrients deficiencies vis-à-vis Food and Nutritional Security of India. Indian J Fertil 10(12):94–112

Shukla AK, Tiwari PK, Pakhare A, Prakash C (2016) Zinc and Iron in soil, plant, animal and human health. Indian J Fertil 12(11):133–149

Sihag SK, Singh MK, Meena RS, Naga S, Bahadur SR, Gaurav YRS (2015) Influences of spacing on growth and yield potential of dry direct seeded rice (*Oryza sativa* L.) cultivars. Ecoscan 9(1–2):517–519

Singh MV (2008) Micronutrients deficiencies in crops in India. Chapter 4. In: Alloway BJ (ed) Micronutrient deficiencies in global crop production. Springer, Dordrecht, pp 93–126

Singh MV (2009) Micronutrient nutritional problems in soils of India and improvement for human and animal health. Indian J Fertil 5:11–26

Singh MV, Goswami V (2013) Efficiency of boron fortified NPK fertilizer in correcting boron deficiency in some cereal and oilseeds crops in India. 17th international plant nutrient colloquium 17–18 August, Istanbul, Turkey

Singh PP, Saini SP, Khurana SM, Matharu GS (2011) Impact of manganese sulphate application on wheat in subtropical soils through on-farm trials. Indian J Fertil 7(11):26–31

Sinha RB, Sakal R, Singh AP, Bhogal NS (1991) Response of some field crops to boron application in calcareous soils. J Indian Soc Soil Sci 39:342–345

Sofi PA, Baba ZA, Hamid B, Meena RS (2018) Harnessing soil rhizobacteria for improving drought resilience in legumes. In: Meena RS et al (eds) Legumes for soil health and sustainable management. Springer. https://doi.org/10.1007/978-981-13-0253-4_8

Sreeja VN (2014) http://www.ibtimes.com/world-population-increase-97b-2050-india-will-overtake-china-mostpopulous-nation-new-study-1413620

Takkar PN (1996) Micronutrient research and sustainable agricultural productivity in India. J Indian Soc Soil Sci 44:562–581

Takkar PN, Nayyar VK (1981) Effect of gypsum and zinc on rice nutrition on sodic soils. Exp Agric 17:49–55

Takkar PN, Singh T (1978) Zinc nutrition of rice as influenced by rates of gypsum and Zn fertilization of alkali soils. Agron J 70:447–450

Takkar PN, Walker CD (1993) The distribution and correction of zinc deficiency In: Robson AD (ed) Zinc in soils and plants. Academic, Dordrecht, 151–165pp

Takkar PN, Chibba IM, Mehta SK (1989) Twenty years of coordinated research on micronutrient in soils and plants. Bulletin 314. Indian Institute of Soil Science, Bhopal, India, 81

Takkar PN, Singh MV, Ganeshmurthy AN (1997) Plant nutrient supply needs, efficiency and policy issues. In: Kanwar JS, Katyal JC (eds) . National Academy of Agricultural Sciences, New Delhi, pp 238–264

Takkar PN, Chhibba IM, Nayyar VK (2004) Micronutrients management in soils and crops. Proceedings of the wrap-up workshop on ITSAP-sustaining agriculture-problems and prospects TIFAC-UNDP, 9–11 November 2004, Punjab Agricultural University, Ludhiana

Thiyagarajan TM (2002) Soil health and its care for sustainable agricultural production. Agriculture, Tamil Nadu Agricultural University, Coimbatore, pp 130–149

Varma D, Meena RS, Kumar S, Kumar E (2017) Response of mungbean to NPK and lime under the conditions of Vindhyan Region of Uttar Pradesh. Legum Res 40(3):542–545

Vempati RK, Loeppert RH (1988) Chemistry and mineralogy of Fe-containing oxides and layer silicates in relation to plant available iron. J Plant Nutr 11:1557–1574

Verma JP, Jaiswal DK, Meena VS, Meena RS (2015a) Current need of organic farming for enhancing sustainable agriculture. J Clean Prod 102:545–547

Verma JP, Meena VS, Kumar A, Meena RS (2015b) Issues and challenges about sustainable agriculture production for management of natural resources to sustain soil fertility and health: a book review. J Clean Prod 107:793–794

Verma SK, Singh SB, Prasad SK, Meena RN, Meena RS (2015c) Influence of irrigation regimes and weed management practices on water use and nutrient uptake in wheat (*Triticum aestivum* L. Emend. Fiori and Paol.). Bangladesh J Bot 44(3):437–442

Welch RM, Graham RD (2004) Breeding for micronutrients in staple food crops from a human nutrition perspective. J Exp Bot 55:353–364

Welch RM, House WA (1984) Factors affecting the bioavailability of mineral nutrients in plant foods. In: Welch RM, Gabelman WH (eds) Crops as sources of nutrients for humans. ASA, Madison, pp 37–54

WHO FAO (2003) Joint WHO/FAO expert consultation on diet, nutrition and the prevention of chronic diseases. World Health Organization, Geneva

WHO (2016) Vitamin and mineral nutrition information system. World Health Organization. www.who.int

Yadav GS, Babu S, Meena RS, Debnath C, Saha P, Debbaram C, Datta M (2017a) Effects of godawariphosgold and single supper phosphate on groundnut (*Arachis hypogaea*) productivity, phosphorus uptake, phosphorus use efficiency and economics. Indian J Agric Sci 87(9):1165–1169

Yadav GS, Lal R, Meena RS, Babu S, Das A, Bhomik SN, Datta M, Layak J, Saha P (2017b) Conservation tillage and nutrient management effects on productivity and soil carbon sequestration under double cropping of rice in North Eastern Region of India. Ecol Indic. http://www.sciencedirect.com/science/article/pii/S1470160X17305617

Yadav GS, Lal R, Meena RS, Datta M, Babu S, Das A, Layek J, Saha P (2017c) Energy budgeting for designing sustainable and environmentally clean/safer cropping systems for rainfed rice fallow lands in India. J Clean Prod 158:29–37

Yadav GS, Das A, Lal R, Babu S, Meena RS, Saha P, Singh R, Datta M (2018a) Energy budget and carbon footprint in a no-till and mulch based rice–mustard cropping system. J Clean Prod 191:144–157

Yadav GS, Das A, Lal R, Babu S, Meena RS, Patil SB, Saha P, Datta M (2018b) Conservation tillage and mulching effects on the adaptive capacity of direct-seeded upland rice (*Oryza sativa*

L.) to alleviate weed and moisture stresses in the North Eastern Himalayan Region of India. Arch Agron Soil Sci. https://doi.org/10.1080/03650340.2018.1423555

Yilmaj A, Ekiz H, Torun B, Gulteki I, Karanlik S, Bagei SA, Cakmak I (1997) Effect of different zinc application methods on grain yield and zinc concentration in wheat grown zinc-deficient calcareous soils in Central Anatolia. J Plant Nutr 20:461–471

Zadeh SS, Begum K (2011) Nutritional supplements and its effect on quality of life and sleep. Am Med J 2:104–110

# Nitrogen Footprint: A Useful Indicator of Agricultural Sustainability

5

Sangita Mohanty, Chinmaya Kumar Swain, Anjani Kumar, and A. K. Nayak

**Abstract**

Nitrogen (N) fertilizer has been identified as a crucial input that has alleviated nitrogen limitation in crop production and substantially enhanced yield. Global N fertilizer consumption in the year 2013 was 107.6 million tons which is approximately ten times that of 1961. However, 60–70% of applied N is lost from the system in the form of reactive N species such as ammonia ($NH_3$), nitrous oxide ($N_2O$), nitric oxide (NO), nitrogen dioxide ($NO_2$), and nitrate ($NO_3$) due to poor N use efficiency of agricultural crop. Intensive agricultural practices therefore are major anthropogenic interference that disrupt natural N cycle, leading to severe environmental hazards in the form of acid rain, smog, eutrophication, ozone depletion, and global warming. Monitoring contribution of agriculture to global N pollution is essential to raise awareness and adopt mitigation measures to ensure environmental sustainability of the production system. Attempts have been made to prepare farm-, region-, and country-specific inventories of N leaching, $N_2O$, and $NH_3$ emission separately, though data on the contribution of agriculture is associated with its inherent uncertainties and biases. Recently, nitrogen footprint approach has been identified as a potential tool to estimate N flow from various sectors such as industry, transport, and agriculture. These tools also provide options of developing strategies for reducing N footprint over a time period. The objectives of this chapter are to comprehensively discuss the contribution of agricultural activity to reactive N flow and analyze the scope of using N footprint tools for comparative assessment of environmental sustainability of various crop management practices.

**Keywords**

Nitrogen · Nitrogen footprint · N cycle · N pollution · Water pollution

---

S. Mohanty (✉) · C. K. Swain · A. Kumar · A. K. Nayak
National Rice Research Institute, Cuttack, Odisha, India


R. S. Meena (ed.), *Nutrient Dynamics for Sustainable Crop Production*,
https://doi.org/10.1007/978-981-13-8660-2_5

## Abbreviations

| | |
|---|---|
| AP | Acidification potential |
| ATP | Adenosine triphosphate |
| EEFs | Enhanced efficiency fertilizers |
| EP | Eutrophication potential |
| Gt | Gigatonnes |
| GWP | Global warming potential |
| MCL | Maximum contaminant level |
| Mt. | Metric tons |
| $N_2O$ | Nitrous oxide |
| $NH_3$ | Ammonia |
| NO | Nitric oxide |
| $NO_2$ | Nitrogen dioxide |
| ODP | Ozone depletion potential |
| POCP | Photochemical ozone creation potential |
| SIA | Secondary inorganic aerosols |
| SSNM | Site-specific nutrient management |
| Tg | Teragrams |

## 5.1 Introduction

Nitrogen is one of the most yield-limiting nutrients in rice production around the world (Samonte et al. 2006). Nitrogen fertilizer is a crucial input for attaining higher crop yield and contributes significantly to total crop production cost (Tirol-Padre et al. 1996). Deficiency of nitrogen adversely affects plant growth and development by restricting the formation of enzymes, chlorophyll, and proteins required for different metabolic activities. Though nitrogen is abundantly available in the atmosphere, it is mainly in nonreactive form ($N_2$). The need for food production led to conversion of nonreactive form of N to reactive form. Before the onset of industrial conversion of $N_2$ to $NH_3$, biological N fixation was the single most important pathway for fixation of atmospheric N, and the N input and output in the terrestrial N cycle were in equilibrium. However, due to excessive use of N fertilizer during the past decades, there are disruptions in natural N cycle leading to severe environmental hazards like eutrophication, acid rain, smog, global warming, ozone depletion, etc. According to earth system researchers, nitrogen is the major factor in biogeochemical pollution and one of four "planetary boundaries" that have been exceeded. World over, approximately 120 million tons of synthetic nitrogen fertilizer is being used each year in agriculture. However, more than half of it is lost from the agroecosystem in the form of reactive N species to the environment due to abysmal low agricultural N use efficiency that shows a declining trend over years. Poisoning of underground water reserves in California, toxic "red tides" off the shores of Florida, and toxic algae spread from river estuaries across the East China Sea are some of the deadly fallout of N pollution over the world, though not enough attention has been given to the problems of N pollution as compared to C pollution. The positive aspect of N use, particularly its role in green revolution, has been recognized widely, but there is little or no awareness related to environmental issues of N fertilizer use. Assessing the environmental impact of agricultural N use is essential to improve understanding of issues related to N pollution. Efforts have been initiated to develop, standardize, and monitor reactive N flow from the agroecosystem and prepare countrywise inventories of reactive N emission; however, available statistics has its own uncertainties and biases due to inherent complexities of the process involved. Since N is an essential input, its use is rationalized to enhance productivity and at the same time minimize its adverse environmental impact to ensure sustainability of crop production world over.

Of late, the concept of ecological footprint has been gaining momentum as indicators of sustainability which aims at measuring the impact a person or community exerts on its environment due to its consumption pattern and the amount of land required to sustain their use of natural resources (Global Footprint Network 2012). Analogous to ecological footprint, several other indicators, e.g., carbon footprint, water footprint, energy footprint, etc., have been developed which are now part of a system of indicators known as footprint family. These indicators estimate the impacts of human activities on the environment in relation to resource consumption waste generation. These indicators provide a quantifiable and rational assessment of efficiency of production processes, limits of resource consumption, and sustainable utilization of global resources. Nitrogen footprint is a recent development in sustainability and footprint research which basically aims at quantifying reactive N load to environment due to anthropogenic activities assessing its impact.

The objective of this chapter is to discuss the contribution of agricultural activities toward N pollution and to analyze different nitrogen footprint indicators for their usefulness as sustainability indicator for agricultural production.

## 5.2 Importance of Nitrogen

Nitrogen is one of the elements essential for performing basic life functions in plants and animals. Being the major constituent of amino acids that are building blocks of proteins and nucleic acids, N is crucial for growth, survival, and subsequent generations of all plants and animals. Protein is an integral component of all animal tissue; hence, growth, replacement, and repairing of tissues require nitrogen. In addition, enzymes which are nothing but proteins play key roles in metabolism of plants and animals. In plants, N constitutes the main structural frame of chlorophyll that is essential for photosynthesis. Being part of adenosine triphosphate (ATP), it plays a significant role in the transfer of energy in both plants and animals. Nitrogen is also a constituent of nonprotein compounds, e.g., "heme" in hemoglobin, which transports oxygen to all parts in human body.

Nitrogen, though the single most abundant element in the earth's atmosphere constituting around 78%, remains mostly inaccessible to animals and plants. A small fraction of N occurring naturally as saltpeter ($KNO_3$, $NaNO_3$) in mineral

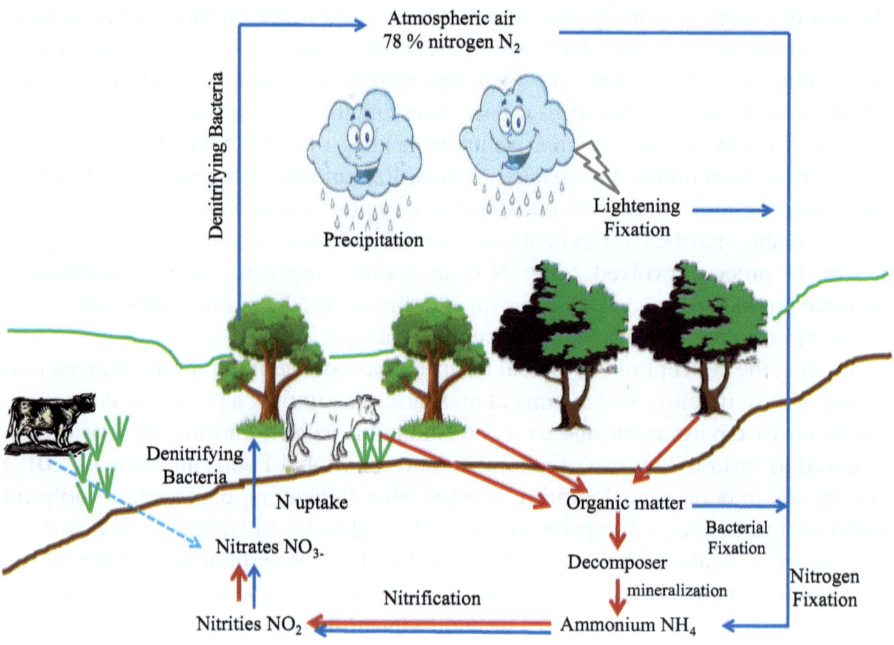


**Fig. 5.1** Schematic representation of natural N cycle

deposits and as bird guano, i.e., excreta of seabirds and bats, is insufficient to meet biological needs. Therefore, N is one of the most limiting elements that significantly influence net primary production in earth. Hence, the cycle of N, i.e., conversion of inert atmospheric N to available N ($NH_4$, $NO_3$ etc.), is of immense ecological significance. Natural N cycle consists of a series of interrelated biogeochemical processes, i.e., fixation, ammonification, nitrification, and denitrification, by which atmospheric $N_2$ is converted to $NH_3$ biologically by N-fixing organisms or chemically (in the presence of lightening) and becomes part of the biosphere, i.e., soil-plant-animal system. Organic N released from plant and animal in the form of residue or manure is converted to inorganic N ($NH_4$-N and $NO_3$-N) through decomposition, ammonification, and nitrification processes. Part of inorganic N is again consumed by plants and microbes and the rest returns back to atmosphere in the form of $NH_3$, $N_2O$, and $N_2$ through volatilization and denitrification (Fig. 5.1).

Recycling of crop residues and manures, application of guano and nitrate deposits, and cultivation of legumes were the main sources of external N supply to sustain plant growth and productivity before the invention of Haber–Bosch process. Haber–Bosch process that involves production of $NH_3$ from $N_2$ and $H_2$ paved the way for fixation of atmospheric N in an industrial scale, and so far, it is considered a breakthrough invention that has far reaching consequences on human civilization. This process is the basis of synthetic nitrogenous fertilizer production which revolutionized agriculture all over the globe by enhancing the productivity. In addition, $NH_3$ has several other industrial uses, i.e., production of plastics, dyes, drugs, soda ash, rocket fuel (hydrazine), explosives (ammonium nitrate), etc. Worldwide, approximately 150 million tons of ammonia is being produced annually, and over 80% of it is used for production of synthetic fertilizers (U.S. Geological Survey, Mineral Commodity Summaries January 2011). The consumption of nitrogen fertilizers in 2013 was 113.2 million tons, and this amount is expected to increase up to nearly 120 million tons in 2018 (https://www.statista.com/statistics/438967/fertilizer-consumption-globally-by-nutrient).

## 5.3 Nitrogen and Food Grain Production

Nitrogen is one of the major essential nutrients of plants and required in larger quantities by almost all food crops. It is also one of the most yield-limiting nutrients for production of major food crops around the world (Ladha and Reddy 2003; Samonte et al. 2006). Availability, uptake, and translocation of N in plants affect many basic physiological functions related to biomass and grain yield (Kaizzi et al. 2012). Apart from the establishment and maintenance of photosynthetic capacity and activity, N also plays an important role in the development and maintenance of sink capacity, more particularly in cereals, by influencing the number and size of the seeds (Dreccer et al. 2000; Fageria et al. 2008; Foulkes et al. 2010). Nitrogen also has varying effects on the quality of produce, e.g., protein content in the grains of cereals (Guarda et al. 2004).

The chlorophyll and N contents of leaves are closely linked (Amaliotis et al. 2004) because N is a structural element of chlorophyll and affects the formation of chloroplasts (Bojovic and Markovic 2009). The study indicated the chlorophyll content in rice seedlings decreased by 8% with a low N supply and increased by 12% with high N supply (Li et al. 2012). Because of its close association with chloroplast, N influences photosynthetic efficiency, carbohydrate accumulation, and subsequent dry matter production. Hence, the availability of N affects the area of leaf, formation of tillers, spikelets, and the percentage of filled grains in rice (Tanaka et al. 2013). Deficiency of N affects reproductive and vegetative phenological development, inhibits growth of a plant, and reduces tiller number and leaf emergence rate (Umehara et al. 2008; Marschner 2012; Meena and Lal 2018). In addition, inadequate N supply to reproductive organs also reduces the number of ear-bearing tillers, number of fertile spikelet, and overall yield (Shahrokhnia and Sepaskhah 2016). Hence, an optimum supply of N is crucial to prevent retardation of plant growth and yield (Tafteh and Sepaskhah 2012). Therefore, nitrogen fertilizers play a crucial role in enhancing agricultural production and ensuring food security for the ever-increasing global population (Spiertz 2010; Qiao et al. 2015). On average, wheat crop with the total biomass of 2.9 t $ha^{-1}$ at maturity contains approximately 95 t $ha^{-1}$ nitrogen in the seed and straw (Stone et al. 1996). The N requirement to produce 1 ton of rice grain, within the yield range of 4–10 t $ha^{-1}$, could vary from 18.0 to 20.0 kg $ha^{-1}$(Guo et al. 2016; Meena et al. 2018). Similarly, maize crop with yield of 40 t $ha^{-1}$ removes 160 kg N $ha^{-1}$; however peak uptake could be as high 210 as kg N $ha^{-1}$ (https://www.pda.org.uk/pda_leaflets/17-forage-maize-fertiliser-requirements/). According to the

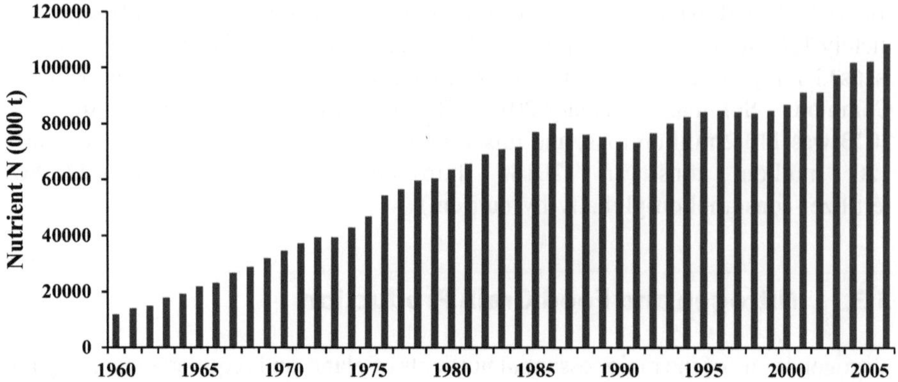


**Fig. 5.2** Graph showing the world consumption of nitrogen till the twenty-first century (IFA 2007)

latest available world fertilizer statistics, cereals including rice, wheat, and maize consume approximately 57.5 TgN, accounting for about 55% of world N fertilizer utilization.

Over the past century, use of N fertilizer is identified as one of the agronomic management options that have alleviated nitrogen limitation in crop production and substantially enhanced yield and soil fertility. According to available statistics, N fertilizer consumption increased from 11.3 Tg N year$^{-1}$ in 1961 to 107.6 Tg N year$^{-1}$ in 2013; during this period, average N consumption per unit cropped area in a year increased from 0.9 to 7.4 g N m$^{-2}$ cropland (Lu and Tian 2017). Between 2000–2001 and 2007–2008, global N fertilizer consumption increased by 23%, from 82.1 to 100.8 Tg N which was further increased from 101 Tg N in 2010, and with this baseline scenario, world N demand is projected to grow by 1.3% per annum to reach a figure of 132 Tg in 2030 (Fig. 5.2). Trends of world cereal production and food supply show strong linear relationship with nitrogenous fertilizer consumption. Based on the relationship between cereal production and fertilizer N consumption, it has been projected that approximately 138–161 Mt. of nitrogenous fertilizer will be required to produce 2.9–3.0 Gt of cereal by 2050 (Ramankutty et al. 2018).

## 5.4 Contribution of Agriculture to N Pollution

Before the discovery of Haber-Bosch process, most of the reactive N input to agroecosystem was through biological N fixation. However, with intensive agricultural practices, there is a dramatic increase of around 9–10 times in global fertilizer N use within the last 4–5 decades. According to an estimate, the total synthetic N fertilizer used since 1985 is approximately more than half of all the synthetic N fertilizers ever used in the world (NRC 2000; Howarth et al. 2002). However, worldwide N use efficiency of applied N is very poor, hardly exceeding 30–40% (Ladha et al. 2005), causing a loss of the rest 60–70% N from the system in the form of reactive

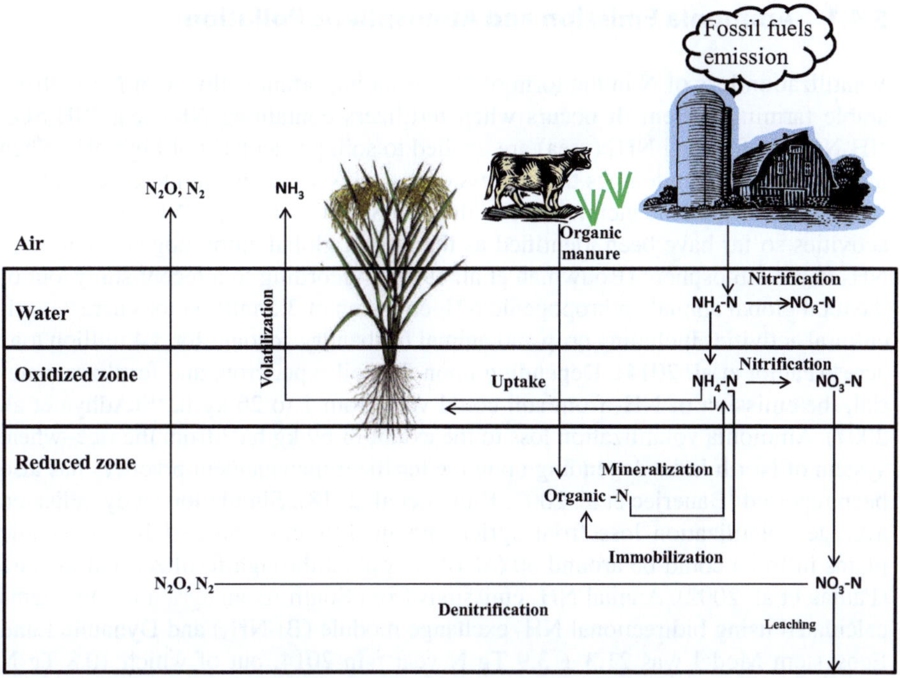


**Fig. 5.3** Schematic diagram of fertilizer-induced reactive N flow in rice ecosystem

N species such as $NH_3$, $N_2O$, NO, $NO_2$, and $NO_3$. Since the N cycle in the agroecosystem is an integral part of the global N cycle, accumulation of reactive N species due to enhanced fertilizer N application has a cascading effect on it. Therefore, intensive farming practice so far has been considered as a major anthropogenic activity that is rapidly altering global N cycle with far-reaching consequences of atmospheric pollution, greenhouse gas emission, and groundwater pollution, thereby raising global sustainability concerns (Smil 1999; Sheldrick et al. 2002) (Fig. 5.3).

Assessment on global nitrogen flows in cropland by Liu et al. (2010) indicated a total nitrogen input of 136.60 million ton of N per year in the year 2000, and almost half of which was from mineral nitrogen fertilizer making it the singlemost source of N (Fig. 5.3). However, there was considerable spatial variability in the contribution of fertilizer N across the regions of the world that ranged from over 55% in Oceania to 25–29% in Africa and South America. The croplands of Europe, Asia, and North America received 48–55% of the N input from mineral fertilizers. The same study calculated that out of the total N output of around 148 Mt. per year, 55% was removed by crops and remaining 45% was lost in leaching (16%), soil erosion (15%), and gaseous emission (14%). However, there were region-specific variations in distribution of loss depending on prevailing climatic and improved fertilizer management technology adopted.

### 5.4.1 Ammonia Emission and Atmospheric Pollution

Volatilization loss of N in the form of $NH_3$ is an important pathway of N loss from arable farming system. It occurs when fertilizers containing $NH_4$ (e.g. $NH_4SO_4$, $NH_4NO_3$) or forming $NH_4$ (urea) are applied to soil, particularly of high pH. When urea is applied, it undergoes the process of hydrolysis causing an increase in pH (> 8.0) of microsite in the vicinity that leads to conversion of $NH_4$ to $NH_3$. Agricultural activities so far have been identified as the largest global anthropogenic source of $NH_3$ to the atmosphere (Bouwman et al. 1997). According to a recent study, out of the total global annual anthropogenic $NH_3$ emission of 3.5 million tons acre$^{-1}$, agricultural activities including crop and animal husbandry account for 3.4 million tons acre$^{-1}$(Paulot et al. 2014). Depending upon the soil type, crop, and fertilizer material, the emission of $NH_3$ from soil could vary from 1 to 26 kg ha$^{-1}$ (Adhya et al. 2007). Ammonia volatilization loss to the extent of 69 kg ha$^{-1}$ from the rice–wheat system of North India depending upon the fertilizer management practices has also been reported (Banerjee et al. 2002; Kumar et al. 2018). Simulation study indicated average volatilization loss from agriculture in different states of Indo-Gangetic plains in India could be around 30.6% of N applied through fertilizer and manure (Pathak et al. 2002). Annual $NH_3$ emissions from South Asian agricultural systems calculated using bidirectional $NH_3$ exchange module (Bi-$NH_3$) and Dynamic Land Ecosystem Model was 21.3 ± 3.9 Tg N year$^{-1}$ in 2014, out of which 10.8 Tg N yearr$^{-1}$ was from synthetic N fertilizer use and 10.4 ± 3.9 Tg N year$^{-1}$ was released from manure production (Xu 2014). In 2000, the total $NH_3$ emission for China was estimated to be 13.6 Tg, 50% of which was from fertilizer applications (Streets et al. 2003).

Enhanced $NH_3$ emissions due to high N inputs could be a potential threat to the environment and human health. Being alkaline in reaction, $NH_3$ undergoes several chemical reactions with the acidic constituents of the atmosphere such as sulfate ($SO_4^{2-}$) and nitrate ($NO_3^-$) forming secondary inorganic aerosols (SIA). Increasing concentration of aerosols in atmosphere has a direct negative relation with the visibility range, climate forcing, and human health (Cheng et al. 2013; He and Dijkstra 2014). Ammonium salts contribute over 50% of the annual light extinction coefficients in air (Tao et al. 2014). Apart from this, increasing concentration of $NH_3$ in atmosphere also leads to dry and wet deposition of $NH_3$ on terrestrial ecosystem causing eutrophication of surface water bodies, loss of biodiversity, and distortion of ecosystem balance due to alterations in N transformation processes, e.g., mineralization, nitrification, etc. in soil (Kim et al. 2011; Sofi et al. 2018).

### 5.4.2 Nitrous Oxide Emission and Global Warming

Nitrous oxide is the by-product of both nitrification and denitrification processes in soil. Under aerobic conditions, nitrification is the main source of $N_2O$, while denitrification dominates under flooded rice fields. Nitrification is the process of microbial conversion of ammonium ion to nitrite and nitrate ions under aerobic condition, and it is also responsible for the production of $N_2O$, possibly from the intermediate compounds $NH_2OH$ or NO, though the exact pathway is not clearly known.

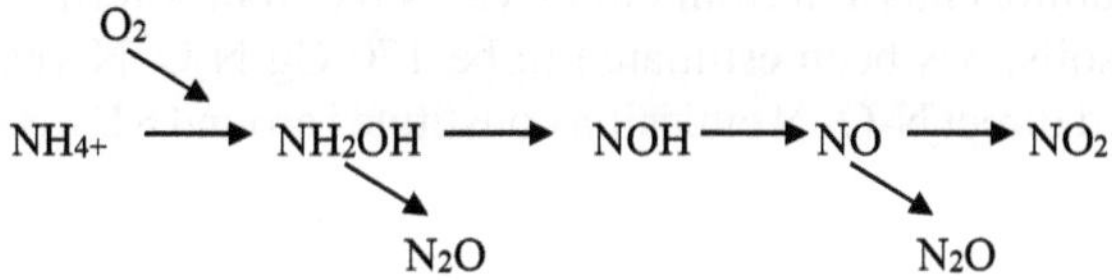


(5.1)

Denitrification is the process of conversion of nitrate or nitrite form of N to dinitrogen or N oxides under anaerobic conditions and presented as

$$NO_3^- \rightarrow NO_2^- \rightarrow NO \rightarrow N_2O \rightarrow N_2 \quad (5.2)$$

$N_2O$ is an important greenhouse gas with atmospheric lifetime of approximately 120 years and radiative forcing 298 times more than that of $CO_2$ on 100 years period. At the global scale, the contribution of $N_2O$ to total radiative forcing has been estimated to be 8%. In addition to its greenhouse effect, $N_2O$ is now also recognized as a major ozone-depleting substance in the stratosphere. On reaction with oxygen, nitrous oxide produces nitric oxide, and this in turn reacts with ozone, regulating its concentration in the stratosphere. The atmospheric concentration of nitrous oxide has risen from a preindustrial value of 270 ppb to 319 ppb in 2005, and during the period 1985–2005, its concentrations continued to increase at a rate of 0.25% per year (IPCC 2007).

Along with industry, transport, and biomass burning, agriculture has been identified as a main source of anthropogenic $N_2O$ emissions. Agricultural activities including application of nitrogenous fertilizer lead to enhanced biogenic production of nitrous oxide directly or indirectly through their impact on the global N cycle. The direct $N_2O$ emissions from fertilized agricultural soils and animal production and indirect emissions from N used in agriculture constitute total agricultural $N_2O$ emission (Mosier et al. 1998). With a value of 2.1 (0.4–3.8) Tg N $year^{-1}$, estimated following the IPCC Phase II methodology (IPCC 1997), the direct $N_2O$ emissions from agricultural soils account for 24% of the total global emission (Mosier et al. 1998). However, the contribution of agriculture to total anthropogenic source of $N_2O$ emission reportedly varies, ranging from 65% to 96% (Mosier et al. 1998; Bouwman et al. 2002; Denman et al. 2007; Yadav et al. 2018). Recently, using the revised emission factor from the IPCC 2006 guidelines (IPCC 2006), Syakila and Kroeze (2011) calculated the share of agriculture to the total anthropogenic source of $N_2O$ is 60% which was lower than the earlier estimation of 80% (Kroeze et al. 1999). Global $N_2O$ emission from agricultural activities involving crop and animal husbandry has been estimated to be 5.3 $N_2O$ Tg N $year^{-1}$ which included both direct and indirect emissions. The direct emission from agricultural soils due to the use of synthetic fertilizer, animal waste, biological $N_2$ fixation, crop residue, and cultivation of histosols was 1.8 $N_2O$ (Tg N $year^{-1}$), and the synthetic fertilizer alone accounted for 0.9 $N_2O$ Tg N $year^{-1}$. Fertilizer-induced $N_2O$ emission factors calculated using meta-analysis of more than 200 observations from 68 studies were 0.68 ± 0.41% and 0.49 ± 0.43% for rice and dry land paddy of China, respectively (Chen et al. 2015). Depending upon crop grown, water management practices, and type and dose of fertilizers, the $N_2O$ emission from agricultural soils in India ranged from 0.06 to 0.93 kg N $ha^{-1}$ (Pathak et al. 2009). The total $N_2O$–N emission from Indian soil, including emission from biological N fixation, N fertilizers, and indirect emission from soils, has been estimated to be 170 Gg $N_2O$–N $year^{-1}$, of which N fertilizer-induced direct $N_2O$–N emission constitutes around 81% (Garg et al. 2001).

### 5.4.3 Nitrate ($NO_3$) Leaching and Water Pollution

Nitrate, the end product of nitrification process that occurs in aerobic soil, is one of the preferable forms of inorganic N by most of the arable plants. However, nitrate ion is highly labile in nature; because of its negative charge, it is loosely attached to the clay surface and hence prone to leaching and runoff losses, particularly in well-drained light-textured soils. Depending upon the soil properties and water and nutrient management practices, loss of N could be as high as 80% of applied N (Watt et al. 1991). Nitrogen loading from agricultural nonpoint sources accounts for more than 50% of the total water pollution in the several countries (Boers 1996; Kronvang and Bruhn 1996) and is one of the nonpoint source of pollution that causes serious threat to water environments (Xie and Zhu 2003; Chen 2005). Nitrate contamination of ground and surface water bodies due to leaching and runoff generally leads to deterioration of water quality and results in eutrophication, algal bloom, and loss of biodiversity in rivers, ponds, and lakes. Drinking of $NO_3$-N polluted water leads to a serious health disorder, i.e., methemoglobinemia (infant cyanosis or blue-baby syndrome) in which nitrite molecules combine with hemoglobin, forming methemoglobin, which prevents the transport of oxygen, leading to suffocation and death of the infant.

The maximum contaminant level (MCL) for nitrate in groundwater has been prescribed as 10 mg $NO_3$-N $l^{-1}$ by World Health Organization, beyond which it is considered unsafe for drinking. However, there are reports of $NO_3$-N concentration exceeding the MCL in many parts of the world, both in developed and developing countries. In the United States, out of 20% of wells surveyed in farmland areas are found to have $NO_3$-N concentration higher than MCL (Galloway et al. 2004). Studies on groundwater nitrate in agricultural regions of China indicated approximately 30% of groundwater in several intensively cultivated regions of China, e.g., Circum-Bohai-Sea Region, middle of the Heishe River irrigation areas, had $NO_3^-$-N concentrations higher than the prescribed limit of WHO for drinking water (Yang and Liu 2010; Ma et al. 2012; Gogoi et al. 2018). There are few reports of enhanced $NO_3$ concentration in groundwater of intensively cultivated Indo–Gangetic states, e.g., Punjab and Haryana in India (Chauhan et al. 2012). A strong connection between higher fertilizer N application and the enhanced nitrate concentrations in ground and surface water has been reported world over (Howarth et al. 1996; Donoso et al. 1999; Sihag et al. 2015). In addition, $NO_3$ enrichment of surface, fresh, and marine aquatic system due to soil erosion and surface runoff causes excessive algal blooms and eutrophication (Howarth 2008). This results in anoxia (no oxygen) or hypoxia (low oxygen) condition and thereby disrupts the food web structure, degrades general habitat, and changes the overall ecosystem function.

### 5.4.4 Uncertainties in Estimation of Reactive N Flow from Agriculture

The processes of N cycling in soil–plant and atmosphere system are intricately related and characterized by large spatial and temporal variability, which lead to a great degree of uncertainties in quantitative understanding of these processes. Standard methodologies for estimation of various N losses, e.g., $NH_3$ emission, denitrification, and $NO_3$ leaching from agricultural fields, are yet to be developed. Existing point data obtained by the following available methodology are with large biases due to the effects of temporal and spatial variability. Agricultural systems are highly dispersed across the region and characterized by diversified climate, crop, and soil conditions. Agroclimatic factors like wind speed, water depth fertilizer, and crop management practices influence N transformation processes, thereby enhancing the uncertainty in the measurement of $NH_3$ emission-reactive N species. Biotic and abiotic factors that affect water storage, transport, and redistribution in soil also influence the extent of N leached from soil, which is generally measured by lysimeters, the measurement of $NO_3$ in drained water by porous ceramic suction. These methods do not consider the preferential flow of water present in cracks and crevices. Accelerated runoff due to presence of hard pan and transport of leached N to runoff enhance the uncertainty associated with the measurement of nitrate leaching. Quantification of spatiotemporal pattern of various pools and fluxes of nitrogen and integrating the data source to a process-based model could help reducing the biases and improving uncertainties in the measurement of N flow (Pathak et al. 2009; Kakraliya et al. 2018). Studies assessing N-related environmental problems in relatively larger areas should consider relevant N inputs, outputs, as well as dynamical processes and scaling aspects.

## 5.5 Management Options to Reduce Reactive N Flow from Agroecosystem

Since there is direct linkage between application of nitrogen fertilizer and flow of reactive N from the agroecosystem, following nutrient management practices to enhance N fertilizer use efficiency is an important strategy to reduce N pollution due to agriculture. Management interventions for enhancing N use efficiency requires a holistic approach to ensure 5 "R" nutrient stewardship, i.e., right source of N, at right dose, right time, at the right place, and with the right method of application. Several approaches for this purpose, e.g., enhanced efficiency fertilizer material, site-specific nutrient management, synchronization of N supply with demand, deep placement of urea super granule, integrated nutrient management, etc., have been devised and evaluated in different agroclimatic conditions with varying effects.

### 5.5.1 Enhanced Efficiency Fertilizer Material

Enhanced efficiency fertilizers (EEFs) are fertilizer products with the coatings of less permeable material and one or more inhibitors (nitrification or urease inhibitors) as extra additives within the formulation or as the coating. The EEFs are generally designed to regulate either nitrification or urea hydrolysis or both to reduce N loss and increase N uptake by plant. Several urease inhibitors (Soares et al. 2012; Verma et al. 2015; Mitran et al. 2018) and nitrification inhibitors are found to regulate urea hydrolysis and nitrification activity both in laboratory and field condition. Meta-analysis study on effects of EEFs indicated over all use of EEFs resulted in 5.7% (95% CI = 3.9–7.7%) and 8.0% (95% CI = 5.2–10.7%) increase in yield and N uptake, respectively, and NBPT [N-(n-butyl) phosphoric triamide] and neem proved effective in increasing yield (Linquist et al. 2013). Urease inhibitors could increase yield and N use efficiency up to 9 and 29%, respectively, and reduce N loss up to 41% in rice–paddy system (Li et al. 2014).

### 5.5.2 Site-Specific Nutrient Management

Response of crops to applied N is highly field-specific and varies with soil condition. The broad-based blanket recommendation on the basis of yield–N response does not consider the availability of N from all possible sources and crop requirement; hence, most of the time excess N is lost from the soil resulting in severe negative economic and environmental repercussions. Site-specific nutrient management (SSNM), on the other hand, takes into consideration climatic yield potential, yield goal, relationship between grain yield and plant nutrient accumulation, indigenous nutrient supply, and recovery efficiency of fertilizer while calculating the dose of N. This approach determines the need of crop according to yield target, effectively utilizes existing source of N, and fills the deficit by applying fertilizers and thereby aims at optimizing N application and reduces N loss. Field experiments conducted in several parts of South Asia showed SSNM-based N application resulted in 10% higher average N uptake in rice as compared to prevailing farmer's practice in a season (Dobermann et al. 2002).

### 5.5.3 Real-Time Nitrogen Management

Application of nitrogen in splits in synchrony with the crop requirement is an important strategy to improve N use efficiency, minimization of N loss, and regulation of $N_2O$ emission from the rice field. Leaf color chart, SPAD meter, etc. can be used to guide farmers in deciding the number of splits, amount of N applied per split, and the time of applications to match the N supply with real-time demand of rice crop. However, there is a need to standardize these tools for critical threshold with respect to cultivars grown and agroclimatic condition. Site-specific nutrient management approach that decides application of nitrogen on the basis of crop requirements, indigenous supply, and the recovery efficiency of applied nitrogen ensures about 30–40% increase in nitrogen use efficiency and hence a potential mitigation option to reduce $N_2O$-N emission from the rice field.

### 5.5.4 Deep Placement of Urea Super Granules

The flooded lowland rice soil consists of a thin oxidized layer overlying a reduced zone, which facilitates simultaneous occurrence of both nitrification and denitrification processes and accelerate the loss of N from various forms. Deep placement of urea super granules, i.e., large particles (1–2 g) of urea at reduced zone, prevents fast conversion of $NH_4$ to $NO_3$ and subsequent losses. Therefore, N availability to the plant lasts for a longer period than the traditional urea fertilizer, which results in significant increases in N uptake and grain yield. Deep placement of USG could save up to 65% of urea fertilizer and increase grain yields up to 50% over that with the same amount of split-applied N as prilled urea (Savant and Stangel 1990). Studies indicated reduction of N loss due to ammonia volatilization and $N_2O$ emission both in field and laboratory conditions (Khalil et al. 2011; Chatterjee et al. 2018; Nayak et al. 2017).

### 5.5.5 Integrated Nutrient Management

The integrated nutrient management (INM) approach that judiciously combines all possible organic, inorganic, and biological sources of N supply has the potential of increasing yield, decreasing nutrient losses, and increasing the efficiency of applied and native nutrients. Panda et al. (2007) observed that at N level of 90 kg $ha^{-1}$, practices involving Dhaincha green manure or *Azolla* dual crop were superior to the chemical source of N.

## 5.6 Nitrogen Footprint: Concept, Scope, and Calculation

Nitrogen footprint is a recent concept that has been used to access the contribution of anthropogenic activities individually or collectively towards loading of reactive N species in the environment. It is defined as accumulated anthropogenic reactive N species released to the environment during the life cycle of an entity (Leach et al. 2012). Nitrogen footprint is the sum total of all forms of nitrogen that are biologically, photochemically, and radiatively active such as $N_2O$, $NO_3^-$, $NO_2^-$, $NH_3$, and ammonium ($NH_4^+$) released to the environment as a result of resource consumption. These reactive forms of nitrogen once enter into the environment have a cascading impact in the form of smog, acid rain, groundwater pollution, biodiversity loss, etc. (Fig. 5.4).

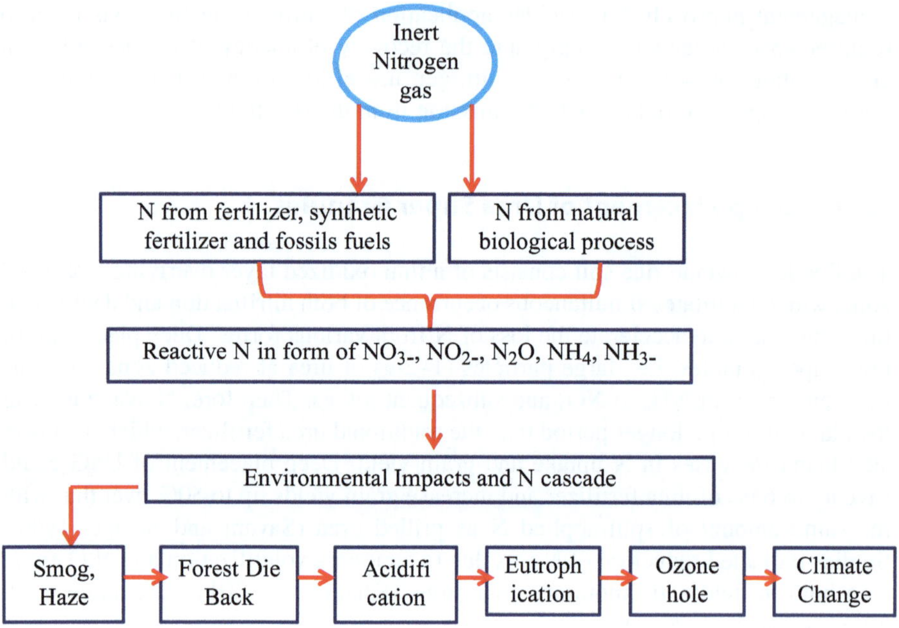


**Fig. 5.4** Schematic flowchart of conversion of inert $N_2$ to reactive N species and other environment impacts causing excessive reactive $N_2$ in the environment. (Source: Andrew Greene, www.sustainableunh.unh.edu)

The N footprint is a useful tool that quantitatively defines the relationship between consumption pattern of an individual or community and extent of N pollution. Conversion of nonreactive N ($N_2$) to reactive N is essential to sustain life on the planet; however, indiscriminate infusion of reactive N to the ecosystem has several cascading harmful effects. Therefore, calculation of N footprint could be a useful indicator for rationalization of N consumption pattern to ensure environmental sustainability.

One of the earliest N footprint models developed by Leach et al. (2012) known as N-Calculator uses per capita data of food and energy consumption, purchase of goods, and use of services. Food nitrogen footprint constituted two parts: food consumption and food production nitrogen footprints. The food consumption footprint is the sum total of the reactive nitrogen present in the food that enters sewage stream through human consumption and excretion. The food production nitrogen footprint is the sum total of reactive N losses that take place at each stage of food production, i.e., fertilizer application, manure storage and application, disposal of food waste, etc. The N footprints calculated for the United States and the Netherlands using the N-Calculator were found to be 41 kg N capita$^{-1}$ year$^{-1}$ and 24 kg N capita$^{-1}$ year$^{-1}$, respectively. Leach et al. (2013) further upscaled the model to N-Institution to calculate the N footprint of University of Virginia. The model calculated the N footprint of the university by tabulating the nitrogen released due to food consumption, food production, food transport and food diversion, N emissions ($NO_x$ and $N_2O$) due to energy usage (both self-generated and purchased), contribution of research animal through animal food consumption and carcass disposal, N emissions due to transportation, and the total amount of nitrogen applied as fertilizers. The results indicated that energy utilities were the largest contributor (48%) to the total N footprint of the university (492 metric tons in 2010) followed by food production (37%). Among the food production categories, meat (22%) was the largest contributor. The scenario analysis for 2025 showed interventions such as improved treatment of sewage, composting of food waste, food donations, reduction of meat consumption, and substitution of chicken for beef, etc. can reduce N footprint by 17% as compared to business as usual.

In addition to this, several other country-specific N footprint models have been developed and used to calculate the reactive N flow for the country and regional level and also for different food items separately. Consumer-based N footprint tool such as N-Calculator calculates a person's N footprint due to consumption of food and other services that involve use of energy, e.g., housing, transportation, goods, services, etc. Thereby, it helps consumers to take informed decision about his or her consumption pattern and bring necessary changes to reduce N footprint. Similarly, N footprint tool like N-Institution calculates the contribution of an institution to the overall N pollution and also provides a quantitative guideline to reduce N footprint by adopting measures like energy saving, choice of low N footprint food, etc.

The overall objective of calculating N footprint of an individual is to provide a quantitative indication of impact of our lifestyle choices, particularly food consumption and energy use pattern on N pollution, which can further be used to raise awareness among all stakeholders, producers, consumers, policy-makers, and government about the impact of anthropogenic activities on N pollution and take necessary measures to curb the same.

## 5.7 Monitoring N Footprint in Agriculture and Allied Sector: Case Studies

Gu et al. (2013) calculated N footprint of China following mass balance approach in the coupled human and natural systems and integrating all anthropogenic reactive N fluxes and sources of anthropogenic reactive fluxes and their contribution. This approach does away with the huge data required in consumer-based approach of N-Calculator. Total national N footprint calculated using this method showed that between 1980 and 2008, the per capita N footprint in China increased from 19 to 32 kg N $ha^{-1}$ and the reactive N loss from the production and consumption of food was the largest component of the N footprint. An integrated nitrogen model MITERRA—Europe was developed to assess the effects of integrated measures in agriculture to reduce $NH_3$ emission on reactive N flow, e.g., $NH_3$ emission, $N_2O$ emission, and $NO_3$ leaching to ground and surface water for country as well as regional level (Velthof et al. 2007; Datta et al. 2017). The model consists of an input module with activity data, emission factors, mitigation measures, and output module. The input data include N input, output, surplus, crop type, topography livestock types, etc. Estimation for the year 2000 showed that denitrification is the largest pathway of N loss in European agriculture followed by $NH_3$ volatilization, N leaching, and emissions of $N_2O$ (Velthof et al. 2009; Layek et al. 2018). Leip et al. (2014) used MITERRA and CAPRI models to calculate the N footprint of food products in the European Union at the country level and for agriculture and found that the N footprint, defined as the total N losses to the environment per unit of product, was substantially higher for livestock products, particularly beef (highest value of 500 g N $kg^{-1}$), as compared to vegetable products. Sugar beet, fruits and vegetables, and potatoes had the lowest N footprint, i.e., 2 g N $kg^{-1}$. Similarly, Xue and Landis (2010) compared eight different food types and found cereals are the food group with low nitrogen footprint while red meat is least environmentally friendly with high nitrogen footprint. Smithwick et al. (2012) followed the life cycle impact analysis (LCIA) method to calculate N footprint during the life cycle of 1 kg Swedish tomato. The life cycle of tomato is divided into fertilizer production, tomato farming, transport to warehouse, transport to retail store, and sewage treatment. Results showed that the treatment of sewage resulted in the highest amount of elementary nitrogen (kg $N_2$ per kg tomato purchased) followed by transport and fertilizer production. The impacts are categorized as global warming potential (GWP), eutrophication potential (EP), acidification potential (AP), photochemical ozone creation potential (POCP), and ozone depletion potential (ODP), and the results showed fertilizer production had the highest nitrous oxide (kg $CO_2$ equivalent) followed by farming. Eutrophication potential measured as kg phosphate equivalent was the highest for sewage treatment due to $NO_3$ and $NO_2$ followed by transport and fertilizer production. Nitrogen oxides were mainly responsible for acidification potential, and the highest value was for transport. Photochemical ozone creation potential due to nitrogen oxides was the highest for transport followed by farming and fertilizer production.

Xue et al. (2016) calculated N footprint of double rice production in Southern China following the life cycle analysis method which included reactive N losses from agricultural inputs and paddy fields at the entire stage of rice production starting from acquisition of agricultural inputs through processes of agricultural production till harvest. The total N emission associated with the entire life cycle of double rice production was calculated as follows:

$$NE_{total} = NE_{inputs} + NVNH_3 + NEN_2O + NLNO_3 + NLNH_4^+ \quad (5.3)$$

where $NE_{input}$ is the total amount of N released due to agricultural input applications, which was obtained by multiplying the total input used with specific emission factors; $NVNH_3$, $NEN_2O$, and $NLNO_3$ are $NLNH_4^+NH_3$ volatilization, $N_2O$ emission, and $NO_3^-$ and $NH_4^{+1}$ leaching during double rice growing periods, respectively. The yield-scale nitrogen footprint calculated for early, late, and double rice were 10.47, 10.89, and 10.68 g N-eq $kg^{-1}$ $year^{-1}$, respectively. The overwhelming majority of nitrogen trifluoride (NF) was due to volatilization of $NH_3$ from paddy fields due to N fertilizer applications for double rice production that contributed maximum to the total N footprint.

## 5.8 Conclusions and Future Perspectives

Nitrogen fertilizers are the integral component of intensive agricultural practices that are instrumental for feeding the ever-increasing population all over the world. Last few decades witnessed tremendous increase in application of reactive N for obvious reasons, but the detrimental environmental impact associated with it is largely ignored. Farmers, consumers, and also policy-makers are mostly unaware or poorly aware about the N-related environmental hazards. In the absence of standard methodology and system for monitoring reactive N flow from agricultural activities, it is almost impossible to assess the environmental impact of agricultural N use. Efforts have been initiated to quantify these losses in larger scale through improved technologies of remote sensing, geographical information system, and simulation modeling. In this context, the concept of N footprint is a welcome development. Though still in nascent stage, it can give an overall quantitative understanding of an individual's contribution toward N pollution through its consumption pattern. Some of the most widely used N footprint models like N-Calculator and N-Institution calculate N footprint in kilograms of $N_2$ per person per year. However, these models do not calculate the accumulation of different N species, e.g., $NO_2$, NOx, $NH_3$, etc., and the associated environmental impact. Some attempts have been made to improve upon these N footprint models by introducing components of environmental impact assessment such as global warming potential, ozone depletion potential, etc; still, a lot of work needs to be done to develop the standard N footprint calculation method for uniform adaptability across the region. Despite their shortcoming, the N-footprint tool is useful to raise awareness among end users about their contribution toward N pollution and also provides alternative options to minimize the impact. The calculation of N footprint of different agricultural operations during the different stages of crop growth will help policy-makers as well as crop growers to adapt environment-friendly measures to reduce N flow from the agroecosystem.

## References

Adhya TK, Pathak H, Chhabra A (2007) N-fertilizers and gaseous–N emission from rice-based cropping systems. Agricultural nitrogen use and its environmental implications. IK International Publishing House Pvt Ltd, New Delhi, pp 459–475

Amaliotis D, Therios I, Karatissiou M (2004) Effect of nitrogen fertilization on growth, leaf nutrient concentration and photosynthesis in three peach cultivars. ISHS, Acta Horticult 449:36–42

Banerjee B, Pathak H, Aggarwal P (2002) Effects of dicyandiamide, farmyard manure and irrigation on crop yields and ammonia volatilization from an alluvial soil under a rice (Oryza sativa L.)-wheat (Triticum aestivum L.) cropping system. Biol Fertil Soils 36(3):207–214

Boers PC (1996) Nutrient emissions from agriculture in the Netherlands, causes and remedies. Water Sci Technol 33(4–5):183–189

Bojovic B, Markovic A (2009) Correlation between nitrogen and chlorophyll content in wheat (Triticum aestivum L). Kragujevac J Sci 31:69–74

Bouwman AF, Lee DS, Asman WA, Dentener FJ, Van Der Hoek KW, Olivier JG (1997) A global high-resolution emission inventory for ammonia. Glob Biogeochem Cycles 11(4):561–587

Bouwman AF, Boumans LJ, Batjes NH (2002) Modeling global annual $N_2O$ and NO emissions from fertilized fields. Glob Biogeochem Cycles 16(4):2

Chatterjee D, Sangita M, Kumar GP, Kumar SC, Tripathi R, Shahid M, Kumar U, Kumar A, Bhattacharyya P, Gautam P, Lal B, Dash PK, Nayak AK (2018) Comparative assessment of urea briquette applicators on greenhouse gas emission, nitrogen loss and soil enzymatic activities in tropical lowland rice. Agric Ecosyst Environ 252:178–190

Chauhan BS, Mahajan G, Sardana V, Timsina J, Jat ML (2012) Productivity and sustainability of the rice–wheat cropping system in the Indo-Gangetic Plains of the Indian subcontinent: problems, opportunities, and strategies. Adv Agron 117:315–369

Chen GQ (2005) Energy consumption of the earth. Ecol Model 184(2–4):363–380

Chen J, Liu X, Li L, Zheng J, Qu J, Zheng J, Zhang X, Pan G (2015) Consistent increase in abundance and diversity but variable change in community composition of bacteria in top soil of rice paddy under short term biochar treatment across three sites from South China. Appl Soil Ecol 91:68–79

Cheng FY, Hsu YC, Lin PL, Lin TH (2013) Investigation of the effects of different land use and land cover patterns on mesoscale meteorological simulations in the Taiwan area. J Appl Meteorol Climatol 52(3):570–587

Datta R, Baraniya D, Wang YF, Kelkar A, Moulick A, Meena RS, Yadav GS, Ceccherini MT, Formanek P (2017) Multi-function role as nutrient and scavenger of free radical in soil. Sustain MDPI 9:402. https://doi.org/10.3390/su9081402

Denman KL, Chidthaisong A, Ciais P, Cox PM, Dickinson RE, Hauglustaine D, Heinze C, Holland E, Lohmann U, Ramachandran S, Dias PL (2007) Couplings between changes in the climate system and biogeochemistry. In: Solomon S, Qin D, Manning M, Chen Z, Marquis M, Averyt KB, Tignor M, Miller HL (eds) Climate change 2007: the physical science basis. Contribution of working group I to the fourth assessment report of the intergovernmental panel on climate change. Cambridge University Press, Cambridge/New York

Dobermann A, Witt C, Dawe D, Gines GC, Nagarajan R, Satawathananont S, Son TT, Tan PS, Wang GH, Chien NV, Thoa VTK, Phung CV, Stalin P, Muthukrishnan P, Ravi V, Babu M, Chatuporn S, Kongchum M, Sun Q, Fu R, Simbahan GC, Adviento MAA (2002) Site-specific nutrient management for intensive rice cropping systems in Asia. Field Crops Res 74:37–66

Donoso G, Cancino J, Magri A (1999) Effects of agricultural activities on water pollution with nitrates and pesticides in the Central Valley of Chile. Water Sci Technol 39(3):49–60

Dreccer MF, Schapendonk AH, Slafer GA, Rabbinge R (2000) Comparative response of wheat and oilseed rape to nitrogen supply: absorption and utilization efficiency of radiation and nitrogen during the reproductive stages determining yield. Plant Soil 220(1–2):189–205

Fageria NK, Baligar VC, Li YC (2008) The role of nutrient efficient plants in improving crop yields in the twenty first century. J Plant Nutr 31(6):1121–1157

Foulkes MJ, Slafer GA, Davies WJ, Berry PM, Sylvester-Bradley R, Martre P, Calderini DF, Griffiths S, Reynolds MP (2010) Raising yield potential of wheat. III. Optimizing partitioning to grain while maintaining lodging resistance. J Exp Bot 62(2):469–486

Galloway JN, Dentener FJ, Capone DG, Boyer EW, Howarth RW, Seitzinger SP, Asner GP, Cleveland CC, Green PA, Holland EA, Karl DM (2004) Nitrogen cycles: past, present, and future. Biogeochemistry 70(2):153–226

Garg BK, Kathju S, Burman U (2001) Influence of water stress on water relations, photosynthetic parameters and nitrogen metabolism of moth bean genotypes. Biol Plant 44(2):289–292

Gogoi N, Baruah KK, Meena RS (2018) Grain legumes: impact on soil health and agroecosystem. In: Meena RS et al (eds) Legumes for soil health and sustainable management. Springer. https://doi.org/10.1007/978-981-13-0253-4_16

Gu H, Gowda GN, Carnevale Neto F, Opp MR, Raftery D (2013). RAMSY: ratio analysis of mass spectrometry to improve compound identification. Anal Chem 85:10771–10779. https://doi.org/10.1021/ac4019268

Guarda G, Padovan S, Delogu G (2004) Grain yield, nitrogen-use efficiency and baking quality of old and modern Italian bread-wheat cultivars grown at different nitrogen levels. Eur J Agron 21(2):181–192

Guo D, Shibuya R, Akiba C, Saji S, Kondo T, Nakamura J (2016) Active sites of nitrogen-doped carbon materials for oxygen reduction reaction clarified using model catalysts. Science 351(6271):361–365

He M, Dijkstra FA (2014) Drought effect on plant nitrogen and phosphorus: a meta-analysis. New Phytol 204(4):924–931

Howarth RW (2008) Coastal nitrogen pollution: a review of sources and trends globally and regionally. Harmful Algae 8(1):14–20

Howarth RW, Billen G, Swaney D, Townsend A, Jaworski N, Lajtha K, Downing JA, Elmgren R, Caraco N, Jordan T, Berendse F (1996) Regional nitrogen budgets and riverine N & P fluxes for the drainages to the North Atlantic Ocean: natural and human influences. In: Nitrogen cycling in the North Atlantic Ocean and its watersheds. Springer, Dordrecht, pp 75–139

Howarth RW, Sharpley A, Walker D (2002) Sources of nutrient pollution to coastal waters in the United States: implications for achieving coastal water quality goals. Estuaries 25(4):656–676

IFA (2007) International fertilizer association. https://www.ifastat.org/dated23/02/2019

IPCC (1997) IPCC (revised 1996) guidelines for national greenhouse gas inventories. Workbook. Intergovernmental Panel on Climate Change, Paris

IPCC (2006) Chapter 11, $N_2O$ emissions from managed soils, and $CO_2$ emissions from lime and urea application. In: Eggleston HS, Buendia L, Miwa K, Ngara T, Tanabe K (eds) IPCC guidelines for national greenhouse gas inventories, prepared by the National Greenhouse Gas Inventories Programme, vol 4. IGES, Hayama

IPCC (2007) Climate change 2007: the physical science basis, contribution of Working Group-I to the fourth assessment report of the Intergovernmental Panel on Climate Change. Cambridge University Press, Cambridge

Kaizzi KC, Byalebeka J, Semalulu O, Alou I, Zimwanguyizza W, Nansamba A, Musinguzi P, Ebanyat P, Hyuha T, Wortmann CS (2012) Maize response to fertilizer and nitrogen use efficiency in Uganda. Agron J 104(1):73–82

Kakraliya SK, Singh U, Bohra A, Choudhary KK, Kumar S, Meena RS, Jat ML (2018) Nitrogen and legumes: a meta-analysis. In: Meena RS et al (eds) Legumes for soil health and sustainable management. Springer. https://doi.org/10.1007/978-981-13-0253-4_9

Khalil SK, Khan F, Rehman A, Muhammad FI, Amanullah KA, Shah MK, Khan H (2011) Dual purpose wheat for forage and grain yield in response to cutting, seed rate and nitrogen. Pak J Bot 43(2):937–947

Kim JH, Kang EJ, Park MG, Lee BG, Kim KY (2011) Effects of temperature and irradiance on photosynthesis and growth of a green-tide-forming species (Ulva linza) in the Yellow Sea. J Appl Phycol 23(3):421–432

Kroeze C, Mosier A, Bouwman L (1999) Closing the global $N_2O$ budget: a retrospective analysis 1500–1994. Glob Biogeochem Cycles 13(1):1–8

Kronvang B, Bruhn AJ (1996) Choice of sampling strategy and estimation method for calculating nitrogen and phosphorus transport in small lowland streams. Hydrol Process 10(11):1483–1501

Kumar S, Meena RS, Lal R, Yadav GS, Mitran T, Meena BL, Dotaniya ML, EL-Sabagh A (2018) Role of legumes in soil carbon sequestration. In: Meena RS et al (eds) Legumes for soil health and sustainable management. Springer. https://doi.org/10.1007/978-981-13-0253-4_4

Ladha JK, Reddy PM (2003) Nitrogen fixation in rice systems: state of knowledge and future prospects. Plant Soil 252(1):151–167

Ladha JK, Pathak H, Krupnik TJ, Six J, van Kessel C (2005) Efficiency of fertilizer nitrogen in cereal production: Retrospect's and prospects. Adv Agron 87:85–156

Layek J, Das A, Mitran T, Nath C, Meena RS, Singh GS, Shivakumar BG, Kumar S, Lal R (2018) Cereal+legume intercropping: an option for improving productivity. In: Meena RS et al (eds) Legumes for soil health and sustainable management. Springer. https://doi.org/10.1007/978-981-13-0253-4_11

Leach AM, Galloway JN, Bleeker A, Erisman JW, Kohn R, Kitzes J (2012) A nitrogen footprint model to help consumers understand their role in nitrogen losses to the environment. Environ Dev 1:40–66

Leach AM, Majidi AN, Galloway JN, Greene AJ (2013) Toward institutional sustainability: a nitrogen footprint model for a university. Sustainability 6(4):211–219

Leip A, Weiss F, Lesschen JP, Westhoek H (2014) The nitrogen footprint of food products in the European Union. J Agric Sci 152(S1):20–33

Li R, Wang JJ, Zhang Z, Shen F, Zhang G, Qin R, Li X, Xiao R (2012) Nutrient transformations during composting of pig manure with bentonite. Bioresour Technol 121:362–368

Li L, Tilman D, Lambers H, Zhang FS (2014) Plant diversity and over yielding: insights from belowground facilitation of intercropping in agriculture. New Phytol 203(1):63–69

Linquist BA, Liu L, Van Kessel C, van Groenigen KJ (2013) Enhanced efficiency nitrogen fertilisers for rice systems: metal-analysis of yield and nitrogen uptake. Field Crops Res 154:246–254

Liu J, You L, Amini M, Obersteiner M, Herrero M, Zehnder AJ, Yang H (2010) A high-resolution assessment on global nitrogen flows in cropland. Proc Natl Acad Sci 107(17):8035–8040

Lu C, Tian H (2017) Global nitrogen and phosphorus fertilizer use for agriculture production in the past half century: shifted hot spots and nutrient imbalance. Earth Syst Sci Data 9(1):181–192

Ma C, Shao X, Cao D (2012) Nitrogen-doped graphene nanosheets as anode materials for lithium ion batteries: a first-principles study. J Mater Chem 22(18):8911–8915

Marschner P (2012) Rhizosphere biology. In: Marschner's mineral nutrition of higher plants, 3rd edn, pp 369–388

Meena RS, Lal R (2018) Legumes and sustainable use of soils. In: Meena RS et al (eds) Legumes for soil health and sustainable management. Springer. https://doi.org/10.1007/978-981-13-0253-4_1

Meena BL, Fagodiya RK, Prajapat K, Dotaniya ML, Kaledhonkar MJ, Sharma PC, Meena RS, Mitran T, Kumar S (2018) Legume green manuring: an option for soil sustainability. In: Meena RS et al (eds) Legumes for soil health and sustainable management. Springer. https://doi.org/10.1007/978-981-13-0253-4_12

Mitran T, Meena RS, Lal R, Layek J, Kumar S, Datta R (2018) Role of soil phosphorus on legume production. In: Meena RS et al (eds) Legumes for soil health and sustainable management. Springer. https://doi.org/10.1007/978-981-13-0253-4_15

Mosier A, Kroeze C, Nevison C, Oenema O, Seitzinger S, Van Cleemput O (1998) Closing the global $N_2O$ budget: nitrous oxide emissions through the agricultural nitrogen cycle. Nutr Cycl Agroecosyst 52(2–3):225–248

National Research Council (NRC) Board OS (2000) Clean coastal waters: understanding and reducing the effects of nutrient pollution. National Academies Press, Washington, DC

Nayak AK, Mohanty S, Chatterjee D, Guru PK, Lal B, Shahid M, Tripathi R, Gautam P, Kumar A, Bhattacharyya P, Panda BB (2017) Placement of urea briquettes in lowland rice: an environment-friendly technology for enhancing yield and nitrogen use efficiency. NRRI Research Bulletin No. 12 ICAR-National Rice Research Institute, Cuttack, Odisha 753006, India, pp 1–26

Network GF (2012) World footprint: do we fit on the planet. See http://www.Footprintnetwork.org/en/index.php/GFN/page/world_footprint

Panda D, Kundu SK, Ghosh A, Prakash NB, Patra DD (2007) Nitrogen use efficiency in rice ecosystems. In: Agricultural nitrogen use and its policy implications, pp 99–120

Pathak H, Bhatia A, Prasad S, Singh S, Kumar S, Jain MC, Kumar U (2002) Emission of nitrous oxide from rice-wheat systems of Indo-Gangetic plains of India. Environ Monit Assess 77(2):163–178

Pathak RK, Wu WS, Wang T (2009) Summertime PM 2.5 ionic species in four major cities of China: nitrate formation in an ammonia-deficient atmosphere. Atmos Chem Phys 9(5):1711–1722

Paulot F, Jacob DJ, Pinder RW, Bash JO, Travis K, Henze DK (2014) Ammonia emissions in the United States, European Union, and China derived by high-resolution inversion of ammonium wet deposition data: interpretation with a new agricultural emissions inventory (MASAGE_$NH_3$). J Geophys Res Atmos 119(7):4343–4364

Qiao C, Liu L, Hu S, Compton JE, Greaver TL, Li Q (2015) How inhibiting nitrification affects nitrogen cycle and reduces environmental impacts of anthropogenic nitrogen input. Glob Chang Biol 21(3):1249–1257

Ramankutty N, Mehrabi Z, Waha K, Jarvis L, Kremen C, Herrero M, Rieseberg LH (2018) Trends in global agricultural land use: implications for environmental health and food security. Annu Rev Plant Biol 69:789–815

Samonte SO, Wilson LT, Medley JC, Pinson SR, McClung AM, Lales JS (2006) Nitrogen utilization efficiency. Agron J 98(1):168–176

Savant NK, Stangel PJ (1990) Deep placement of urea supergranules in transplanted rice: principles and practices. Fertil Res 25(1):1–83

Shahrokhnia MH, Sepaskhah AR (2016) Effects of irrigation strategies, planting methods and nitrogen fertilization on yield, water and nitrogen efficiencies of safflower. Agric Water Manag 172:18–30

Sheldrick WF, Syers JK, Lingard J (2002) A conceptual model for conducting nutrient audits at national, regional, and global scales. Nutr Cycl Agroecosyst 62(1):61–72

Sihag SK, Singh MK, Meena RS, Naga S, Bahadur SR, Gaurav YRS (2015) Influences of spacing on growth and yield potential of dry direct seeded rice (Oryza sativa L) cultivars. Ecoscan 9(1–2):517–519

Smil V (1999) Nitrogen in crop production: an account of global flows. Glob Biogeochem Cycles 13(2):647–662

Smithwick EA, Naithani KJ, Balser TC, Romme WH, Turner MG (2012) Post-fire spatial patterns of soil nitrogen mineralization and microbial abundance. PloS One 7(11):Ne50597

Soares JR, Cantarella H, de Campos Menegale ML (2012) Ammonia volatilization losses from surface-applied urea with urease and nitrification inhibitors. Soil Biol Biochem 52:82–89

Sofi PA, Baba ZA, Hamid B, Meena RS (2018) Harnessing soil rhizobacteria for improving drought resilience in legumes. In: Meena RS et al (eds) Legumes for soil health and sustainable management. Springer. https://doi.org/10.1007/978-981-13-0253-4_8

Spiertz H (2010) Food production, crops and sustainability: restoring confidence in science and technology. Curr Opin Environ Sustain 2(5–6):439–443

Stone ML, Solie JB, Raun WR, Whitney RW, Taylor SL, Ringer JD (1996) Use of spectral radiance for correcting in-season fertilizer nitrogen deficiencies in winter wheat. Trans ASAE 39(5):1623–1631

Streets DG, Bond TC, Carmichael GR, Fernandes SD, Fu Q, He D, Klimont Z, Nelson SM, Tsai NY, Wang MQ, Woo JH (2003) An inventory of gaseous and primary aerosol emissions in Asia in the year 2000. J Geophys Res 108:8809

Syakila A, Kroeze C (2011) The global nitrous oxide budget revisited. GHG Measure Manage 1(1):17–26

Tafteh A, Sepaskhah AR (2012) Yield and nitrogen leaching in maize field under different nitrogen rates and partial root drying irrigation. Int J Plant Prod 6(1):93–114

Tanaka N, Uraguchi S, Saito A, Kajikawa M, Kasai K, Sato Y, Nagamura Y, Fujiwara T (2013) Roles of pollen-specific boron efflux transporter, OsBOR4, in the rice fertilization process. Plant Cell Physiol 54(12):2011–2019

Tao M, Chen L, Xiong X, Zhang M, Ma P, Tao J, Wang Z (2014) Formation process of the widespread extreme haze pollution over northern China in January 2013: implications for regional air quality and climate. Atmos Environ 98:417–425

Tirol-Padre A, Ladha JK, Singh U, Laureles E, Punzalan G, Akita S (1996) Grain yield performance of rice genotypes at suboptimal levels of soil N as affected by N uptake and utilization efficiency. Field Crop Res 46(1–3):127–143

Umehara M, Hanada A, Yoshida S, Akiyama K, Arite T, Takeda-Kamiya N, Magome H, Kamiya Y, Shirasu K, Yoneyama K, Kyozuka J (2008) Inhibition of shoot branching by new terpenoid plant hormones. Nature 455(7210):195

Velthof GL, Oudendag DA, Oenema O (2007) Development and application of the integrated nitrogen model MITERRA-EUROPE. Task 1 service contract "integrated measures in agriculture to reduce ammonia emissions". Alterra, Wageningen

Velthof GL, Oudendag D, Witzke HP, Asman WA, Klimont Z, Oenema O (2009) Integrated assessment of nitrogen losses from agriculture in EU-27 using MITERRA-EUROPE. J Environ Qual 38(2):402–417

Verma JP, Meena VS, Kumar A, Meena RS (2015) Issues and challenges about sustainable agriculture production for management of natural resources to sustain soil fertility and health: a book review. J Clean Prod 107:793–794

Watt W, Tulinsky A, Swenson RP, Watenpaugh KD (1991) Comparison of the crystal structures of a flavodoxin in its three oxidation states at cryogenic temperatures. J Mol Biol 218(1):195–208

Xie HM, Zhu B (2003) Research progress on non-point source pollution of nitrogen in agroecosystem. Ecol Environ 12:349–352

Xu K (2014) Electrolytes and interphases in Li-ion batteries and beyond. Chem Rev 114(23):11503–11618

Xue X, Landis AE (2010) Eutrophication potential of food consumption patterns. Environ Sci Technol 44(16):6450–6456

Xue JF, Pu C, Liu SL, Zhao X, Zhang R, Chen F, Xiao XP, Zhang HL (2016) Carbon and nitrogen footprint of double rice production in southern China. Ecol Indic 64:249–257

Yadav GS, Das A, Lal R, Babu S, Meena RS, Patil SB, Saha P, Datta M (2018) Conservation tillage and mulching effects on the adaptive capacity of direct-seeded upland rice (Oryza sativa L.) to alleviate weed and moisture stresses in the North Eastern Himalayan region of India. Arch Agron Soil Sci 64(9):1254–1267. https://doi.org/10.1080/03650340.2018.1423555

Yang R, Liu W (2010) Nitrate contamination of groundwater in an agroecosystem in Zhangye Oasis, Northwest China. Environ Earth Sci 61(1):123–129

# Strategies for Identification of Genes Toward Enhancing Nitrogen Utilization Efficiency in Cereals

6

Alka Bharati and Pranab Kumar Mandal

**Abstract**

Global cereal demand will increase up to 38% by 2025, and to achieve it in a sustainable way, 60% increase in global nitrogen (N) use will be necessary. In cereals ~30 to 50% of the applied N is taken up by the crop, and the rest is lost in the environment causing pollution. Hence, improvement of N use efficiency (NUE) in cereals is really important. The NUE is the total biomass or grain yield produced per unit of applied N fertilizer. Soil and plant management practices play a key role toward enhancing N recovery, but again it greatly depends on environmental conditions. Another option for improvement of NUE is the genetic strategy. Broadly, NUE has two components, N uptake efficiency (NUpE), which is N acquisition by the plant per unit of available N in the soil, and N utilization efficiency (NUtE), which is yield per unit of acquired N by the plant. As NUtE is directly related to the crop yield, it depends on subcomponent N assimilation, remobilization, and finally efficient utilization of assimilated N for starch biosynthesis in the grain. Understanding the mechanisms and gene regulating of these processes, exploiting genotypic variant in each subcomponent (N uptake, assimilation, and remobilization) to find genes and superior alleles is crucial for the improvement of NUE in crop plants. In addition, the studies on starch metabolism during grain filling are an important factor for N utilization. To study this, genotypes with similar background of uptake and assimilation but differing in grain filling should be taken into consideration. Global metabolomic profiling of these genotypes, transcriptome profiling, identification, and mapping of quantitative trait loci (QTLs) in combination with marker-assisted selection (MAS), analyzing mutants defective in their normal response to N limitation, and studying plants that show better growth under N-limiting conditions are different options to study the N-utilization efficiency and gene identification. In the first topic, we have highlighted the N application and its effect on yield in cereals.

A. Bharati · P. K. Mandal (✉)
ICAR-National Research Center on Plant Biotechnology, New Delhi, India



R. S. Meena (ed.), *Nutrient Dynamics for Sustainable Crop Production*,
https://doi.org/10.1007/978-981-13-8660-2_6

Introduction of N-responsive genotype during green revolution has enhanced yield, but indiscriminate use of fertilizer mainly N fertilizer has caused severe damage to environment. In the subsequent topic, we have defined NUE as a whole; later the main focus was on biological NUE and their different components. Thereafter we described strategies for genetic improvement to reduce N use without much compromising yield. Primarily we tried to highlight candidate genes and their role in NUE reported in cereals as well as model plant system. We have also described the advance molecular techniques to identify the gene in strategic manner. As a part of molecular breeding, QTL identification and its introgression are described in one of the topics at the last part.

**Keywords**
Candidate genes · Cereals · Nitrogen use efficiency

## Abbreviation

| | |
|---|---|
| AAP | Amino acid permease |
| ABA | Abscisic acid |
| ADP | Adenosine diphosphate |
| AE | Agronomic efficiency |
| AFG | Auxin signaling F-box |
| AGPase | ADP glucose pyrophosphorylase |
| AlaAT | Alanine aminotransferase |
| AMT | Ammonium transporter |
| ARE | Apparent recovery efficiency |
| AS | Asparagine synthase |
| ATF | Amino acid transporter |
| BE | Branching enzyme |
| BNF | Biological nitrogen fixation |
| C | Carbon |
| CaMV | Cauliflower mosaic virus |
| cDNA | Complimentary DNA |
| CGs | Candidate genes |
| CLC | Chloride channel family |
| CPSase | Carbamoyl phosphate synthase |
| CRISPR-Cas9 | Clustered regularly interspaced short palindromic repeats |
| DNA | Deoxyribonucleic acid |
| Dof | DNA binding with one zinc finger |
| EMS | Ethyl methanesulfonate |
| G-1-P | Glucose-1-phospate |
| GBSS | Granule-bound starch synthase |
| GDH | Glutamate dehydrogenase |
| GMPase | GDP mannose pyrophosphorylase |
| GOGAT | Glutamine-2-oxoglutarate aminotransferase or glutamate synthase |

| | |
|---|---|
| GS | Glutamine synthetase |
| $H^+$ | Hydrogen |
| HAT | High-affinity transport system |
| HSN1 | Hypersensitive to $NH_4^+$ |
| HYVs | High-yielding varieties |
| ISA | Isoamylase |
| LATS | Low-affinity transport |
| LHT | Lysine/histidine transporter |
| MAS | Marker-assisted selection |
| miR | micro RNA |
| MPSS | Massively parallel signature sequencing |
| mQTLs | Metabolic QTLs |
| N | Nitrogen |
| $N_2O$ | Nitrogen oxide |
| NAD(P)H | Nicotinamide adenine dinucleotide phosphate |
| NAGK | N-acetyl glutamate kinase |
| $NH_3$ | Ammonia |
| NIL | Near-isogenic lines |
| NiR | Nitrite reductase |
| NO | Nitric oxide |
| $NO_3^-$ | Nitrate |
| NPF | Peptide transporter family |
| NpUE | N physiological use efficiency |
| NR | $NO_3^-$ reductase |
| NRA | $NO_3^-$ reductase activity |
| NRT | $NO_3^-$ transporter |
| NUE | Nitrogen use efficiency |
| NUpE | N uptake efficiency |
| NUtE | N utilization efficiency |
| PEP | Partial factor productivity |
| PEPC | Phosphoenolpyruvate carboxylase |
| PNB | Partial nutrient balance |
| PTST | Protein targeting to starch |
| QTL | Quantitative trait loci |
| RNA | Ribonucleic acid |
| RNAi | RNA interference |
| RS | Root system |
| RSA | Root system architecture |
| SAGE | Serial analysis of gene expression |
| SAV | Senescence associated vacuoles |
| SBE | Soluble starch branching enzymes |
| SG | Starch granules |
| SLAC/SLAH | Slow anion-associated channel homolog |
| SS | Starch synthase |
| SSH | Suppression subtractive hybridization |

| | |
|---|---|
| T-DNA | Transfer DNA |
| TALEN | Transcription activator-like effector nucleases |
| TCA | Tricarboxylic acid |
| TIP | Tonoplast intrinsic protein |
| TOR | Target of rifampicin |
| UI | Usage index |
| WUE | Water use efficiency |
| ZFN | Zinc finger nucleus |

## 6.1 Introduction

During last 50 years, the global human population has increased from 5 billion to 9 billion, which primarily compelled to increase the staple foodstuff production. Population in India has increased from 0.36 billion to 1.21 (fourfold). The major two cereals (rice and wheat) production has increased from 21 to 103 million tonnes and from 6 to 88 million tonnes in India (GOI 2012). These two cereals occupy the major cultivated area and are used as staple food throughout the world. Wheat alone provides nearly 55% of the carbohydrate and 20% of the calories consumed globally (Breiman and Graur 1995).

Nitrogen (N) is a key element involved in various life processes of plants like biosynthesis of amino acid, nucleic acid, chlorophyll, cofactors, etc. Plants uptake on average 40–50 mg N to produce 1 kg of dry weight, whose limited supply in soil affects its availability to plant making it a critical limiting element for growth and development (Robertson and Vitousek 2009). External supply of N in the form of chemical fertilizers has become necessarily a need to sustain food production. In the last five decades, cereal production has boosted up to 260%, but this increase in production was at the cost of sevenfold increase in N fertilizer application (Ladha and Chakraborty 2016). As per the input survey data of 2006–2007, the average N fertilizer use of wheat in India is 122 kg $ha^{-1}$ which makes it the third most N fertilizer consumable crop after sugarcane and cotton. Presently, India stands second in N fertilizer use with 17.2 million tonnes after China. The global N fertilizer demand is expected to increase from a total of 105.3 million tonnes in 2011 to 119 million tonnes in 2018 (FAO 2015). At the same time, global cereal demand is also expected to increase up to 38% by 2025, and to achieve this, 60% increase in global N use will be needed till that time (Dobermann 2005). In case of the wheat crop, the nitrogen use efficiency (NUE) is 40%, whereas in rice, it varies from 20–40% (under submergence condition) to 40–60% (under upland condition) (Raghuram et al. 2007); the remaining N gets lost in the environment and cause pollution. Intensive use of N fertilizers in agricultural production is continuing, leaving the negative consequence to environment. Because of these consequences, the diversity and functioning of nonagricultural system are facing detrimental effect, i.e., eutrophication of freshwater and marine water due to leaching of N (Hirel et al. 2007). In addition to environmental effect, N fertilization also increases input cost because the Haber-Bosch process of N production consumes natural gas as high as 873 $m^3$ for producing 1 metric ton of fertilizer N (Xu et al. 2012). One percentage increase in NUE for cereal production could save around 1731 crore rupees ($234) fertilizer cost (Mosier and Syers 2004). Hence, developing cultivar which can efficiently use N is important to render these problems.

Nitrogen use efficiency has several definitions, but for plant biologists and breeder's, definition and expression given by Moll et al. (1982) is mostly used. The NUE is calculated by dividing grain weight to total available soil N. Nitrogen use efficiency is the multiplication of two components, N uptake efficiency (NUpE) and N utilization efficiency (NUtE). The NUpE is calculated by dividing N in plant harvest to total available N in soil, whereas NUtE is obtained by dividing grain dry mass to total N in plant harvest. In totality, NUE is defined as the total biomass or grain yield produced per unit of available N, which mostly comes from applied fertilizer in case of cereals.

Genetics and biochemical mechanisms involved in each component of N use have been widely studied in model plants as well as cereal crops to identify many key genes and regulators which are potential targets for improving crop NUE and also understanding genetic basis of NUE.

## 6.2 Nitrogen Application and Yield Enhancement in Cereals

Nitrogen derives from atmosphere which is subsequently transformed and transported to pedosphere and hydrosphere through the process of biological nitrogen fixation (BNF). Therefore, N is unique among major nutrient elements which mostly form by weathering of rocks and hence also present in less amount in soil. Atmosphere contains a large, well-mixed, biologically non-available N pool, a relatively small part of which is converted into a biologically available or reactive N pool primarily through BNF. Industrial N fixation has, however, become more important in agriculture, since taking care of the developing demand for sustenance has resulted in huge increments in the utilization of N fertilizer.

Nitrogen is one of the nutrients that most often limits crop production, particularly in the major staple cereals. In the tropics, lowland rice produces an approximate yield of 10–15 $qha^{-1}$ using naturally available N derived from processes like BNF by diazotrophs and mineralization of soil N and wet and dry depositions from the atmosphere. Similarly, wheat and maize yields of 10–15 $qha^{-1}$ were obtained without any fertilizer application (Janssen et al. 1990; Witt et al. 1999; Pathak et al. 2003). Such agro-systems have been sustained, though with low yields, for numbers of years without external N supplement (Fischer 2000). Whether it is good or bad, environmentally friendly or not, that is a matter of debate, but additional N application is necessary to increase the yield to three- to fourfold to feed the ever-increasing population (Lassaletta et al. 2014).

## 6.3 Green Revolution and Effect of Nitrogen Fertilizer in Environment

During Green Revolution the major emphasis was on genetic improvement of crop to produce high-yielding varieties (HYVs). Another aspect was intended to shortening of maturity time in cereal to ultimately achieve greater cropping intensity. Rice-wheat system adopted in the Indo-Gangetic Plain is contribution of shorter crop growing period. And lastly improved inputs, including fertilizer, irrigation, and, to a certain extent, pesticides, also contributed significantly in the Green Revolution.

Therefore, this increased N fertilization resulted in many negative impacts on the environment. Added N allows farmers to simplify plant community by displacing need of N fixing plant. Addition of N to soil system also interferes with their inhabiting microbial community and their associated processes like decomposition, nitrification, denitrification, etc. (Robertson and Vitousek 2009). Other than this, losses of fertilizer N to the soil and atmospheric environment cause serious pollution which is the fact why the plant use efficiency of N fertilizers is lower and thus higher cost of cultivation. Environmental pollution includes eutrophication of freshwater (London 2005) and marine ecosystems (Beman et al. 2005) which leads to augmentation of algal growth. Subsequently, after the death and decomposition of algae in water bodies, the content of organic matter in the water increases. In turn, this consumes higher amount of oxygen, causing drop in its level and creating conditions of hypoxia or dead zone. In absence of adequate oxygen availability, the aquatic organisms like fish, crabs, and others die (Rabalais et al. 2002). Some soluble portion of N leaches down as nitrate ($NO_3^-$) and pollute groundwater reserve (Powlson et al. 2008). In addition to leaching losses, reactive N can also lose to the atmosphere in the form of N-containing gases like ammonia ($NH_3$), nitrogen oxide ($N_2O$), nitric oxide (NO), etc. causing buildup of greenhouse gases (Robertson and Vitousek 2009).

## 6.4 Definitions of Nitrogen Use Efficiency (NUE)

The simplest definition of NUE is "the ratio of yield either in terms of grain or biomass to the total N (soil N and fertilizer applied)" (Good et al. 2004). However, NUE is a complex concept, and its expression and measurement must be context dependent. Complexity in meaningful definition and expression of NUE is dependent on many factors: (i) nutrient source including soil N, manure/fertilizer, or atmosphere and (ii) factor influencing crop nutrient demand like environmental factor, crop management practices, or genotypic makeup of plant. In addition to these factors, variation in NUE expression also depends on (iii) kind of data available and (iv) scale of interest (for plant breeder, interest may be a single plant, and for policy-making it may be as large as a country). Some of the measurements of NUE are partial factor productivity (PEP), agronomic efficiency (AE), partial nutrient balance (PNB), apparent recovery efficiency (RE), and N physiological use efficiency (NpUE) (Dobermann 2005, 2007; Fixen et al. 2015) which mainly denote efficiency of fertilizer input and are used by agronomist and policy-makers (these are explained briefly below). Plant biologist and breeders are mostly interested in another concept of NUE where it is mainly presented as plant phenotypic trait. Plant NUE (Xu et al. 2012) is a combination of two plant physiological components NUpE and NUtE. This book chapter is mainly focused on plant NUtE, and it is elaborated later in the section.

***Partial Factor Productivity (PEP)*** It is the expression of production efficiency calculated as amount of crop yield produced per unit of nutrient added. The PEP can be calculated at farm, regional, or national level if proper statistical record of fertilizer input as well as crop yields is available. It is also dependent on cropping system, which mainly explains productivity of a cropping system within a region in comparison to its N input and is a long-term indicator of trend.

***Agronomic Efficiency (AE)*** It is calculated as unit of grain yield increased per unit of nutrient added. It denotes degree of gain in productivity by nutrient input. Usually short-term impact of applied nutrient on productivity is indicated by AE, but if long-term trials are conducted, then contribution of fertilizer input to crop yield can also be indicated by AE.

***Partial Nutrient Balance (PNB)*** It is denoted as nutrient output per unit of nutrient input. It expresses the amount of nutrient being removed from the system in relation to amount of application. So basically, it expresses nutrient recovery rate. If PNB is close to 1, then it is assumed that fertility of soil will be sustained at steady state. However, there is lacuna in this assumption because removal of N from soil by means of losses like leaching and erosions is not considered. It indicates trends in long term and is more advantageous when used along with soil fertility record.

***Apparent Recovery Efficiency (RE) or Apparent N Recovery Rate (ANR)*** It is the relative amount of nutrient uptake in aboveground parts of fertilized and unfertilized crop plants as a function of quantity of applied nutrient. It tells what amount of fertilizer N is acquired by the plant, hence also indicating potential nutrient loss from that cropping system and management practice.

***N Physiological Use Efficiency (NpUE)*** It is the ratio of net increased grain weight to net increased N uptake in aboveground plant parts with and without application of fertilizer N. It is more like uptake efficiency denotes ability of plants to transform nutrients acquired by soil and fertilizer to economic yield. This expression also used to denote plant NUE.

***NUE as Phenotypic Trait*** Many people presented different definitions and evaluation methods for NUE (reviewed in Good et al. 2004; Fageria et al. 2008), but the appropriate way to estimate it depends on the crop, its harvest product, and whether the researcher wants to analyze specific physiological processes involved in NUE. The most common and widely used definition among plant breeders is the definition given by Moll et al. (1982). He defined it as grain production per unit of N available in the soil, and its expression is as follows:

$$Gw / Ns = (Nt / Ns)(Gw / Nt)$$

where *Gw*, grain weight; *Ns*, *N* supply (gplant$^{-1}$); and *Nt*, total *N* in plant at maturity. Ratio of Gw and *Ns* denotes NUE which is the multiplication of its two components NUpE (*Nt*/*Ns*) and NUtE (*Gw*/*Nt*).

Another definition is usage index (UI). Usage index is calculated by multiplying total plant biomass to the ratio of the total plant biomass and total plant N. For cereals, NUE is to be described as NUEg, which is the grain production per unit of N available and is a more appropriate presentation than UI plant could produce a lot of biomass for every unit N (high UI) without changing over the procured N to seed production; therefore, in spite of having high UI, it should have a low NUEg. Crop improvement mainly focused on the improvement of cereal yield is the main driver to bring economic prosperity in developing countries like India from the last five decades (Conant et al. 2013). Scientific approach to plant breeding accelerated its rate during this period, but this increase in crop production per unit area also owed intensive use of N nutrient input in the form of synthetic N fertilizers. All the selection for high-yielding, lodging-resistant, short-stature cultivars of cereals are carried out in high N environment which results in high N-responsive varieties, which means giving more yields with increased N input. Although cereal boosted the food production by about 260%, the contribution of N fertilizer to N input increased by 45% (Ladha et al. 2016). Annually 100 million tonnes of N fertilizers are applied in cropland and pastures, globally. Approximately half of this N input is taken by three major cereal crops, namely rice, wheat, and maize (Ladha and Chakraborty 2016). By seeing the rate of increase in N fertilizer input in these crops for the last 50 years, it is assumed that to meet 3 million tonnes of global requirement of cereal by 2050 and with 7% increase in crop land area, fertilizer N input will increase by 65% in these three crops with no change in NUE. Synthetic fertilizer production would also increase two times by 2050 (Ladha et al. 2016).

## 6.5 Components of Nitrogen Use Efficiency (NUE)

### 6.5.1 Nitrogen Uptake

The factors like soil type, pH, temperature, precipitation, and wind affect the N availability and its ionic form in soil both spatially and temporarily. Therefore, the N form to be taken by plant will depend on its adaptability to soil condition. Forest soils are mostly acidic; hence plant adapted to this condition prefers ammonium and amino acids, similarly plants adapted to more aerobic soil and high pH prefer $NO_3^-$ form as in the case of most of crop plants except lowland rice because it grows in anaerobic condition. For N uptake transporters and root system architecture plays a crucial role.

#### 6.5.1.1 Nitrate Transporters

Nitrate form taken by the plants through $H^+ / NO_3^-$ (hydrogen/nitrate) cotransports mechanism (Ullrich 1987). Many experiments showed $NO_3^-$ uptake causes alkalization of medium due to strong ionic difference (Mistrik and Ullrich 1996); this occurs due to 2:1 stoichiometry between $H^+ / NO_3^-$ cotransporter. Nitrogen availability in soil fluctuates, and to counteract this plants have developed two types of transporter systems (Epstein 1972). They are called low-affinity transport system (LATS) and high-affinity transport system (HATS). For both the forms, inducible and constitutive types coexist and act coordinately to uptake nutrient from soil through roots and distribute it all over the plants. Low-affinity uptake system acquires nutrients in high external substrate concentration (>0.5 mM for $NO_3^-$), while the high-affinity system uptakes substrate at low external concentrations (< 0.2 for $NO_3^-$). Along with these, some dual-affinity transporters are also reported which act as both LATS and HATS (Liu et al. 1999). In model plant *Arabidopsis thaliana*, transport system for $NO_3^-$ is well studied. Four families, namely, nitrate transporter 2 (NRT2) transporters, NRT 1/peptide transporter family (NPF) transporters, chloride channel family (CLC) transporters, and slow anion-associated channel homolog (SLAC/SLAH) (Orsel et al. 2002a; Negi et al. 2008; Barbier-Brygoo et al. 2011; Krapp et al. 2014; Léran et al. 2014), are responsible for $NO_3^-$ uptake, distribution, and storage. Among these, transporters from NPF and NRT2 family are responsible for $NO_3^-$ uptake from root. In addition to it, some members in NPF and NRT2 are also involved in $NO_3^-$ sensing and signaling. The NRT2 family consists of high-affinity $NO_3^-$ transporters (Krapp et al. 2014). A total of seven members are known from *Arabidopsis* NRT2 family. The NRT2.1 and NRT2.2 exclusively express in root, while the NRT2.5 and NRT 2.6 express in leaves along with root, whereas NRT2.7 expresses in leaves (Orsel et al. 2002b). Many NRT2 gene expressions are regulated by availability of $NO_3^-$ and some other factors (Zhuo et al. 1999; Orsel et al. 2002b, 2006). AtNRT2.1 interacts with another protein NAR2 and forms two component $NO_3^-$ uptake systems (Orsel et al. 2006). The ZmNRT2.1 and ZmNRT2.2 are the main transports controlling high-affinity $NO_3^-$ uptake in maize plant (Garnett et al. 2013). Lupini et al. (2016) showed ZmNRT2.1 expression and localization influenced by ZmNAR2.1; expression of these two genes was also correlated by $NO_3^-$ influxes. Five members of NRT2 family are characterized in rice (Feng et al. 2011), unlike *Arabidopsis* each showing different affinities to $NO_3^-$ and N supply dependent regulation also differs. Three NRT2 transporters (NRT2.1, NRT2.2, and NRT2.3a) interact with OsNAR2.1 at messenger RNA (mRNA) as well as protein levels to influence $NO_3^-$ uptake over both high- and low-concentration ranges (Yan et al. 2011).

For low-affinity uptake, one family of $NO_3^-$ transporters, NPF (NRT1), is responsible, but the exception is NRT 1.1/NPF6.3/CHL1 which is a dual-affinity transporter (Wang et al. 1998; Liu et al. 1999); along with it, NRT1.1 also act as $NO_3^-$ sensor (Ho et al. 2009). Fifty-three NRT genes have been reported till date in *Arabidopsis*, of which 51 gene families may have unique functions of each because its expression is found to be limited to a specific tissue (Tsay et al. 2007). The NRT 1.1 initially discovered as low-affinity transporter by Tsay et al. (1993) was later found to act as high-affinity transporter at low $NO_3^-$ concentration after phosphorylation of Thr101 (Liu and Tsay 2003). The protein complex CIPK23-CBL9 (CIPK, CBL-interacting protein kinase; CBL, calcineurin B-like protein) phosphorylates Thr101 at low $NO_3^-$ condition which results in switching of NRT1.1 to HAT (Liu and Tsay 2003; Ho et al. 2009). Another protein ABI2, a phosphatase, also regulates NRT1.1 and enhances rate of transport by inhibiting CIPK23-CBL1 complex phosphorylation (Léran et al. 2015). The location of NRT1.1 is plasma membrane of epidermis and root tip, because gene expression is observed in these locations along with it in mature part of root; it is also located in the cortex and endodermis (Huang et al. 1999). Another member NRT1.2/NPF4.6/AIT1 is also involved in soil $NO_3^-$ uptake, and it constitutively is expressed in root epidermis, but it comes under LATS (Liu et al. 1999). Interestingly NRT1.2 also transports abscisic acid (ABA) and plays an important role in transpiration and seed dormancy (Kanno et al. 2012). After uptake from soil, $NO_3^-$ is transported and distributed to all tissue. For root-to-shoot transport of $NO_3^-$ across several cell membranes, it is carried out by NRT1.5, NRT1.8, and NPF2.3 (Lin et al. 2008; Li et al. 2010; Taochy et al. 2015). Another NRT1 family member, NPF2.7/NAXT1, involves in $NO_3^-$ efflux to maintain $NO_3^-$ homeostasis (Segonzac et al. 2007).

#### 6.5.1.2 Ammonium Transporter

Nitrogen is also taken up as ammonium ion ($NH_4^+$) by plants growing under anaerobic condition. The ammonium transporters (AMT) are responsible for uptake and transport of $NH_4^+$. In *Arabidopsis,* six genes from this family have been reported (Gazzarrini et al. 1999), whereas in rice, which is a species well-adapted to $NH_4^+$ nutrition, ten genes are reported (Sonoda et al. 2003). Analysis of single and multiple mutants from this gene family in *Arabidopsis* revealed AMT1.1 and AMT1.3 as main transporters conferring 30–35% transport followed by AMT1.2 which confers 18–25%. The AMT 1.5 has lower $K_m$ than that of the previous two transporters (AMT 1.1 and 1.3), but in spite of that, it is a low-capacity transporter (Yuan et al. 2007). Spatial organizations of these AMT transporters are also very interesting. Outer root cells and root hairs contain transporters with highest $NH_4^+$ affinity (AMT 1.3, AMT1.5) to take $NH_4^+$ from soil. The electrochemical gradient between the vacuole and cytosol is responsible for $NH_3$ import in exchange of $NH_4^+$ exported to the external side of the vacuole. An intrinsic protein TIP in tonoplast plays an important role in $NH_3$ import to vacuole (Loque et al. 2005).

#### 6.5.1.3 Amino Acid Transporters

Plant also takes up amino acid from soil in very trace amount. Rentsch et al. (2007) reported that at least 5 gene families comprised of a total of 65 genes are considered as putative transporter of amino acid. A member of ATF (amino acid transporter) family, LHT1 (lysine/histidine transporter), is important for uptake of neutral and acidic amino acid by roots. Under high external concentration of amino acids, the uptake of uncharged amino acids is taken care of by AAP1 (amino acid permease 1) (Lee et al. 2007). The AAP5 is responsible for uptake of cationic amino acids (Svennerstam et al. 2008).

#### 6.5.1.4 Root System Architecture

Root system architecture is an important trait which affects the performance of crop in various stresses like drought, nutrient, and mineral toxicity and has implication in providing tolerance from these stresses (Manske and Vlek 2002). Tendency of plants to capture limited resources in soil is greatly influenced by the capacity of its root system to explore and forage. The root system architecture (RSA) mainly comprises of primary roots, lateral roots, and accessory roots. These are the key determinants of NUE and water use efficiency (WUE). At minor scale, root hairs are also included in root system (RS). It alleviates uptake of water and nutrient by increasing surface area. Many factors influence RSA, one of which is the form and concentration of N (Marschner 1995). Localized supply of $NH_4^+$ and $NO_3^-$ is responsible for initiation and elongation of lateral roots, respectively (Zhang and Forde 1998; Lima et al. 2010). Nitrate-dependent root elongation is regulated by miR393 (micro RNA 393) and AFG3 (auxin signaling F-box 3), and AFG3 is induced by $NO_3^-$ itself and miR399 by N metabolites (Vidal et al. 2010). In addition to these, NRT1.1 also regulates lateral root proliferation. It senses external N and also induces N signaling pathway by activating ANR1 (MAD box gene) (Remans et al. 2006; Ho et al. 2009). The $NH_4^+$-regulated root growth is governed by some AMTs like AMT1.3 and a GMPase (GDP mannose pyrophosphorylase) which is encoded by *HSN1* (*hypersensitive to* $NH_4^+$) (Qin et al. 2008; Lima et al. 2010).

### 6.5.2 Nitrogen Assimilation

Nitrogen assimilation is a vital process in controlling growth and development of plants. Plants take N from soil in the form of $NO_3^-$, ammonium ($NH_4$), and in small amount as amino acid. Ultimate substrate for amino acid biosynthesis is $NH_3$; therefore $NO_3^-$ form has to first reduce to $NH_4^+$. Reduction of $NO_3^-$ to $NH_4^+$ takes place in two steps. In the first step, $NO_3^-$ reduced to form $NO_2^-$ (nitrite) in cytosol by the action of enzyme nitrate reductase (NR) (Meyer and Stitt 2001). After that, $NO_2^-$ is transported to chloroplast where it gets reduced to $NH_4^+$ by nitrite reductase enzyme (NiR). Inorganic $NH_3$ is then assimilated to amino acid glutamine and glutamate which serve to translocate organic N from source to sink (Peoples and Gifford 1993; Roche et al. 1993; Lam et al. 1996). Enzyme NR is homodimer, and each monomer is attached to three prosthetic groups: (i) flavin adenine dinucleotide [cytochrome b reductase, binding NAD(P)H] (nicotinamide adenine dinucleotide phosphate), (ii) a heme (cytochrome b), and (iii) a molybdenum cofactor (MoCo) (site for $NO_3^-$ binding and reduction). The nitrate reductase activity (NRA) is considered as the rate-limiting step in the $NO_3^-$ assimilating pathway, and different genotypes of a species differ in NRA (Gniazdowska-Skoczek 1997; Bussi et al. 1997; Marwaha 1998). The NRA is induced by light in plants (Li and Oaks 1994; Kronzucker et al. 1995), but effect of light can be replaced by glucose, sucrose, or acetate (Galvan et al. 1996; Pajuelo et al. 1997; Sivasankar et al. 1997; Gniazdowska et al. 1998). Two classes of genes, Nia and Cnx, code for NR apoenzyme and MoCo cofactor, respectively. Most of the reports showed NR localized in cytoplasm, but there is some evidence of its localization in plasma membrane also (maize root and barley) (Ward et al. 1989). The next enzyme of pathway, NiR, is located in chloroplast and encoded by *Nii* gene whose number varies from one to two in different species (Meyer and Stitt 2001).

Ammonium is further assimilated to form amino acid. Nitrate reduction, photorespiration, or breakdown of amino acid generates $NH_4^+$ which is assimilated in plastid/chloroplast by GS (glutamine synthetase)/GOGAT (glutamine-2-oxoglutarate aminotransferase or glutamate synthase) cycle. The first step of fixation of $NH_4^+$ to amino acid is catalyzed by GS. It fixes $NH_4^+$ group to the δ-carboxylic group of glutamate to form glutamine. This step consumes an ATP. The second enzyme GOGAT transfers δ-amide group of glutamines to 2-oxoglutarate and forms two molecules of glutamate. Therefore, the net outcome of GS/GOGAT cycle is glutamate which further forms another amino acid by transferring its amino group to other carbon (C) skeletons with the help of aminotransferases or transaminases (Forde and Lea 2007). All N-containing molecules like chlorophyll, protein, secondary metabolites, nucleic acid, cytochrome/phytochrome, etc. are subsequently synthesized by a specific amino acid precursor. As we see, N metabolism requires energy and C skeletons, which comes from C-metabolism, so there must be crosstalk between these two pathways. Evidences in support of it are as follows: (a) Reduction of $NO_3^-$ also requires parallel C oxidation to form 2-oxoglutarate through respiratory pathway (Foyer et al. 2011). (b) Correlation between starch and protein contents has been always found in plants (Sulpice et al. 2009). (c) N availability affects partitioning of assimilated C between carbohydrates and organic acids (Foyer et al. 2011).

Enzymes GS and GOGAT exist in the form of different isoenzymes, and individual isoenzymes have been proposed to play a role in three major $NH_3$ assimilation processes: primary N assimilation, re-assimilation of photorespiratory $NH_3$, and re-assimilation of recycled N (Lam et al. 1996). Traditional assignment of GS/GOGAT isoenzyme functions is based on their organ-specific distribution. Chloroplastic GS2 and Fd-GOGAT (ferredoxin-GOGAT) are predominant GS-GOGAT enzymes located in leaves therefore proposed to involve in primary assimilation of $NH_3$ to glutamine and glutamate. Photorespiratory mutants also show they are specifically defective in enzyme GS2, hence the also highlighted role of these two isoforms in photorespiratory $NH_3$ re-assimilation. Catabolic processes include protein breakdown, deamination of amino acid, and biosynthetic pathways which release ammonia (met hionine, isoleucine, phenylpropanoid, and lignin biosynthesis) (Miflin and Lea 1976; Lea 1993). These processes are highest during seed germination and senescence of leaf. Cytosolic GS1 and NADH-GOGAT are involved in these processes (Stewart et al. 1980; Lea et al. 1990). In higher plants, chloroplastic GS2 is coded by a single nuclear gene *GLN2* and cytosolic GS1 by multiple *GLN1* genes (Peterman and Goodman 1991; Oliveira and Coruzzi 1999). Expression of *GLN2* is primarily observed in leaves and is regulated by light via phytochrome (Peterman and Goodman 1991; Oliveira and Coruzzi 1999). By contrast, the *GLN1* genes encoding cytosolic GS1 isoenzymes are expressed at higher levels in roots (Oliveira and Coruzzi 1999).

In addition to glutamine synthase and GOGAT, three more enzymes playing important roles in $NH_4^+$ assimilation are cytosolic asparagine synthase (AS), carbamoyl phosphate synthase (CPS), and mitochondrial NADH-glutamate dehydrogenase (GDH). The AS catalyzes ATP-dependent transfer of amide group of glutamines to aspartate and generates glutamate and asparagine. Some evidences also showed utilization of$NH_3$ as substrate (Lam et al. 2003; Masclaux-Daubresse et al. 2006). Three genes *ASN1*, *ASN2*, and *ASN3* encode AS. Asparagine is suitable for long-distance transport and storage of fixed N because it has high N/C ratio than glutamine (Rochat and Boutin 1991; Lam et al. 2003), but all four molecules glutamate, glutamine, aspartate, and asparagine have shown to be used for translocation from source to sink in different plant parts (Lea and Miflin 1980; Peoples and Gifford 1993).

The CPSase synthesizes carbamoyl phosphate, using $NH_4^+$ or amide group of glutamine, bicarbonate, and ATP. Synthesis of carbamoyl phosphate is carried out within plastid which further acts as a precursor of arginine and citrulline. The CPSase is a heteromeric enzyme, and its small and large subunits are encoded by genes *CarA* and *CarB*, respectively (Potel et al. 2009). Mitochondrial NADH-glutamate dehydrogenase (GDH) transaminates glutamate reversibly in response to high $NH_4^+$ concentration under stress (Skopelitis et al. 2006). However, glutamate deamination has been shown as major catalytic activity of this enzyme in plant cells (Masclaux-Daubresse et al. 2006) (Fig. 6.1).

### 6.5.3 Nitrogen Remobilization

Recycling of N by-product of various catabolic processes is necessary for efficient utilization of primary assimilated N (Lam et al. 1996). These catabolic processes embody protein catabolism, amino acid deamination, and specific synthesis reactions related to methionine, isoleucine, phenylpropanoid, and lignin biosynthesis (Miflin and Lea 1980; Lea 1993). In plant's entire growth period, there are two foremost times when re-assimilation of cast-off $NH_3$ to glutamine and glutamate is maximum. First is germination, when seed storage proteins are catabolized and amino acids are transferred to growing seedling (Lea and Miflin 1980), and second is during senescence of leaves where amino acids are transported to developing seed (Miflin and Lea 1976). Increased activity of cytosolic GS1, NADH-GOGAT, AS, and GDH during this process supports involvement of these enzymes in remobilization (Stewart et al. 1980; Lea et al. 1990). Enormous N is obtained during leave senescence due to extensive degradation of photosynthetic proteins of leaves. Plants can tap this N to enhance the nourishment of developing organs, for example, new leaves and seeds. Up to 95% of seed protein is taken from amino acids that are traded to the seed after the degradation of existing proteins in leaves (Taylor et al. 2010). Glutamine and asparagine play key roles in rendering N available for remobilization from the senescing leaves (Masclaux-Daubresse et al. 2008). These amino

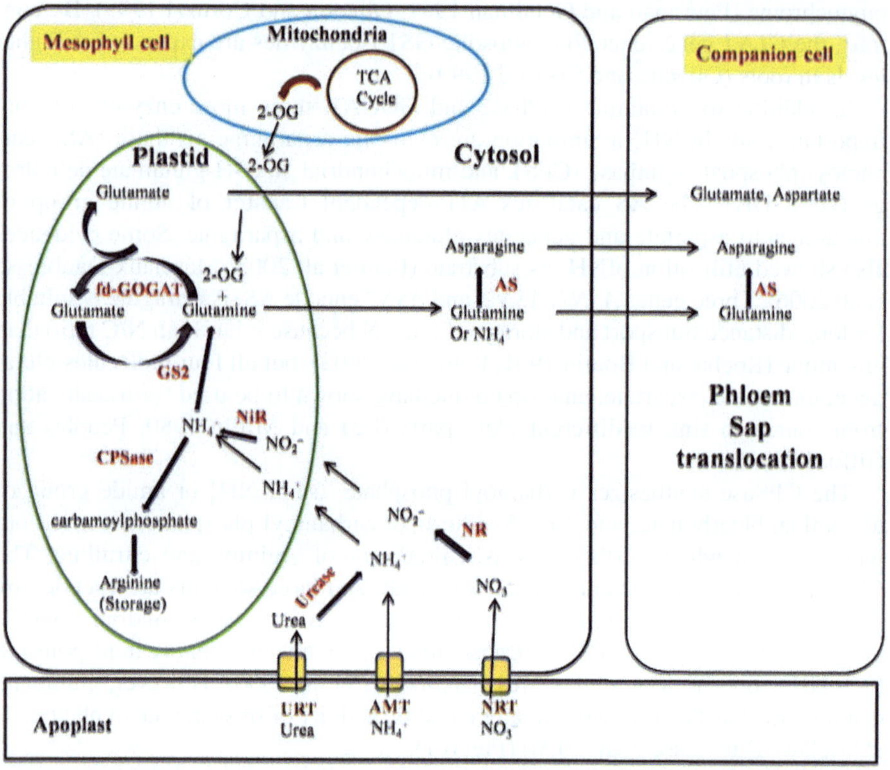


**Fig. 6.1** Schematic presentation of key enzymes involved in primary N assimilation. Various transporter imports $NO_3^-$, $NH_3$, and trace amount of urea into cytoplasm. Cytoplasm located in NR reduces $NO_3^-$ to $NO_2^-$, and urease converts urea into $NH_3$. $NO_2^-$ and $NH_3$ are transported to plastid where enzyme NiR GS2, fd-GOGAT CPSase, incorporates the $NH_3$ into amino acids. Carbon skeleton for amino acid biosynthesis is obtained by mitochondrial respiratory pathway. Amino acids are transported in the form of glutamine and asparagine through phloem

acids are carried from source tissue to sink via phloem. Rate of phloem loading is determined by activity of amino acid transporters present in sieve element (Tilsner et al. 2005). Phloem loading of amino acids is taken care of by members of AAP which are the nonspecific amino acid transporters (Koch et al. 2003). Uptake of amino acids by root is facilitated by LHT1 transporter which is lysine/histidine transporter with very less Km. This xylem-derived amino acids are supplied to mesophyll cell by the same transporter (Hirner et al. 2006). Induced expression of *LHT1* gene during leaf senescence highlights its role in N remobilization. One $NH_4^+$ transport AMT1.1 and a $NO_3^-$ transporter NRT2.5 also showing enhanced expression during leaf senescence suggests that inorganic N might also be mobilized in senescing tissues. A key step for economical N remobilization is phloem loading; however, whether or not it's limiting for NUE remains to be evidenced. Other processes, like sink strength, may be limiting steps for economical N remobilization from senescing leaves (Fig. 6.2).

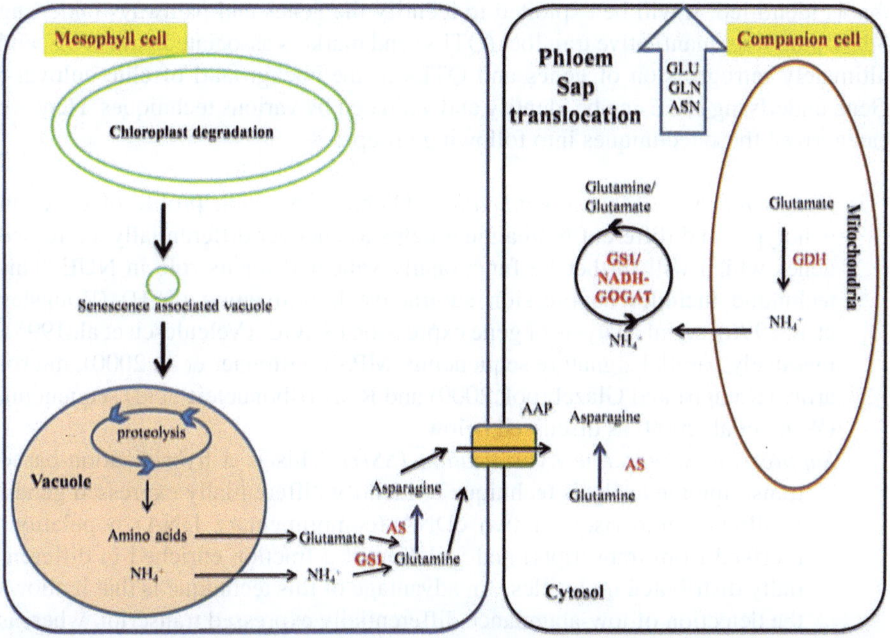


**Fig. 6.2** Schematic representation of N remobilization process and key enzymes involved in it. During senescence degradation of chloroplast and other proteins occurs, degraded products are transported to vacuole for further degradation via SAV (senescence-associated vacuoles). Similarly recycling of $NH_3$ occurs in different organelles (mitochondria, cytosol) of mesophyll and companion cell. These events generate $NH_3$ which is reduced to glutamine glutamate and asparagine by the action of enzyme GDH, GS1, NADH-GOGAT, and AS. Schematic presentation of key enzymes involved in N assimilation remobilization

## 6.6 Strategies for Genetic Improvement

### 6.6.1 Playing with Candidate Genes (CGs)

Nitrogen use efficiency in crop plants is genetically controlled, and therefore natural variation in NUE in different genotype of a crop species is common phenomena and reported in many crops. This variation arises either due to one or more components of NUE including total N uptake, post-anthesis N uptake, N remobilization, and N assimilation. Therefore, genotypes must be screened for its performance for individual traits. Some NUE components are developmental-specific like N uptake, so screening for uptake efficiency must be at different developmental stages. Another important aspect is the screening of NUE variant in different N regimes as different genotypes behave differently in low- and high-N doses. Under present scenario there is a need to find out genotypes which are efficient in N use under low-N dose and also give acceptable yield. Maintaining acceptable yield is important because plant will still be called N use-efficient if it requires less N and also gives less yield (as NUE is ratio between yield and available N). Once genotypes/germplasm for this is identified, it will be exploited to identify the genes and pathways underling NUE, mapping quantitative trait loci (QTLs) and markers associated with NUE, and ultimately introgression of genes and QTLs in the background of elite cultivars. Gene underlying NUE can be identify and validated by various techniques. Here we categorized these techniques into following groups:

(i) *Expression-based gene identification*: Global expression profile of different genotypes and different N treatment helps to uncover differentially expressed genes which will further be functionally validated for its role in NUE. This technique includes suppression subtractive hybridization (SSH) (Rounsley et al. 1996), serial analysis of gene expression (SAGE) (Velculescu et al. 1995), massively parallel signature sequencing (MPSS) (Brenner et al. 2000), microarray (Katagiri and Glazebrook 2009) and RNA (ribonucleic acid) sequencing (Wang et al. 2009) as discussed below.

*Suppression subtractive hybridization (SSH)*: This is a hybridization-based transcriptome analysis technique to identify differentially expressed genes. It allows comparison of two cDNA (complimentary DNA) populations (derived from transcripts) and isolation of a fraction enriched in differentially distributed molecules. An advantage of this technique is that it allows the detection of low-abundance differentially expressed transcript, whereas limitation includes false-positives and possibility of only pairwise comparison.

*SAGE (serial analysis of gene expression)*: The main principle is a short sequence tag (10–14 bp) developed from individual mRNA which contains sufficient information to uniquely identify a transcript. Sequence tags are then linked together to form long serial molecules that can be cloned and sequenced. The number of times a particular tag is observed provides the information about expression level of the corresponding transcript.

*Microarray*: Microarray technology is based on the fact that complementary sequences of cDNA can be used to hybridize immobilized cDNA molecules. Here in this technique, probe-target hybridization is usually detected and quantified by detection of fluorophore or chemiluminescence-labeled targets to determine relative abundance of nucleic acid sequences in the target. An advantage of this technique includes large data generation in a single experiment. Its limitation includes cost, escape of rare allele detection, need for sequence information to generate microarray chip, do not detect unique genes, and analysis is tedious.

*RNA sequencing*: RNA sequencing is a comparatively new and high-throughput technique of NGS-based sequencing cDNA. It works by sequencing every RNA molecule and profiling the expression of a particular gene by counting the number of its transcript, thereby correlating phenotype with CGs. The advantages of this technique are high reproducibility, prior sequence information may not be required, discovery of rare alleles, and unique genes are possible, although this technique has complexity of analysis and cost.

(ii) *Discovery of gene and its functional validation by mutation and transgenic studies*: Artificial mutant population and natural variants both can be used to discover a gene which has a role in NUE. In addition, the functional validation of putative NUE-governing gene is also possible by targeted mutation of that gene. After determining a phenotype which is associated with NUE, the very first step is creating mutagenized population and then screening for the plant showing mutated phenotype of desired trait. Mutation can be point or insertion. The insertion mutation can be further divided into insertion leading to loss of function and insertion resulting into gain of function mutation. At last the gene is recovered through map-based cloning approach. The mutagenized population may be created by EMS (ethyl methanesulfonate), T-DNA, or transposon tagging/activation tagging. However, NUE is a complex trait defining a single phenotype for it will be misleading.

(iii) *T-DNA/transposon tagging*: Insertional mutagenesis is useful since flanking site information; in other words the disrupted gene, can be retrieved easily but limited by gene redundancy, lethal knockouts, and inability to target the inserted element to a specific gene.

(iv) ***Activation tagging***: Through activation tagging, gain of function phenotype can be obtained because it contains four copies of enhancer elements of CaMV 35S (cauliflower mosaic virus 35S) which can mediate transcriptional activation of nearby genes. It is mainly useful for genes having redundant function. However, activation tagging may fail because every gene may not have an overexpression phenotype. After identification of genes, its function should be validated either by disruption of its function or by overexpression. Disruption of gene can be carried out by the following techniques:

*RNA interference (RNAi):* RNA interference (abbreviated RNAi) is a mechanism for RNA-guided regulation of gene expression in which double-stranded ribonucleic acid inhibits the expression of genes with complementary nucleotide sequences. It has several specific advantages over forward genetics. Targeted, requiring only a few transformants per target gene. It facilitates the study of essential genes whose inactivation would lead to lethality. But there are some disadvantages too, like it requires prior sequence information and has leaky expression.

(v) Other methods for gene disruption are techniques such as site-directed mutagenesis, gene knockout through homologous recombination, T-DNA insertion, targeted genome modification through TALEN (transcription activator-like effector nucleases), ZFN (zinc finger nucleus), and CRISPR-Cas9 (Cas9-clustered regularly interspaced short palindromic repeats) systems. The overexpression is carried out by expressing the gene with a strong promoter (CaMV 35S).

#### 6.6.1.1 Gene for Nitrogen Acquisition

Many CGs for N uptake and RSA are known which are exploited to improve NUE in the past decades either by overexpressing them or by knockout mutation. The *CKX1* gene code for cytokinin oxidase/dehydrogenase is responsible for degradation of cytokinin (negative regulator of root growth). Root-specific overexpression of this gene in *Arabidopsis* and tobacco leads to increase in root length, branching, and root-to-shoot ratio without affecting shoot growth and development (Werner et al. 2010). N accumulation is not estimated in this study, which highlights the sufficiency of a single dominant gene to alter complex trait like root growth. The ANR1 is a transcription factor involved in $NO_3^-$ signaling pathway to induce $NO_3^-$ -stimulated lateral root growth. The ANR1 overexpression can stimulate lateral root growth, but presence of $NO_3^-$ and posttranscriptional modification of ANR1 is prerequisite (Walch-Liu and Forde 2008). Root-based traits can provide exquisite opportunities for future improvements in NUE for cereals, but direct evidence in terms of gene manipulation is still lacking. Overexpression of $NO_3^-$ and $NH_4^+$ transporter is also documented. The *NRT2.7* overexpression in *Arabidopsis* enhanced $NO_3^-$ accumulation and improved seed germination (Chopin et al. 2007). In rice, *OsNRT2.1* overexpression showed improved seedling growth without any effect on N uptake (Katayama et al. 2009). This may be due to absence of required amount of OsNAR2.1, as we know NRT2.1 interacts with NAR2.1 to form functional transporter. The expression of *OsNRT2.3b* may increase rice yield and total N uptake (Xu et al. 2012). Overexpression of *AMT1*enhances $NH_4^+$ uptake capacity, but probably due to its toxicity, the plant biomass gets reduced. Therefore, *AMT1* can be a potential gene under low-$NH_4^+$ condition (Hoque et al. 2006). Transporter and RSA controlling gene may be the potential candidate to increase NUE, provided utilization of acquired N should be efficient. So, introgression of these genes in the genotypic background efficient in N utilization might be useful to meet the need of NUE genotype.

#### 6.6.1.2 Genes for Nitrogen Utilization

The improvement of N utilization can improve grain yield per unit of N acquired. The first step of $NO_3^-$ reduction by NR has been long known as rate-limiting step. The utility of NR/NiR overexpression is to improve how NUE is limited. Overexpression of *NR* in tobacco showed retention of NR activity for a longer period and some advantage during water stress (Ferrario-Mery et al. 1998). Similar result showed with *NiR* overexpression in *Arabidopsis*, potato, and tobacco. It reduces $NO_3^-$ concentration in plant tissues but does not show effect on seed or tuber yield. This may be due to regulation of these enzymes in posttranscriptional level (Pathak et al. 2009). The *GS1* overexpression studies are more numerous than *GS2*, and in many cases it resulted in higher growth yield in biomass in low N supply (Habash et al. 2001; Oliveira et al. 2002). Overexpression of *ASN1*in *Arabidopsis* enhances plant fitness and growth in low N condition and also increases total protein content (Lam et al. 2003). Under anaerobic condition (i.e., flooding), alanine is the major storage amino acid. The *AlaAT* (alanine aminotransferase) catalyzes the synthesis of alanine and 2-oxoglutarate from pyruvate and glutamate. Expression of barley *AlaAT* in rice with the help of rice tissue-specific promoter showed improved NUpE, biomass, and yield and hence NUE (Shrawat et al. 2008). In addition to enzymes of N assimilation and amino acid metabolism, the attempts have been made to generate plants modified for expression of transcription factors. This is of particular importance because one transcription can regulate more than one gene in a metabolic pathway, and hence modification in TF expression may achieve modification of more than one gene. The Dof1 (DNA binding with one zinc finger 1) is reported to be one of the regulators for N metabolism by coordinated gene expression involved in TCA (tricarboxylic acid) cycle and hence C-skeleton production, viz., pyruvate kinase, PEP carboxylase (phosphoenolpyruvate carboxylase), citrate synthase, and isocitrate dehydrogenase. In *Arabidopsis*, expression of *ZmDof1* results in enhanced N assimilation and better adaption of transgenic plants under N stress (Yanagisawa 2004). In rice, transgenic plant overexpressing *ZmDof1* showed enhanced N assimilation (Kurai et al. 2011). Similarly, in wheat and sorghum, the constitutive expression of *ZmDof1* leads to increase in PEPC expression (Peña et al. 2017). These effects suggest that NUE could also be improved by manipulating carbon metabolism pathways. Other potential regulatory proteins which could be subjected to further research are P-II, NPL7, and TOR (target of rifampicin). The P-II is a nuclear-encoded plastid protein which is homologous to bacterial P-II signaling proteins known to be involved in regulation of N metabolism by regulating key enzyme of arginine biosynthesis pathway. N-acetyl glutamate kinase (NAGK. P-II) knockout mutant in *Arabidopsis* showed accumulation of reduced ornithine, citrulline, and arginine accumulation in response to $NH_4^+$ supply after N starvation (Ferrario-Mery et al. 2006). The NLP7 (NIN-like protein 7) is an important element of the $NO_3^-$ signal transduction pathway. New regulatory protein specific for N assimilation in non-nodulating plants *Arabidopsis NLP7*-knockout mutants constitutively showed several features of N-starved plants (Castaings et al. 2009). Target of rifampicin (TOR) kinase showed positive regulation of growth in *Arabidopsis* under environmental stress (Deprost et al. 2007) (Fig. 6.3 and Table 6.1).

#### 6.6.1.3 Genes for Starch Metabolism

Although N utilization components includes N assimilation and N remobilization, if we see the definition of NUE, it explains grain yield per unit of N captured, i.e., how efficiently captured N is utilized to ultimately produce yield. So, in case of cereal, better N utilization means efficient grain filling by starch biosynthesis and accumulation because starch is the main component of grain in cereals. Good N utilization in cereals does not mean better partitioning of N to the grain. Better N partitioning may enhance protein quality of grain, but grain protein and cereal yield have inverse relation, so more protein/N accumulation in grain will actually lead to less grain filling due to less starch biosynthesis which results in low NUE. Therefore, the role of N in grain filling and enhancement of yield is more important in cereal to achieve NUE. The effect of different levels of N fertilization on dry weight partitioning, grain filling, and starch metabolism activities has been investigated in wheat. Results indicated increase of N input within a certain extent could increase dry matter weights of stem and sheath at the heading and harvest stages. Activities of key enzymes of starch synthesis, namely, soluble starch synthase (SS), ADP (adenosine diphosphate) glucose pyrophosphorylase (AGPase), and soluble starch branching enzymes (SBE), are also influenced by N dose during grain filling. The study highlighted higher export and transform percentages of stem and sheath and

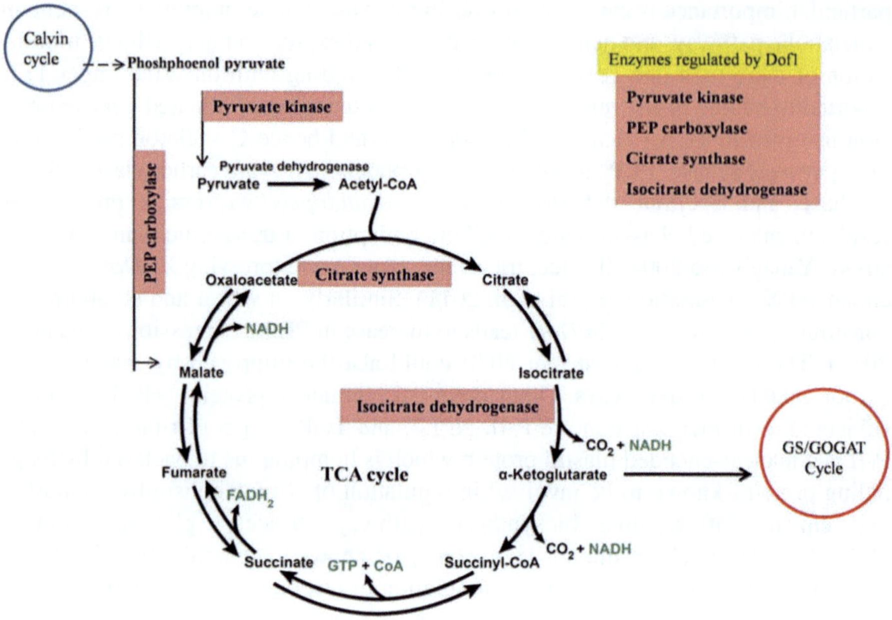


**Fig. 6.3** Schematic presentation of enzymes of carbon metabolism regulated by Dof1

**Table 6.1** Potential CGs for NUE

| NUE subcomponent | | Potential CGs | Phenotype | References |
|---|---|---|---|---|
| Uptake | RSA | *CKX1*, codes for cytokinin oxidase/dehydrogenase | Root-specific overexpression in *Arabidopsis* and *Nicotiana tobacum* leads to increased root length, branching, and root-to-shoot ratio | Werner et al. (2010) |
| | | *AtANR1* | Overexpressions stimulate lateral root growth in presence of $NO_3^-$ | Walch-Liu and Forde (2008) |
| | Transporters | *AtNRT 2.7* | Overexpression in *Arabidopsis* enhanced $NO_3^-$ accumulation and improved seed germination | Chopin et al. (2007) |
| | | *OsNRT2.1* | Overexpression showed improved seedling growth in rice | Katayama et al. (2009) |
| | | *OsAMT1* | Overexpression of *AMT1* enhances $NH_4^+$ uptake capacity in rice; potential CG in low $NH_4^+$ condition | Hoque et al. (2006) |

**Table 6.1** (continued)

| NUE subcomponent | | Potential CGs | Phenotype | References |
|---|---|---|---|---|
| Utilization | N assimilation | *NR* | Overexpression in tobacco showed retention of NR activity for longer period | Ferrario-Mery et al. (1998) |
| | | *NiR* | Overexpression in *Arabidopsis*, potato, and tobacco reduced $NO_3^-$ concentration in plant tissues | Pathak et al. (2009) |
| | | *GS1* | Overexpression in wheat and *Nicotiana* resulted in higher growth and yield in biomass in low N supply | Habash et al. (2001), Oliveira et al. (2002) |
| | | *ASN1* | Overexpression in *Arabidopsis* enhanced plant fitness, growth, and total protein content in low-N condition | Lam et al. (2003) |
| | | *AlaAT* | Expression of barley *AlaAT* in rice showed improved NUE | Shrawat et al. (2008) |
| | Carbon metabolism | *Dof1* | Expression of *ZmDof1* in *Arabidopsis* and rice enhanced N assimilation and better adaption in N stress | Yanagisawa (2004), Kurai et al. (2011) |
| | Signaling factors | *P-II* | Knock out mutant in *Arabidopsis* showed accumulation of reduced ornithine, citrulline, and arginine accumulation in response to $NH_4^+$ supply after N starvation | Ferrario-Mery et al. (2006) |
| | | *NLP7* | *Arabidopsis NLP7* knockout mutants constitutively showed features of N-starved plants | Castaings et al. (2009) |
| | | TOR | Positive regulation of growth in *Arabidopsis* in environmental stress | Deprost et al. (2007) |

rate of grain filling, and activities of key enzymes of wheat grain are the physiological basis for higher yield under an appropriate N level (Wang et al. 2013). Xiong et al. (2014) provided visual evidence of N effect on starch granules (SGs) in wheat endosperm. The results suggest that increased N fertilizer application mainly increased the numbers of small SGs and decreased the numbers of large SGs but that the results varied in different regions of the wheat endosperm. These observations give some hint of N effect on starch, but very less study is done so far. N assimilation is estimated only in terms of protein content, and amino acid content, parameters like yield and grain N, is and indicator of plant NUE so far. But in the cereals, to estimate utilization efficiency of N, it is more appropriate to focus on the factors which are responsible for starch synthesis and grain filling to give more yield

**Table 6.2** Types of starch synthetase and their function

| Enzyme | Function |
|---|---|
| SS1 (starch synthase 1) | Amylopectin synthesis; to produce the short single cluster-filling chains [degree of polymerization (DP) -8] |
| SS2 (starch synthase2) | Amylopectin synthesis; to produce the short single cluster-filling chains (DP-18) |
| SS3 (starch synthase 3) | Amylopectin synthesis; synthesize longer cluster-spanning B chains; role in granule initiation, at least in absence of SS4 |
| SS4 (starch synthase 4) | Starch granule initiation |
| GBSS (granule-bound starch synthase) | Amylose synthesis |

per unit of plant N. To identify these factors and genes underlying efficient grain filling and starch biosynthesis, it is important to screen out the genotype which does not differ in terms of N uptake, assimilation, and remobilization capacity, but they differ in grain filling, mainly due to starch accumulation. By comparing genes and alleles related to starch transport and metabolism, regulatory genes for starch metabolic in these genotypes will help to uncover few underlying CGs which will be responsible for efficient grain filling and therefore yield per-unit N in cereals.

The process of starch biosynthesis and genes involved is explained here. Phosphoglucose isomerase and plastidial phosphoglucomutase forms glucose-1-phospate (G-1-P) from fructose-6-phosphate derived from the Calvin cycle. The G-1-P forms ADP glucose by the action of AGPase enzyme. The ADP glucose is the substrate of starch synthases (SSs) and granule-bound starch synthase (GBSS). The GBSS synthesizes amylose, while soluble SSs (different types of starch synthetase are summarized in Table 6.2), branching enzymes (BEs), and isoamylase-type debranching enzyme (ISA), pollunase collectively synthesize amylopectin (Pfister and Zeeman 2016).

The SS4 is proposed to play a role in generating glucan primers required to initiate starch granule synthesis (Roldán et al. 2007). The SS4 loss-of-function mutant showed a smaller number of granules per plastid but larger in size. Protein targeting to starch (PTST1) is a protein which functions to localize GBSS to starch granules for normal amylose synthesis. PTST2 and PTST3 control starch granule initiation (Seung et al. 2017). Along with these other enzymes like D-enzyme, phosphorylase, glucan water dikinase is also important. Studies focusing on these genes and its regulator including miRNA should be taken up in the genotypes differing only in grain filling under the same background (uptake and assimilation).

### 6.6.2 Discovery of QTLs and Genes by Mapping Studies

Mapping approaches include (a) linkage analysis in biparental mapping population and (b) association mapping in naturally existing population. They may not need any sequence information, but based on markers, they identify the position of the gene. Linkage analysis involves creating biparental populations, genotyping and phenotyping of segregating progeny in the populations, and testing if sequence variations in the CGs/marker co-segregate or co-localize with the loci controlling the trait in the populations. The main advantage of biparental mapping through linkage analysis is that it offers high level of confidence. Among disadvantages, it needs to construct mapping population, so it requires more time and resource and also cannot offer high resolution. Association mapping, also known as "linkage disequilibrium mapping," is a method of mapping QTLs taking the advantage of linkage disequilibrium to link phenotype to genotype. It actually discovers linked markers associated with gene controlling the trait by exploiting diverse lines from natural population or germplasm collection. It also has own advantages as well as disadvantages. The prominent advantages over biparental mapping is (1) higher resolution, (2) lesser time, and that (3) it can discover additional genes or rare alleles for trait of interest. But it has somewhat lesser confidence because it may show association with spurious marker.

The QTL generally falls in two groups of genes; one is a major gene having large effect. These QTLs contribute to larger variation in highly heritable trait and called major QTLs, whereas minor QTLs have lesser effect. In this group, each QTL explains a small portion of total trait variation. Many agronomically important quantitative traits have small number of moderate effect QTLs and very large number of small effect QTLs to derive its genetic variation (Robertson 1967; Kearsey and Farquhar 1998). For a given study, the number of QTLs detected also depends on various factors like mapping population size and type, trait of interest, and effect of environment on that trait, in which environment phenotyping has been done as well as genome coverage. Major gene effect can be studied by means of segregation analysis, as properly as evolutionary history; however, when several genes with minor effect decide a trait (like in case of NUE), it is a good deal of great challenge, as they commonly can't be investigated individually. Despite of these challenges, NUE trait has been mapped in many crop plants. Obara et al. (2001) investigated QTL association with NUE in rice. The main focus of their investigation is to see co-segregation of NUE with GS1 and NADH-GOGAT and identify seven and six loci for co-segregation with GS1 and NADH-GOGAT, respectively. Also, in GS2 spanning chromosomal region in wheat and rice has been mapped for number of QTLs for NUE, yield, and other agronomic traits (Han et al. 2015). Therefore, introgression of this chromosomal region to NUE-inefficient background in rice and wheat may be useful to develop cultivar with improved NUE and agronomic traits. The QTL association with NUE in maize has been carried out in segregating maize population in numerous studies (Gallais and Hirel 2004). A meta-analysis of QTLs for yield and yield components was carried out in low and optimum N regimes to identify meta-QTLs and characterize its map position. It reveled 22 QTLs under low N condition (Liu et al. 2012). But many researchers do not consider it the analysis for NUE as the selection was for yield at low N. Due to multiple QTLs for the trait and very low contribution of each QTL for variation, these studies point out many challenges. Genome-wide association mapping in 196 accession of wheat for yield and yield components reveled 23 N-responsive genomic regions which may be useful for breeding for N responsiveness (Bordes et al. 2013). Traits such as enzyme activity for N metabolism and assay of N metabolites are incorporated with QTL studies. It gave some metabolic QTLs (mQTLs) (Habash et al. 2007) advantage from mQTL as it provides evidence to link agronomic traits with potential gene(s) underlying the QTLs (Hill et al. 2013). But a disadvantage is that such traits are vulnerable to environmental cues. Nitrogen use efficiency and grain protein content in wheat and barley are controlled by a common QTL known as *Gpc-B1*. These loci control N remobilization, hence resulting in significant difference in NUE. The NILs (near-isogenic lines) with *Gpc-B1* showed increase GPC without decreasing grain yield (Uauy et al. 2006; Heidlebaugh et al. 2008).

As we discussed the main objective of a mapping studies is to identify CG responsible for the trait, an identification of marker which can further be used to track trait of interest and cloning of favorable allele for the trait. Identification of CGs for complex trait like NUE faces challenges in spite of precision of these mapping studies because QTLs identified have many other genes that also co-segregate. Hence, further narrowing of the mapping interval to a place of the chromosome that carries a practical wide variety of genes may constrain. Therefore, identification of genes that would possibly have an effect on NUE is nonetheless based totally on our information of the gene's characteristic like gene for N uptake, assimilation, remobilization, etc. But there are many evidences that other types of genes completely unrelated to N metabolism are also associated with NUE. For example, various studies describe involvement of phenology gene with NUE traits, which are semi-dwarf gene *Rht* and *Ppd*, and *Vrn* (Habash et al. 2007; Laperche et al. 2007).

## 6.7 Conclusion

Nitrogen use efficiency is a complex trait regulated by many environmental factors. The manipulation of NUE is a difficult process in spite of genetic potential. Recent years have seen a tremendous increase in the number of genes found to be involved in the mechanisms of inorganic N uptake and utilization in plants, but maximum study is limited to model plant, and there is need to extend it to crop plants. The first component of NUE is uptake; many genes of transports are identified in *Arabidopsis*, and some of it is also utilized to develop transgenic, but none of it resulted in NUE phenotype so far. Improvement of N uptake may be helpful in counteracting N loss from the environment, but it will not necessarily generate NUE phenotype. Nitrogen utilization is utilizing of the acquired N to produce grain yield. Traditionally it consists of N assimilation and remobilization, but for cereal, utilization of N to produce starch is also important. Several genes in N assimilation and remobilization have been reported, and some are used to overexpress in crop or model plant to achieve NUE phenotype. Genes of starch metabolism, which potentially affect grain filling in N limitation, need to identify in cereals which may further be exploited in creating NUE cereal.

## References


Barbier-Brygoo H, DeAngeli A, Filleur S, Frachisse JM, Gambale F, Thomine S et al (2011) Anion channels/transporters in plants: from molecular bases to regulatory networks. Annu Rev Plant Biol 62:25–51

Beman JM, Arrigo K, Matson PM (2005) Agricultural runoff fuels large phytoplankton blooms in vulnerable areas of the ocean. Nature 434:211–214

Bordes J, Ravel C, Jaubertie JP, Duperrier B, Gardet O et al (2013) Genomic regions associated with the N limitation response revealed in a global wheat core collection. Theor Appl Genet 126:805–822

Breiman A, Graur D (1995) Wheat evolution. Isr J Plant Sci 43:85–98

Brenner S, Johnson M, Bridgham J, Golda G, Lloyd DH, Johnson D et al (2000) Gene expression analysis by massively parallel signature sequencing (MPSS) on micro bead arrays. Nat Biotechnol 18:630–634

Bussi C, Gojon A, Passama L (1997) In situ nitrate reductase activity in leaves of adult peach trees. J Hortic Sci 72:347–353

Castaings L, Camargo A, Pocholle D et al (2009) The nodule inception-like protein 7 modulates nitrate sensing and metabolism in Arabidopsis. Plant J 57:426–435

Chopin F, Orsel M, Dorbe MF, Chardon F, Truong HN et al (2007) The *Arabidopsis* ATNRT2.7 nitrate transporter controls nitrate content in seeds. Plant Cell 19:1590–1602

Conant RT, Berdanier AB, Grace PR (2013) Patterns and trends in N use and N recovery efficiency in world agriculture. Glob Biogeochem Cycles 27:558–566

Deprost D, Yao L, Sormani R et al (2007) The Arabidopsis TOR kinase links plant growth, yield, stress resistance and mRNA translation. EMBO Rep 8:864–870

Dobermann A (2005) Nutrient use efficiency – state of the art. In: IFA International Workshop on Enhanced-Efficiency Fertilizers. International Fertilizer Industry Association (IFA), Frankfurt/Paris, pp 1–16

Dobermann A (2007) Nutrient use efficiency – measurement and management. In: Fertilizer best management practices. General principles, strategy for their adoption and voluntary initiatives vs regulations. IFA International Workshop on Fertilizer Best Management Practices. International Fertilizer Industry Association (IFA), Brussels; Paris, pp 1–28

Epstein E (1972) Mineral nutrition of plants: principles and perspectives. Wiley, New York

Fageria NK, Baligar VC, Li YC (2008) The role of nutrient efficient plants in improving crop yields in the twenty first century. J Plant Nutr 31:1121–1157

FAO (2015) Current world fertilizer trends and outlook to 2015. Food and Agriculture Organization of the United Nations, Rome

Feng HM, Yan M, Fan XR, Li BZ, Shen QR et al (2011) Spatial expression and regulation of rice high-affinity nitrate transporters by N and carbon status. J Exp Bot 62:2319–2332

Ferrario-Mery S, Valadier M, Foyer C (1998) Overexpression of nitrate reductase in tobacco delays drought-induced decreases in nitrate reductase activity and mRNA. Plant Physiol 117:293–302

Ferrario-Mery S, Besin E, Pichon O, Meyer C, Hodges M (2006) The regulatory PII protein controls arginine biosynthesis in Arabidopsis. FEBS Lett 580:2015–2020

Fischer KS (2000) Frontier project on N fixation in rice: looking ahead. In: Ladha JK, Reddy PM (eds) The quest for N fixation in rice. International Rice Research Institute, Los Banos, pp 25–31

Fixen P, Brentrup F, Bruulsema TW, Garcia F, Norton R, Zingore S (2015) Nutrient/fertilizer use efficiency: measurement, current situation and trends. In: Drechsel P, Heffer P, Magen H, Mikkelsen R, Wichelns D (eds) Managing water and fertilizer for sustainable agricultural intensification. Copyright 2015 IFA, IWMI, IPNI and IPI, Paris, pp 8–38

Forde BG, Lea PJ (2007) Glutamate in plants: metabolism, regulation and signalling. J Exp Biol 58:2339–2358

Foyer CH, Noctor G, Hodges M (2011) Respiration and N assimilation: targeting mitochondria associated metabolism as a means to enhance N use efficiency. J Exp Bot 62:1467–1482

Gallais A, Hirel B (2004) An approach to the genetics of N use efficiency in maize. J Exp Bot 55:295–306

Galvan A, Quesada A, Fernandez E (1996) The use of mutants to study nitrate assimilation in green microalgae. Sci Mar 60:191–194

Garnett T, Conn V, Plett D, Conn S, Zanghellini J, Mackenzie N et al (2013) The response of the maize nitrate transport system to N demand and supply across the lifecycle. New Phytol 198:82–94

Gazzarrini S, Lejay L, Gojon A, Ninnemann O, Frommer WB, von Wiren N (1999) Three functional transporters for constitutive, diurnally regulated, and starvation induced uptake of ammonium into Arabidopsis roots. Plant Cell 11:937–947

Gniazdowska A, Mikulska M, Rychter AM (1998) Growth, nitrate uptake and respiration rate in bean roots under phosphate deficiency. Biol Plant 41:217–226

Gniazdowska-Skoczek H (1997) Properties of nitrate reductase from seedling leaves of selected barley genotypes. Acta Physiol Plant 19:137–145

GOI (2012) Press information bureau, government of India. Ministry of Agriculture & Farmers Welfare. http://pib.nic.in/newsite/PrintRelease.aspx?relid=80083

Good AG, Shrawat AK, Muench DG (2004) Can less yield more? Is reducing nutrient input into the environment compatible with maintaining crop production? Trends Plant Sci 9:597–605

Habash DZ, Massiah AJ, Rong HI, Wallsgrove RM, Leigh RA (2001) The role of cytosolic glutamine synthetase in wheat. Ann Appl Biol 138:83–89

Habash DZ, Bernard S, Schondelmaier J, Weyen J, Quarrie SA (2007) The genetics of N use in hexaploid wheat: N utilisation, development and yield. Theor Appl Genet 114:403–419

Han M, Okamoto M et al (2015) The genetics of N use efficiency in crop. Plants Annu Rev Genet 49:269–289

Heidlebaugh NM, Trethewey BR, Jukanti AK, Parrott DL, Martin JM, Fischer AM (2008) Effects of a barley (*Hordeum vulgare*) chromosome 6 grain protein content locus on whole-plant N reallocation under two different fertilisation regimes. Funct Plant Biol 35:619–632

Hill C, Taylor J, Edwards J, Mather D (2013) Whole-genome mapping of agronomic and metabolic traits to identify novel quantitative trait loci in bread wheat grown in a water-limited environment. Plant Physiol 162:1266–1281

Hirel B, Le Gouis J, Ney B, Gallais A (2007) The challenge of improving N use efficiency in crop plants: towards a more central role for genetic variability and quantitative genetics within integrated approaches. J Exp Bot 58:2369–2387

Hirner A, Ladwig F, Stransky H, Okumoto S, Keinath M, Harms A, Frommer WB, Koch W (2006) *Arabidopsis* LHT1 is a high-affinity transporter for cellular amino acid uptake in both root epidermis and leaf mesophyll. Plant Cell 18:1931–1946

Ho CH, Lin SH, Hu HC, Tsay YF (2009) CHL1 functions as a nitrate sensor in plants. Cell 138:1184–1194

Hoque MS, Masle J, Udvardi MK, Ryan PR, Upadhyaya NM (2006) Over-expression of the rice *OsAMT1-1* gene increases ammonium uptake and content, but impairs growth and development of plants under high ammonium nutrition. Funct Plant Biol 33:153–163

Huang NC, Liu KH, Lo HJ, Tsay YF (1999) Cloning and functional characterization of an Arabidopsis nitrate transporter gene that encodes a constitutive component of low-affinity uptake. Plant Cell 11:1381–1392

Janssen BH, Guiking FCT, Van Der Eijk D, Smaling EMA, Wolf J, Reuler H (1990) A system for quantitative evaluation of the fertility of tropical soils (QUEFTS). Geoderma 46:299–318

Kanno Y, Hanada A, Chiba Y, Ichikawa T, Nakazawa M, Matsui M et al (2012) Identification of an abscisic acid transporter by functional screening using the receptor complex as a sensor. Proc Natl Acad Sci 109:9653–9658

Katagiri F, Glazebrook J (2009) Overview of mRNA expression profiling using DNA microarrays. Curr Protoc Mol Biol 22:22

Katayama H, Mori M, Kawamura Y, Tanaka T, Mori M, Hasegawa H (2009) Production and characterization of transgenic rice plants carrying a high-affinity nitrate transporter gene (*OsNRT2.1*). Breed Sci 59:237–243

Kearsey MJ, Farquhar AGL (1998) QTL analysis in plants; where are we now? Heredity 80:137–142

Koch W, Kwart M, Laubner M, Heineke D, Stransky H, Frommer WB, Tegeder M (2003) Reduced amino acid content in transgenic potato tubers due to antisense inhibition of the leaf $H^+$/amino acid symporter StAAP1. Plant J 33:211–220

Krapp A, David LC, Chardin C, Girin T, Marmagne A, Leprince AS et al (2014) Nitrate transport and signaling in *Arabidopsis*. J Exp Bot 65:789–798

Kronzucker H, Glass ADM, Siddiqi MY (1995) Nitrate induction in spruce: an approach using compartmental analysis. Planta 196:683–690

Kurai T, Wakayama M, Abiko T, Yanagisawa S, Aoki N, Ohsugi R (2011) Introduction of the ZmDof1 gene into rice enhances carbon and N assimilation under low-N conditions. Plant Biotechnol J 9:826–837

Ladha JK, Chakraborty D (2016) Solutions to improve N use efficiency for the world. In: Proceedings of the international N initiative conference, pp 4–8. December 2016, Melbourne, Australia

Ladha JK et al (2016) Agronomic improvements can make future cereal systems in South Asia far more productive and result in lower environmental footprint. Glob Chang Biol 22:1054–1074

Lam HM, Coschigano KT, Oliveira IC, Melo-Oliveira R, Coruzzi GM (1996) The molecular-genetics of N assimilation into amino acids in higher plants. Annu Rev Plant Physiol Plant Mol Biol 47:569–593

Lam HM, Wong P, Chan HK et al (2003) Overexpression of the ASN1 gene enhances N status in seeds of Arabidopsis. Plant Physiol 132:926–935

Laperche A, Brancourt-Hulmel M, Heumez E, Gardet O, Hanocq E et al (2007) Using genotype × N interaction variables to evaluate the QTL involved in wheat tolerance to N constraints. Theor Appl Genet 115:399–415

Lassaletta L, Billen G, Grizzetti B, Anglade J Garnier J (2014) 50 year trends in N use efficiency of world cropping systems: the relationship between yield and N input to cropland. Environ Res Lett 9:105011

Lea PJ (1993) N metabolism. In: Lea PJ, Leegood RC (eds) Plant biochemistry and molecular biology. Wiley, New York, pp 155–180

Lea PJ, Miflin B (1980) Transport and metabolism of asparagine and other N compounds within the plant, vol 40, pp 569–607

Lea PJ, Robinson SA, Stewart GR (1990) The enzymology and metabolism of glutamine, glutamate, and asparagine. In: Miflin BJ, Lea PJ (eds) The biochemistry of plants, vol 16. Academic, New York, pp 121–159

Lee YH, Foster J, Chen J, Voll LM, Weber AP, Tegeder M (2007) AAP1 transports uncharged amino acids into roots of Arabidopsis. Plant J 50:305–319

Léran S, Varala K, Boyer JC, Chiurazzi M, Crawford N, Daniel-Vedele F et al (2014) A unified nomenclature of NITRATE TRANSPORTER 1/PEPTIDE TRANSPORTER family members in plants. Trends Plant Sci 19:5–9

Léran S, Edel KH, Pervent M, Hashimoto K, Corratge-Faillie C, Offenborn JN et al (2015) Nitrate sensing and uptake in *Arabidopsis* are ABI2, a phosphatase inactivated by the stress hormone abscisic acid. Sci Signal 8:43. https://doi.org/10.1126/scisignal.aaa4829

Li XZ, Oaks A (1994) Induction and turnover of nitrate reductase in *Zea mays*: influence of light. Plant Physiol 106:1145–1149

Li JY, Fu YL, Pike SM, Bao J, Tian W, Zhang Y et al (2010) The *Arabidopsis* nitrate transporter NRT1.8 functions in nitrate removal from the xylem sap and mediates cadmium tolerance. Plant Cell 22:1633–1646

Lima JE, Kojima S, Takahashi H, von Wiren N (2010) Ammonium triggers lateral root branching in *Arabidopsis* in an AMMONIUM TRANSPORTER1;3-dependent manner. Plant Cell 22:3621–3633

Lin SH, Kuo HF, Canivenc G, Lin CS, Lepetit M, Hsu PK et al (2008) Mutation of the *Arabidopsis* NRT1.5nitrate transporter causes defective root-to-shoot nitrate transport. Plant Cell 20:2514–2528

Liu HS, Tsay YF (2003) Switching between the two action modes of the dual-affinity nitrate transporter CHL1 by phosphorylation. EMBO J 22:1005–1013
Liu KH, Huang CY, Tsay YF (1999) CHL1 is a dual-affinity nitrate transporter of *Arabidopsis* involved in multiple phases of nitrate uptake. Plant Cell 11:865–874
Liu R, Zhang H, Zhao P, Zhang Z, Liang W et al (2012) Mining of candidate maize genes for N use efficiency by integrating gene expression and QTL data. Plant Mol Biol Rep 30:297–308
London JG (2005) N study fertilizes fears of pollution. Nature 433:791
Loque D, Ludewig U, Yuan L, von Wiren N (2005) Tonoplast intrinsic proteins AtTIP2;1 and AtTIP2;3 facilitate NH3 transport into the vacuole. Plant Physiol 137:671–680
Lupini A, Mercati F, Araniti F, Miller AJ, Sunseri F, Abenavoli MR (2016) NAR2.1/NRT2.1 functional interaction with $NO3^-$ and $H^+$ fluxes in high-affinity nitrate transport in maize root regions. Plant Physiol Biochem 102:107–114
Manske GG, Vlek PL (2002) Root architecture–wheat as a model plant. In: Waisel Y, Eshel A, Beeckman T, Kafkafi U (eds) Roots: the hidden half. CRC, Boca Raton, pp 249–259
Marschner H (1995) Mineral nutrition of higher plants, 2nd edn. Academic, London
Marwaha RS (1998) Nitrate assimilation in potato cultivars during plant growth. Indian J Plant Physiol 3:147–151
Masclaux-Daubresse C, Reisdorf-Cren M, Pageau K et al (2006) Glutamine synthetase-glutamate synthase pathway and glutamate dehydrogenase play distinct roles in the sink-source N cycle in tobacco. Plant Physiol 140:444–456
Masclaux-Daubresse C, Reisdorf-Cren M, Orsel M (2008) Leaf N remobilisation for plant development and grain filling. Plant Biol 10:23–36
Meyer C, Stitt M (2001) Plant N. In: Lea PJ, Goudry JFM (eds) Nitrate reduction and signaling. Springer, Berlin/Heidelberg, pp 37–59
Miflin BJ, Lea PJ (1976) The pathway of N assimilation in plants. Phytochemistry 15:873–885
Miflin BJ, Lea PJ (1980) Ammonia assimilation. In: Miflin BJ (ed) The biochemistry of plants: amino acids and derivatives, vol 5. Academic, New York, pp 169–202
Mistrik I, Ullrich CI (1996) Mechanism of anion uptake in plant roots: quantitative evaluation of $H+/NO_{3-}$ and $H+/H_2PO_{4-}$ stoichiometries. Plant Physiol Biochem 34:629–636
Moll RH, Kamprath EJ, Jackson WA (1982) Analysis and interpretation of factors which contribute to efficiency of N utilisation. Agronomy 74:562–564
Mosier A, Syers JK (2004) Agriculture and the N cycle: assessing the impacts of fertilizer use on food production and the environment, vol 65. Island Press, Washington, DC
Negi J, Matsuda O, Nagasawa T, Oba Y, Takahashi H, Kawai-Yamada M et al (2008) $CO_2$ regulator SLAC1 and its homologues are essential for anion homeostasis in plant cells. Nature 452:483–486
Obara M, Kajiura M, Fukuta Y, Yano M, Hayashi M et al (2001) Mapping of QTLs associated with cytosolic glutamine synthetase and NADH-glutamate synthase in rice (*Oryza sativa* L.). J Exp Bot 52:1209–1217
Oliveira IC, Coruzzi GM (1999) Carbon and amino acids reciprocally modulate the expression of glutamine synthetase in Arabidopsis. Plant Physiol 121:301–310
Oliveira IC, Brears T, Knight TJ, Clark A, Coruzzi GM (2002) Overexpression of cytosolic glutamate synthetase. Relation to N, light, and photorespiration. Plant Physiol 129:1170–1180
Orsel M, Filleur S, Fraisier V, Daniel-Vedele F (2002a) Nitrate transport in plants: which gene and which control? J Exp Bot 53(370):825–833
Orsel M, Krapp A, Daniel-Vedele F (2002b) Analysis of the NRT2 nitrate transporter family in *Arabidopsis*. Structure and gene expression. Plant Physiol 129:886–896
Orsel M, Chopin F, Leleu O, Smith SJ, Krapp A, Daniel-Vedele F, Miller AJ (2006) Characterization of a two-component high-affinity nitrate uptake system in Arabidopsis. Physiology and protein protein interaction. Plant Physiol 142(3):1304–1317
Pajuelo PE, Pajuelo B, Forde G, Marquez AJ (1997) Regulation of the expression of ferredoxin-glutamate synthase in barley. Planta 203:517–525

Pathak H, Aggarwal PK, Roetter R, Kalra N, Bandyopadhaya SK, Prasad S, Keulen HV (2003) Modeling the quantitative evaluation of soil nutrient supply, nutrient use efficiency, and fertilizer requirements of wheat in India. Nutr Cycl Agroecosyst 65:105–113

Pathak RR, Ahmad A, Lochab S, Raghuram N (2009) Molecular physiology of plant N use efficiency and biotechnological options for its enhancement. Curr Sci 94:1394–1403

Peña PA, Quach T, Sato S et al (2017) Expression of the maize Dof1 transcription factor in wheat and sorghum. Front Plant Sci 8:434. https://doi.org/10.3389/fpls.2017.00434

Peoples MB, Gifford RM (1993) Plant physiology, biochemistry and molecular biology. In: Dennis DT, Turpin DH (eds) Long distance transport of carbon and N from sources to sinks in higher plants. Wiley, New York, pp 434–447

Peterman TK, Goodman HM (1991) The glutamine synthetase gene family of Arabidopsis thaliana: light regulation and differential expression in leaves, roots and seeds. Mol Gen Genet 230:145–154

Pfister B, Zeeman SC (2016) Formation of starch in plant cells. Cell Mol Life Sci 73:2781–2807

Potel F, Valadier MH, Ferrario-Mery S et al (2009) Assimilation of excess ammonium into amino acids and N translocation in Arabidopsis thaliana – roles of glutamate synthases and carbamoylphosphate synthetase in leaves. FEBS J 276:4061–4076

Powlson DS, Addiscott TM, Benjamin N, Cassman KG, DeKok TM (2008) When does nitrate become a risk for humans? J Environ Qual 37:291–295

Qin C, Qian W, Wang W, Wu Y, Yu C et al (2008) GDP-mannose pyrophosphorylase is a genetic determinant of ammonium sensitivity in *Arabidopsis thaliana*. Proc Natl Acad Sci U S A 105:18308–18313

Rabalais NN, Turner RE, Wiseman WJ (2002) Gulf of Mexico hypoxia, aka "the dead zone.". Annu Rev Ecol Syst 33:235–263

Raghuram N, Sachdev MS, Abrol YP (2007) Towards an integrative understanding of reactive N. In: Agricultural N use and its environmental implications, pp 1–6

Remans T, Nacry P, Pervent M, Filleur S, Diatloff E et al (2006) The *Arabidopsis* NRT1.1 transporter participates in the signaling pathway triggering root colonization of nitrate-rich patches. Proc Natl Acad Sci 103:19206–19211

Rentsch D, Schmidt S, Tegeder M (2007) Transporters for uptake and allocation of organic N compounds in plants. FEBS Lett 581:2281–2289

Robertson A (1967) The nature of quantitative genetic variation. In: Brink R (ed) Heritage from Mende, Univ. Wisc. E Styles. University of Wisconsin Press, Madison, pp 265–280

Robertson GP, Vitousek PM (2009) N in agriculture: balancing the cost of an essential resource. Annu Rev Environ Resour 34:97–125

Rochat C, Boutin JP (1991) Metabolism of phloem-borne amino acids in maternal tissues of fruit of nodulated or nitrate-fed pea plants (*Pisum sativum* L.). J Exp Bot 42:207–214

Roche D, Temple SJ, Sengupta GC (1993) Two classes of differentially regulated glutamine synthetase genes are expressed in the soybean nodule: a nodule specific and a constitutively expressed class. Plant Mol Biol 22:971–983

Roldán I, Wattebled F, Mercedes Lucas M, Delvallé D, Planchot V, Jiménez S, Pérez R et al (2007) The phenotype of soluble starch synthase IV defective mutants of Arabidopsis thaliana suggests a novel function of elongation enzymes in the control of starch granule formation. Plant J 49:492–504

Rounsley SD, Glodek A, Sutton G, Adams MD, Somerville CR, Venter JC et al (1996) The construction of Arabidopsis expressed sequence tag assemblies. A new resource to facilitate gene identification. Plant Physiol 112:1177–1183

Segonzac C, Boyer JC, Ipotesi E, Szponarski W, Tillard P, Touraine B et al (2007) Nitrate efflux at the root plasma membrane: identification of an *Arabidopsis* excretion transporter. Plant Cell 19:3760–3777

Seung D, Boudet J, Monroe J, Schreier TB, David LC et al (2017) Homologs of protein targeting to starch control starch granule initiation in Arabidopsis leaves. Plant Cell 29:1657–1677

Shrawat AK, Carroll RT, DePauw M, Taylor GJ, Good AG (2008) Genetic engineering of improved N use efficiency in rice by the tissue-specific expression of *alanine aminotransferase*. Plant Biotechnol J 6:722–732

Sivasankar S, Rothstein S, Oaks A (1997) Regulation of the accumulation and reduction of nitrate by N and carbon metabolites in maize seedlings. Plant Physiol 114:583–589

Skopelitis D, Paranychianakis N, Paschalidis K, Pliakonis E, Delis I, Yakoumakis D, Kouvarakis A, Papadakis A, Stephanou E, Roubelakis-Anfelakis K (2006) Abiotic stress generates ROS that signal expression of anionic glutamate dehydrogenases to form glutamate for proline synthesis in tobacco and grapevine. Plant Cell 18(10):2767–2781

Sonoda Y, Ikeda A, Saiki S, Yamaya T, Yamaguchi J (2003) Feedback regulation of the ammonium transporter gene family AMT1 by glutamine in rice. Plant Cell Physiol 44:1396–1402

Stewart GR, Mann AF, Fentem PA (1980) Enzymes of glutamate formation: glutamate dehydrogenase, glutamine. In: Miflin BJ (ed) The biochemistry of plants: comprehensive treatise, vol. 5. Amino acids and derivatives. Academic, New York, pp 271–327

Sulpice R, Pyl ET, Ishihara H, Trenkamp S, Steinfath M et al (2009) Starch as a major integrator in the regulation of plant growth. Proc Natl Acad Sci 106:10348–10353

Svennerstam H, Ganeteg U, Nasholm T (2008) Root uptake of cationic amino acids by Arabidopsis depends on functional expression of amino acid permease 5. New Phytol 180:620–630

Taochy C, Gaillard I, Ipotesi E, Oomen R, Leonhardt N, Zimmermann S et al (2015) The *Arabidopsis* root stele transporter NPF2.3 contributes to nitrate translocation to shoots under salt stress. Plant J 83:466–447

Taylor L, Nunes-Nesi A, Parsley K, Leiss A, Leach G, Coates S, Wingler A, Fernie AR, Hibberd JM (2010) Cytosolic pyruvate, orthophosphatedikinase functions in N remobilization during leaf senescence and limits individual seed growth and N content. Plant J 62(4):642–652

Tilsner J, Kassner N, Struck C, Lohaus G (2005) Amino acid content and transport in oilseed rape (*Brassica napus* L.) under different N conditions. Planta 221:328–338

Tsay YF, Schroeder JI, Feldmann KA, Crawford NM (1993) The herbicide sensitivity gene CHL1of *Arabidopsis* encodes anitrate-inducible nitrate transporter. Cell 72:705–713

Tsay YF, Chiu CC, Tsai CB, Ho CH, Hsu PK (2007) Nitrate transporters and peptide transporters. FEBS Lett 581:2290–2300

Uauy C, Brevis JC, Dubcovsky J (2006) The high grain protein content gene *Gpc-B1* accelerates senescence and has pleiotropic effects on protein content in wheat. J Exp Bot 57:2785–2794

Ullrich WR (1987) Nitrate and ammonium uptake in green algae and higher plants: mechanism and relationship with nitrate metabolism. In: Ullrich WR, Aparicio PJ, Syrett PJ, Castillo F (eds) Inorganic N metabolism. Springer, Berlin/Heidelberg/New York, pp 32–38

Velculescu VE, Zhang L, Vogelstein B, Kinzler KW (1995) Serial analysis of gene expression. Science 270:484–487

Vidal EA, Araus V, Lu C, Parry G, Green PJ et al (2010) Nitrate-responsive miR393/*AFB3* regulatory module controls root system architecture in *Arabidopsis thaliana*. Proc Natl Acad Sci 107:4477–4482

Walch-Liu P, Forde BG (2008) Nitrate signaling mediated by the NRT1.1 nitrate transporter antagonizes l-glutamate-induced changes in root architecture. Plant J 54:820–828

Wang R, Liu D, Crawford NM (1998) The Arabidopsis CHL1 protein plays a major role in high-affinity nitrate uptake. Proc Natl Acad Sci 95:15134–15139

Wang Z, Gartein M, Snyder M (2009) RNA-Seq: a revolutionary tool for transcriptomics. Nat Rev Genet 10(1):57–63

Wang H et al (2013) Regulating effect of N fertilization on grain filling and activities of enzymes involved in starch synthesis of Yumai49-198. Plant Nutr Fertil Sci 19:288–296

Ward M, Grimes H, Huffaker R (1989) Latent nitrate reductase activity is associated with the plasma membrane of corn roots. Planta 177:470–475

Werner T, Nehnevajova E, Kollmer I, Novak O, Strnad M, Kramer U, Schmulling T (2010) Root specific reduction of cytokinin causes enhanced root growth, drought tolerance, and leaf mineral enrichment in *Arabidopsis* and tobacco. Plant Cell 22:3905–3920

Witt CA, Dobermann S, Abdulrachman HC, Gines W, Guanghuo R, Nagarajan S, Satawathananont TT, Son PS, Tan L, Van Tiem G, Simbahan Olk DC (1999) Internal nutrient efficiencies in irrigated lowland rice of tropical and subtropical Asia. Field Crop Res 63:113–138

Xiong et al. (2014) Effect of N fertilizer on distribution of starch granules in different regions of wheat endosperm. Crop J 46:54

Xu G, Fan X, Miller AJ (2012) Plant N assimilation and use efficiency. Annu Rev Plant Biol 63:153–182

Yan M, Fan XR, Feng HM, Miller AJ, Shen QR, Xu GH (2011) Rice OsNAR2.1 interacts with OsNRT2.1, OsNRT2.2 and OsNRT2.3a nitrate transporters to provide uptake over high and low concentration ranges. Plant Cell Environ 34:1360–1372

Yanagisawa S (2004) Dof domain proteins: plant-specific transcription factors associated with diverse phenomena unique to plants. Plant Cell Physiol 45:386–391

Yuan L, Loque D, Ye F, Frommer WB, von Wiren N (2007) N-dependent posttranscriptional regulation of the ammonium transporter AtAMT11. Plant Physiol 143:732–744

Zhang H, Forde BG (1998) An *Arabidopsis* MADS box gene that controls nutrient-induced changes in root architecture. Science 279:407–409

Zhuo D, Okamoto M, Vidmar JJ, Glass AD (1999) Regulation of a putative high-affinity nitrate transporter (AtNrt2.1) in roots of Arabidopsis thaliana. Plant J 17(5):563–568

# Improving the Nitrogen Cycling in Livestock Systems Through Silvopastoral Systems

# 7

Lucero Sarabia, Francisco J. Solorio, Luis Ramírez, Armin Ayala, Carlos Aguilar, Juan Ku, Camila Almeida, Rafael Cassador, Bruno J. Alves, and Robert M. Boddey

**Abstract**

Conventional livestock are monoculture systems where the major species are native grasses or African grasses, with biomass production being limited by seasonality of rainfall and low soil fertility. In animal production systems, the pasture degradation is associated with the nitrogen (N) cycle. Therefore, if farmer applied no subsequent fertilizer, milk production or live weight gains have been gradually reduced. As animals slowly gain weight, they produce more methane ($CH_4$) and nitrous oxide ($N_2 0$). This has led to the search for strategies to help minimize the impacts of livestock, and the excessive application of fertilizers, on the environment and natural resources. One strategy with promising results that has been developed in Latin America is the conversion of traditional livestock systems to silvopastoral systems (SPS), which include the establishment of shrub legumes at high densities and forage grasses aimed at increasing livestock profitability. With the association of legumes and forage grasses, forage quality can increase, more than 100%, compared to monoculture-based pastures and, consequently, production costs related to the purchase of imported cereal grains and nitrogen fertilizers are reduced. On the other hand, changes in climate and graz-

L. Sarabia · C. Almeida · R. Cassador
Department of Soils, Federal University Rural of Rio de Janeiro (UFRRJ), Seropédica, RJ, Brazil

F. J. Solorio (✉) · L. Ramírez · A. Ayala · C. Aguilar · J. Ku
Agricultural and Biological Sciences Campus, Autonomous University of Yucatán, Merida, Yucatán, Mexico
e-mail: ssolorio@correo.uady.mx

B. J. Alves
Brazilian company of agricultural research – National Research Center in Agrobiology, Seropédica, RJ, Brazil

R. M. Boddey
Embrapa Agrobiologia, Seropédica, Rio de Janeiro, Brazil



R. S. Meena (ed.), *Nutrient Dynamics for Sustainable Crop Production*,
https://doi.org/10.1007/978-981-13-8660-2_7

ing pressure to increase stocking rate have resulted in extensive degradation of existing vulnerable pastures, which favour poorly palatable, perennial species, affecting directly livestock production and enhancing greenhouse gas (GHG) emissions and the loss of soil carbon and nitrogen stock severely affecting soil fertility. The importance of the association of species of legumes with grasses and *Leucaena* (*Leucaena leucocephala* L., (Lam.) de Wit) is an environmentally friendly proposal of positive interactions to improve soil fertility and animal productivity. Overall, improving forage quality and N efficiency of dietary nutrients is an effective way of decreasing GHG. Silvopastoral systems (SPS) are used successfully in many regions around the world, and there is considerable evidence that SPS can increase production efficiency, increase carbon sequestration and improve N cycling on land used for livestock production.

**Keywords**
Animal urine · Grass N uptake · Greenhouse gasses · Silvopastoral systems

## Abbreviations

| | |
|---|---|
| BNF | Biological nitrogen fixation |
| $CH_4$ | Methane |
| $CO_2$ | Carbon dioxide |
| CP | Crude protein |
| DE | Digestible energy |
| GHG | Greenhouse sasses |
| IFA | International Fertilizer Industry Association |
| ISPS | Intensive silvopastoral systems |
| N | Nitrogen |
| $N_2$ | Atmospheric nitrogen |
| $N_2O$ | Nitrous oxide |
| NDF | Neutral detergent Fibre |
| $NH_4^+$ | Ammonium ion |
| $NO_3^-$ | Nitrate |
| OM | Organic matter |
| SPS | Silvopastoral systems |

## 7.1 Introduction

Global continuous increase of livestock production (milk, eggs and meat) demands more forage to feed the animals; consequently, more grazing areas would be necessary to reach the animal food intake; this could increase the deforestation, replacing forest areas for pastures, if the pastures are managed wrong, for example, overgrazing it enters in a process of degradation, so in a few years, more areas are necessary. Lately, overgrazing is one of the major causes of grassland degradation and represents the main cause of degradation among the major biomes. Steinfeld et al. (2006) estimated that approximately 73% of the pastures have been degraded; the same authors estimated that in the Amazon, the introduction of pastures is responsible for 70% of the deforestation.

It has been estimated that approximately 5% of soil organic carbon has been lost from overgrazing and during the dry season, ruminants are usually fed low-quality forages, which are characterized by their low concentrations of crude protein (CP), digestible energy (DE) and their high contents of neutral detergent fiber (NDF) and lignin, which induce a higher emission of methane ($CH_4$). Degradation of grazing biomes not only has a negative and direct effect on livestock production but also effects on the soil and the environment. Under tropical grassland conditions, cattle generally loses weight, and the milk production per cow is severely affected. Grass-based diet can lead to a daily weight gain of less than 300 g or a daily milk production of 4 kg/day (Ku et al. 2014; Meena and Meena 2017; Ashoka et al. 2017).

These grassland pastures are one of the ecosystems most vulnerable to climate change. In the tropics, livestock in extensive mixed systems suffer from permanent or seasonal nutritional stress (Ku et al. 2012; Yadav et al. 2018b). Poor nutrition is one of the major production constraints in small holder systems, particularly in tropical areas. Additionally, in the last decade, climate change and human population growth began to threaten the productivity of those grasslands due to changes in vegetation, mainly due to variability in rainfall along the year, frequently raising temperatures (IPCC 2007), including incorrect grazing management practices.

Two options to solve these problems have relied on the use of nitrogen-based fertilizer and the use of imported supplemental feed concentrates with social, economic and environmental negative effects. The dependence on grain as feed animal component has created a competition ground between humans and animals for the same source of food (Thornton 2009; Meena et al. 2015d; Kumar et al. 2017b; Meena and Lal 2018a, b). The abuse of chemical fertilizer, mainly those based on nitrogen, leads to environment and soil (nitrification and denitrification process) contamination. Both of these two options, however, have considerably increased animal production.

In the tropical regions, environmental temperature and relative humidity are high, and frequently above the physiological capacity of livestock to dissipate body heat, causing enormous economic damages. Under these conditions, silvopastoral systems (SPS) are an important tool to increase livestock production and enhance resilience to drought and in reducing the contribution of cattle to climate change. Greenhouse gas (GHG) emissions are reduced due to fewer applications of nitrogen-based synthetic fertilizers that improved forage quality and production ($CH_4$ emission reductions estimated at 15–20% and nitrous oxide ($N_2O$) emission reduction at 25–30%). Given the prevalence of many leguminous species, closer integration of trees and shrub with grasses can rise to increased productivity and increased soil fertility (atmospheric nitrogen fixation) including animal welfare. Integrating local leguminous trees and shrubs species is usually well adapted to water stress of the tropical climate by storing nutrients and carbohydrates in perennial belowground organs, thus improving the capacity to store carbon in the soil and in the aboveground biomass due to trees and shrubs integrating with grass. The objective of this chapter is to discuss the role of SPS as a strategy to improve the N cycling in tropical and subtropical livestock systems.

## 7.2 Livestock and the Environment

Climate change is the result of the accumulation of GHG emissions in the atmosphere, caused mainly by human activities. The most important GHG directly emitted by humans include carbon dioxide ($CO_2$), methane ($CH_4$) and nitrous oxide ($N_2O$). Since the Industrial Revolution, the amount of GHGs emitted exceeds the capture capacity of the biosphere, and the net result is the constant increase in GHG concentrations, which prevent heat from escaping the atmosphere (IPCC 2007). This global warming is the most obvious manifestation of climate change and refers to the rise in average surface temperatures.

Human activities such as agriculture and deforestation contribute to the rise of GHG emissions. Within this, the livestock sector is considered one of the main activities with the biggest impact on climate change through the emission of greenhouse gases (Herrero et al. 2013; Varma et al. 2017a; Buragohain et al. 2017). According to the FAO (2013), the livestock sector contributes 18% of the total gases emitted into the atmosphere. Large amounts of $CO_2$ are emitted from the burning of fossils fuels to make fertilizers which are used to grow grain to fed animals, including the deforestation to grassland expansion (Fig. 7.1).

Livestock activities contribute to global warming due to the large release of GHG into the atmosphere, originated from the enteric fermentation. Ruminant animals are the major emitters of $CH_4$, due to the digestive process in which microbes ferment the food consumed.

**Fig. 7.1** Deforestation and soil preparation for grassland expansion in the Mexican tropics

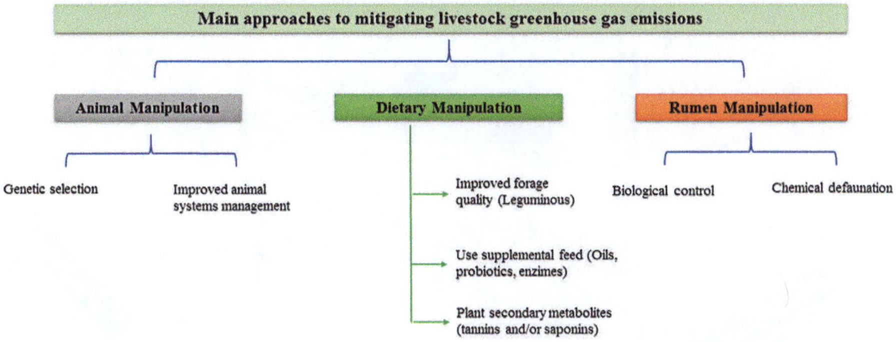


**Fig. 7.2** Specific solutions for the mitigation of GHG emissions in the livestock sector

Ruminants grazing low-quality fodder increase $CH_4$ emissions, with a bigger negative impact on climate change. Excreted urine and faeces by ruminants significantly contributed to $CH_4$ and $N_2O$ emissions. Although the amount of nitrogen (N) excreted depends on the feed quality ingested, low-quality feeds result in low N-excreted factors like temperature, and soil moisture can facilitate the release of GHG emissions on pastures based livestock production (Lessa et al. 2014; Meena and Yadav 2015; Dhakal et al. 2015; Datta et al. 2017b). Although GHG emissions have been increased markedly in the last five decades, many options exist that can mitigate GHG emissions from the livestock sector (Fig. 7.2).

## 7.3 Nitrogen Cycling in Livestock Systems

In the tropics and subtropics, millions of people have no food security; about 60% of rural communities are permanently affected by the decline in household food production, with sub-Saharan Africa and parts of Latin America, the Caribbean and Central Asia suffering worst (Stocking 2003; Meena et al. 2016a; Verma et al. 2015c). From 33 Latin American countries, Mexico occupies the second place in deforestation just below Brazil. Current estimates of deforestation rates in Mexico range from 400,000 to 1,500,000 hectares per year, the largest area located in the southern Mexico including the Peninsula of Yucatan (Cairns et al. 1995). The consequence of forest deforestation is the decline in soil productivity.

The progressive deterioration of natural resources in the tropics has led to the need of alternative methods to sustain crop production. Optimizing nutrient cycling and improving biological nitrogen fixation as a source of nitrogen have been proposed as the better strategy for achieving improved levels of production without further damage to the natural resource (Greenland 1975; Meena and Lal 2018a, b; Yadav et al. 2017c; Dadhich and Meena 2014; Kumar et al. 2018b). In the tropics, fast-growing trees, particularly nitrogen-fixing trees, are increasingly being recommended for land restoration where soil has been degraded (Franco and De Faria 1997; Dubeux et al. 2015; Varma et al. 2017b; Meena et al. 2015e), for fallow

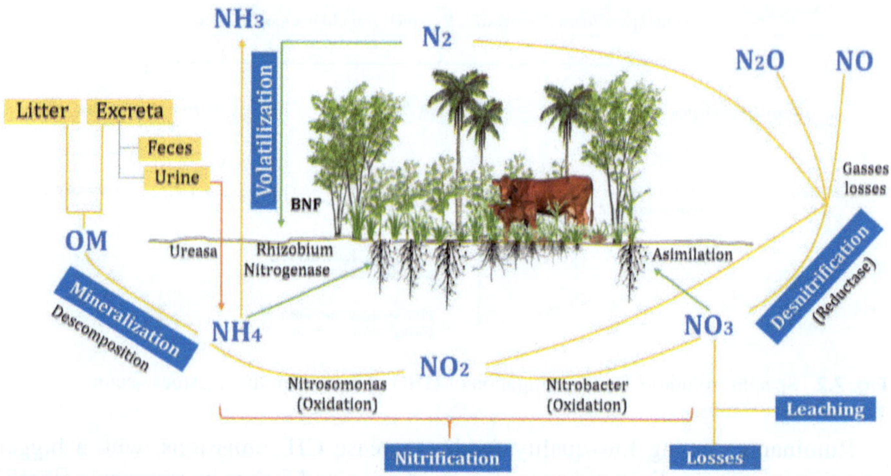


**Fig. 7.3** Approach to outline for synthesizing understanding of N dynamics in livestock SPS

improvement and for erosion control; examples of benefits in soil improvement are well-documented by Kamara et al. (2000), Giller (2001) and Schroth and Sinclair (2003). In tropical agroecosystems, the leaf litter decomposition and subsequent nutrients released represent a good tool for the poor farmer in order to reduce external input and maintain or increase agricultural products (Mthembu et al. 2018; Kumar et al. 2018a). About 70–90% of the nutrients required for growing plants could be proportioned by leaf litter decomposition of a companion plants (Waring and Schlesinger 1985). Using this system, fodder trees, apart from being a source of foliage rich in nutrients to animal feed, could also be a very important source of nutrient to the soil (Fig. 7.3).

## 7.4 Agroecological Strategies to Improve N Cycling

In tropical countries the gap between food production and population growth is widening (Stocking 2003). This increased demand for more effective food production generally means increased application of industrially produced fertiliser nitrogen, because N is the most important nutrient for crop growth. Nevertheless, research has shown that biological nitrogen fixation by legumes is an efficient way to supply the large amounts of nitrogen needed to produce high-yielding crops with high protein content (Crews and Peoples 2004; Boddey et al. 2015; Meena et al. 2018a; Sihag et al. 2015). Introducing legumes into farming systems can provide a continuous supply of N for plant growth and provide good-quality organic matter to be incorporated into the soil; thus, legume species should play an important role in developing new strategies to increase food production (CGIAR 2004). Biological $N_2$ fixation contributes to enhanced production directly by increasing the yield of grain or other food crops for human or animal consumption or indirectly by contributing to the maintenance of soil fertility (Giller and Cadisch 1995; Graham and

Vance 2003; Meena et al. 2016b; Yadav et al. 2017b). In the latter case, $N_2$ fixation by leguminous trees is most likely to constitute a relevant input to farming systems when the soil is low in N and when N fertiliser is scarce (Schroth et al. 2001; Verma et al. 2015b). Additionally, trees or shrubs survive in most dry seasons and can contribute significantly to animal feed by providing nutritional foliage, in places where grassland productivity depends largely on the rainfall.

## 7.5 Silvopastoral Systems

At present, livestock production in tropical areas face serious problems related to climate change and the prevailing model of extensive production (FAO 2012). These problems are characterized by the transformation of natural ecosystems into large monoculture pastures, low productivity and with strong demand for fertilizers mainly nitrogen (Meena and Lal 2018a, b; Meena and Yadav 2014). This has a negative impact on agricultural production and plant biodiversity, which leads to a high dependence on external inputs and little integration between the agricultural, livestock and forestry sectors (Williams et al. 2017; Dadhich et al. 2015). In addition, extensive livestock systems have low levels of efficiency and profitability (Ku et al. 2014; Sofi et al. 2018) and are more vulnerable to extreme climatic conditions such as droughts or floods (Cuartas et al. 2014; Meena et al. 2015c).

The availability of nitrogen is one of the main constraints of tropical animal production (Ku et al. 2014). The application of nitrogenous inorganic fertilizers is a frequent practice, to correct the problem as well as to increase the productivity and quality of the pastures (Silveira et al. 2013; Yadav et al. 2018a). This causes considerable quantities of fertilizers to be imported annually to be applied to the soil (FAOSTAT 2014).

Additionally, the indiscriminate application of fertilizers causes irreversible damage to the environment by emissions of gases, mainly $N_2O$ and contamination of groundwater (IFA 2002). In addition, the high costs of nitrogen N fertilizers increase the costs of livestock production (Pelletier and Tyedmers 2010; Verma et al. 2015a). In Mexico, for example, the price of agrochemicals doubled in recent years, to such a degree that in 1 year more than 60 million pesos were invested in their purchase; just to cite one example, urea has a cost of approximately $ 500 per ton (FAO 2008) and, 1 hectare of pasture requires approximately 120 kg of N/ha/year. This has led to the search for strategies to minimize the environmental impact of livestock production and the excessive application of fertilizers. Efforts have also been made with particular emphasis on increasing forage production and quality within these livestock production systems, which may contribute to a more efficient use of N throughout the production system.

One of the strategies with promising results that has arisen in Latin American tropical livestock is the reconversion of traditional monoculture systems with intensive silvopastoral systems (ISPS) (Murgueitio et al. 2011; Bacab et al. 2012; Solorio et al. 2012; Meena et al. 2015b), including shrub legumes at high densities associated with forage grasses which would increase both yield and quality of forage and

**Fig. 7.4** Shrub forage legume for effective reconversion grassland to silvopastoral systems

promote the N atmospheric fixation and recycling nutrients (Fig. 7.4). With the association of legumes and forage grasses, forage quality could increase more than 100% in comparison to monoculture-based pastures (Sturludóttir 2011; Datta et al. 2017a), and, consequently, the production costs related to the purchase of nitrogen fertilizers and the use of feed concentrates will be reduced (Murgueitio et al. 2015; Meena et al. 2015a).

Intensive silvopastoral systems is an agroecological model of agricultural production, in which perennial woody trees (multipurpose trees and high-forage shrub densities) interact with pastures and animals under an integrated management system (Fig. 7.5). These systems have been designed and proposed based on research results, which have evaluated the yield and quality of forage (Murgueitio et al. 2015; Meena et al. 2014), animal productivity (e.g. production and quality of meat and milk) and carbon capture (Solorio et al. 2016; Dhakal et al. 2016), as well as improved the microclimate of grazing animals and biodiversity (Broom et al. 2013; Kakraliya et al. 2018).

With the implementation of SPS, it is feasible to reduce the environmental impact of traditional extensive livestock production systems (Murgueitio et al. 2011; Solorio et al. 2012; Ram and Meena 2014). Silvopastoral systems are developed as sustainable animal production strategies. The main structure of the SPS includes the association of grasses with shrub legumes and represents additional advantages, including legumes in the paddocks that obtain benefits in the production and quality

**Fig. 7.5** ISPS with multipurpose trees, edible shrubs and the integration of animal-grass

of forage biomass, as a consequence of the biological fixation and transfer of N (Casanova et al. 2014). Several studies have demonstrated the potential benefits of silvopastoral systems and their effect on the profitability of livestock systems including environmental benefits, soil and their benefits in the productivity and sustainability of agroecosystems (Goh et al. 1996; Mercado et al. 2011; Alvarez et al. 2014; Meena et al. 2017c).

Other positive interactions among trees or shrub species may maximize above- and belowground resource utilization for growth (Cadisch et al. 2002; Yadav et al. 2017a; Kumar et al. 2017a). Intercropping legumes and nonlegumes increases the opportunity for complementary N use (van Kessel and Hartley 2000). Mixing species may also improve resilience, by improving nutrient cycling and enhancing resistance to pests or diseases. Additionally, mixing species can exploit interactions in which one species enhances the biological performance of another (Khanna 1997; Cadisch et al. 2002; Gathumbi et al. 2003; Meena et al. 2017a); perhaps the most important feature is the direct transfer of N from a $N_2$-fixing plant to a non-fixing plant, which can be better exploited in a mixed system.

## 7.6 Challenges in the Tropical Livestock Production Systems

Livestock activity has been associated with problems of deforestation, soil degradation, loss of biodiversity, environmental pollution and low productivity (Broom et al. 2013). These livestock systems often remove all the vegetation (as erroneous, considered competition for water and soil nutrients) in order to allow the growth of native or introduced forage grasses, (Grande et al. 2010; Jiménez et al. 2011).

Most Latin American beef cattle production systems use tropical grasses as the basal forage source. However, extensive livestock production that is mainly pasture-based results in lower yield (milk or beef) per unit of land and a negative environment impact (Ku et al. 2014). Its development and profitability are based on the extension of the grazing area with large areas that have been deforested for the production of milk and meat with dual-purpose cattle, specialized breeds and their crosses (Murgueitio et al. 2011). These problems can result in livestock weight losses and hence greatly restrict farmer's income. Fluctuations in forage availability, throughout the year, restrict the ability to achieve sustainable levels of animal production (Ku et al. 2014; Solorio et al. 2012), mainly due to the low quality of the grass which will consequently limit pasture intake and digestibility. Mismanagement of the grazing systems, the seasonality and low quality of forages, consequently, further decreases animal production (Murgueitio et al. 2011) and contributed to increase the $CH_4$ emissions from the digestive process of ruminants.

Mismanagement of the grazing systems has also contributed to the pasture degradation, due to overgrazing that has extracted nutrients from the soil without restitution (Murgueitio et al. 2013). The conservation and maintenance of ecosystems are currently under threat by the intensification of livestock systems. In order to reduce the deficiency of pasture-based animal production, farmers frequently rely on large quantities of imported concentrate feed (e.g. grains and cereals), and in the use of fertilizers, trying to increase livestock production, but highly dependent of external inputs and polluting the environment (Solorio et al. 2009).

In order to resolve the feed deficit in the dry season and meet the competing demands of increasing animal intensity while maintaining sustainable grain production, alternative forage sources need to be identified (Fig. 7.6). One of the options increasing the profitability of the livestock farmers includes the renovation of degraded pastures. The identification of the grass nutritional deficiencies can be reduced by improving livestock and grassland management with the incorporation of tropical forage legumes into their farming system. Figure 7.6 shows the effects of tropical forage legumes on the animal nutrition.

Another very important issue related to animal production is water. Water is a scarce and valuable resource essential to human and animal life. Water for livestock

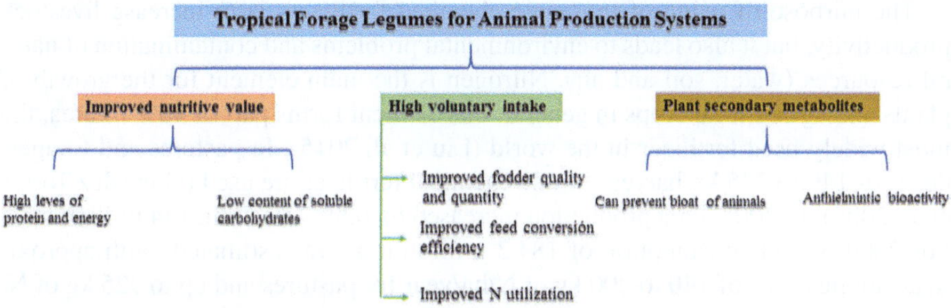


**Fig. 7.6** Forage legumes for animal production and the GHG mitigation

production is used for drinking, irrigation and growing crops/pasture. The water required to produce feed is the major factor behind the water footprint of animal products (Mekonnen and Hoekstra 2012). Animals can negatively affect water quality by having free access to water sources where animals can excrete faeces. Waste from animals can be dangerous because it carries harmful bacteria. Bacteria can enter water sources during heavy rainfalls that might result in an overflow of the manure catchment basin or from manure that has been put on fields as fertilizer (McAllister et al. 2012).

The challenge for livestock production therefore is to improve environmental sustainability and reduce greenhouse gas emissions. Challenges remain to reverse the economic losses from grassland degradation while accommodating growing demand and simultaneously avoiding the conversion of ecosystems.

## 7.7 Fertilizers and Livestock

Livestock production, as mentioned previously, is constrained by many factors. However, feed shortages during the dry season constitute the greatest challenge in terms of quantity and quality given by the seasonality of rainfall. Other factors include low soil fertility for forage production. Of the 17 chemical elements that are essential for plant growth, N is the nutrient that most often limits grass growth. N is very mobile in the soil and can become limiting in areas with high rainfall or irrigation, in coarse or shallow soils and in soils with low organic matter.

Climate change together with inadequate grassland management and the soil's low fertility is the main constraint to increasing livestock productivity. The lack of good-quality livestock feed, produced at a competitive cost, in the dry season, can jeopardize food security. Improvements in forage production through improved soil fertility practices have the potential to increase income and reduce livestock production costs. As pasture soil do not usually contain sufficient amounts of N for high and sustained fodder production, frequently farmers rely on the use of chemical fertilizer. The most common sources of commercial fertilizer N are urea. However, urea application is highly susceptible to volatilization (leaching to the atmosphere and water pollution).

The purpose of using of inorganic chemical fertilizers is to increase livestock productivity, but it also leads to environmental problems and contamination of natural resources (water, soil and air). Nitrogen is the main element for the growth of plants and agricultural crops in general; this element forms part of 46% of urea, the most widely used fertilizer in the world (Liu et al. 2015). In pastures and forages, doses of 140 to 325 kg/ha/year of this chemical fertilizer are used (González Torres et al. 2009). In 2007, urea production increased by 6.6%, reaching 144 million tons. For 2009, world consumption of 184.3 million tons was estimated, with approximate applications of 140 to 200 kg of N/ha/year for pastures and up to 325 kg of N/ha/year in crops such as forage maize (González Torres et al. 2009).

A common problem with the indiscriminate use of chemical fertilizers is related to the infiltration and consequent contamination of aquifers (approximately 10% of the fertilizer applied to the soil is infiltrated) and emissions of gases to the atmosphere (approximately 5% of the applied fertilizer is lost as a gas) (Vendramini et al. 2007). In addition, livestock contribute approximately 40% of global ammonia emissions mainly from animal excreta and from the use of fertilizers in pastures (IFA 2002). According to the International Fertilizer Industry Association (IFA), between 2002 and 2007, there was an increasing trend in world fertilizer production. The global supply grew by 3.4%, reaching an average of 165.3 million tons of nutrients: nitrogen, phosphorus and potassium, mainly (FAO 2008).

However, agroecological opportunities for improving the nutrition of livestock do exist, for instance, multipurpose legume trees can provide high-quality feed and improve soil fertility (Lenné and Thomas 2006). Intercropping legumes with grasses, which are an excellent source of N, improves forage quality. In Queensland, Australia, *Leucaena* (*Leucaena leucocephala* L., -Lam- de Wit)-grass mixes had higher live weight gain. A steer must consume a diet containing 35–40% *Leucaena* (4 kg/day for a 450 kg steer) to gain more than 1 kg/day (Dalzell et al. 2006). The main value of *Leucaena* is as a much-needed protein supplement to cattle grazing tropical grass pastures. Cattle require about 13% CP in their diets to produce good weight gains; they cannot get this from grass alone. When cattle are allowed to graze in *Leucaena* paddocks, their intake of protein immediately increases.

Intercropping *Leucaena* pastures can also enhance the environment by revitalizing the fertility of degraded soils by contributing biologically fixed nitrogen (BNF). *Leucaena* pastures offer the opportunity to intensify production in an environmentally sustainable manner. *Leucaena*-grass pastures are persistent and productive at higher stocking rates. Beef production is 4–6 times higher than from the best native pastures. Most crop-livestock production relies directly on rainfall, and adverse changes in quantity and temporal patterns of rainfall are a major risk to production. The drought tolerance of deep-rooted *Leucaena* can protect the land against the worst effects of drought. Also these may increase soil organic matter (OM), aggregation, nutrient availability, plant resistance to stresses and yield.

According to Ku et al. (2014), for improving meat and milk production and quality in tropical regions, different options have been created for manipulating the energy metabolism of ruminants. Silvopastoral systems, based on *Leucaena* and Tanzania grass *(Megathyrsus maximus (jacq) B.K. Simon & S.W.L. Jacobs)* association, can provide live weight gains of 770 g/d in growing cattle. For milk, it is possible to increase the concentration of unsaturated fatty acids from tannins in foliage that are beneficial effect on human health. This addition could provide aggregate value to the cow's milk. Improved feeding practices are required to decrease $CH_4$ emissions from the rumen. This can be done by feeding animals with foliage and fruits which contain secondary metabolites which are capable of affecting ruminal fermentation.

## 7.8 Silvopastoral Systems Benefits

### 7.8.1 Environmental Issue

The benefits of ISPS are associated with the integration of the multipurpose trees and shrubs. The leguminous component viz; *Leucaena*, *Gliricidia sepium (Jacq.) Kunth ex Walp.*, *Sesbania grandiflora L. Pers.*, *Cratylia argentea*, (*Desv.*) O. Kuntze, of these species has the capacity to fix atmospheric nitrogen through symbiotic association with bacteria of the genus Rhizobium (Peoples and Herridge 1999; Peoples et al. 2009). For example, *Leucaena* has the capacity to fix between 70 and > 285 kg atmospheric N/ha/year (Goh et al. 1996; Giller 2001; Sarabia 2013; Meena et al. 2017b). Therefore, the leguminous component contributes to the reduction of the excessive use of nitrogen fertilizers required by grasses in monocultures to improve the production and quality (Sierra et al. 2007; Peoples et al. 2009; Layek et al. 2018).

In addition, this type of shrub legume species presents greater tolerance to droughts, besides having a wide range of adaptation to diverse climatic conditions (climate change mitigation) and a great capacity to regrow in short periods or to resist frequent defoliations. SPS also integrates multipurpose trees that contribute to improving environmental conditions, reducing temperatures, improving animal behaviour and generating other products (e.g. wood, fruits, fodder, etc.). In addition, the inclusion of trees in association with pastures helps carbon sequestration, reduce water loss through evaporation and increase OM content in soils (Don 2012; Casanova Lugo et al. 2014).

### 7.8.2 Production and Quality of Forage

Silvopastoral systems are characterized by the diversity of species that can be incorporated into the system to increase animal production (Solorio et al. 2016). It highlights the importance of the association between shrub legumes, such as *Leucaena*, with grasses, since a variety of positive interactions occur, such as increased nitrogen availability, reduced solar radiation impact and temperature, improves animal comfort, shrubs also improve forage quality (increase protein content of the whole grassed forage), which contributes to improved animal productivity (Solorio et al. 2016). An important aspect of these interactions is the fixation and transfer of N, of which the associated pastures are directly benefited (Tessema and Baars 2006). Due to the high production of forage of high nutritional value, shrub legumes contribute significantly to animal feed, increasing the production and quality of the ingested forage.

Leguminous shrubs can continuously fix nitrogen and due to their deep rooting system act as a 'pumps'; they can also bring up nutrients from lower soil horizons and return them to the surface in the litter (Young 1997). Proper management of trees will ensure that the foliage is available for animal feeding during the critical period of food scarcity. For example, the association of leguminous with non-leguminous trees in a mixture may increase rates of N cycling. On nitrogen-deficient sites, mixed stands present an ecological opportunity for increasing both total stand growth and the growth of non-fixing trees (Binkley et al. 1992; Dakora and Keya 1997; Parrota 1999).

Evidence suggests that in addition to the positive advantages to agriculture from the leguminous species, they also play a major role in the growth of non-leguminous plants if they are planted in close contact with them (Parrota 1999; Rothe and Binkley 2001; Forrester et al. 2004). Reports on mixtures of leguminous trees with non-leguminous trees or annual crops show that the N concentration tends to increase in the leaves of the non-leguminous species in comparison to that of monocrop stands of the non-fixing plants (Khanna 1997; Chirwa et al. 2003). New evidence suggests that non-leguminous crops benefit from the direct transfer of N fixed by the plants (Fagbola et al. 1998; Graham and Vance 2000). The roots of nitrogen-fixing species have more nodules when they grow in close contact with roots of non-nitrogen-fixing plants (Van Noordwijk and Dommergues 1990; Sanchez 1995; Young 1997). This increased nodulation may lead to the direct transfer of nitrogen to the non-nodulating plant. Symbiotic activity in some intercropped legume species can be stimulated if the associated plants in the mixture exert intense competition for soil N, forcing the legume to rely more on symbiosis for its N nutrition (Eaglesham et al. 1981; Rerkasem et al. 1988). Since legumes usually do not have to compete with non-leguminous plants for soil N uptake, legumes grown in mixture with non-leguminous plants usually derive a higher percentage of their N from symbiosis (Graham and Vance 2000).

### 7.8.3 Importance of $N_2$ Fixation

Nitrogen is a key element in soil fertility and in the development of food production systems, being one of the most important elements for plant growth (Mafongoya et al. 2004). The N content of the soil is maintained thanks to natural processes, such as the BNF and the application of organic fertilizers and mineral fertilizers (Giller et al. 1997). However, BNF is the most important source for sustaining soil fertility (Stockdale et al. 2001; Unkovich et al. 2008; Meena et al. 2018b).

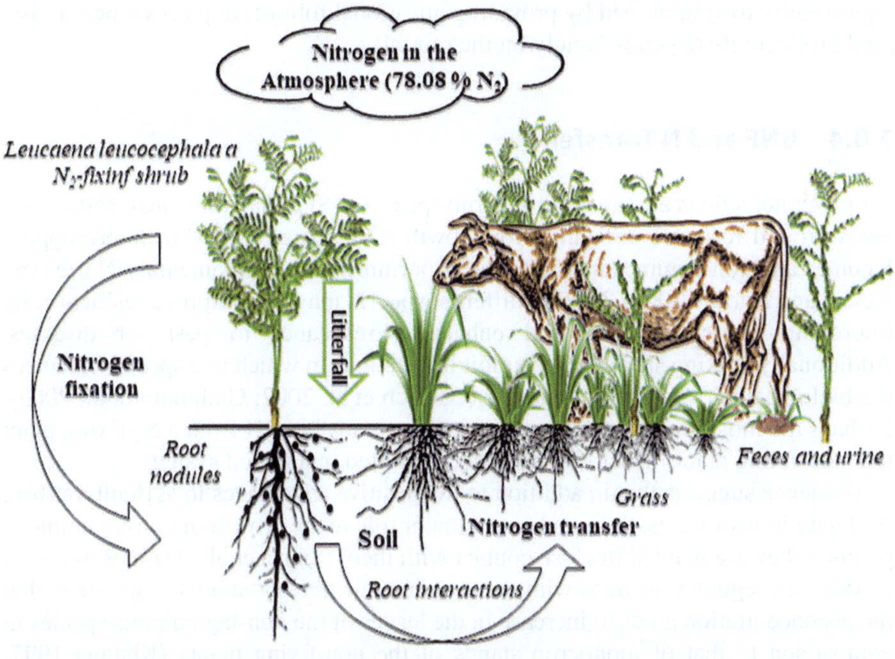


**Fig. 7.7** The proposed model of the N fixation in SPS for animal production

Generally, the N absorbed by the plants comes from the soil, which must be mineralized in the form of nitrate ($NO_3^-$) or ammonium ($NH_4^+$). On the other hand, 78% of the air in the atmosphere is nitrogen gas ($N_2$), which is not readily available to plants; only some species are able to directly use $N_2$ by symbiosis with soil bacteria that are $N_2$-binding agents (Fig. 7.7). In this case, there are different genera of $N_2$-fixing bacteria. The genus *Rhizobium* is one of the most important for leguminous plants, since they play a very important role in some N transformations through the biological fixation process.

The amount of nitrogen required by the plants may be greater than that provided by the soil; in most crops, N fertilization is necessary, since the nitrogen cycle has been altered by the removal of trees (particularly legumes) from the system and the excessive use of grassland in monoculture.

Introducing legumes into livestock farming systems can provide a continuous supply of N for animals and for plant growth (Fig. 7.7). Also, providing OM of good quality to be incorporated in the soil, legume species should play an important role in developing new strategies to increase animal production. Biological $N_2$ fixation contributes to enhanced production directly by increasing forage biomass and quality or indirectly by contributing to the maintenance of soil fertility (Giller and Cadisch 1995; Murgueitio et al. 2015). In the latter case, $N_2$ fixation by leguminous trees is most likely to constitute a relevant input to farming systems when the soil is low in N and when N fertilizer is scarce (Schroth et al. 2001; Meena et al. 2018c). Additionally, trees or shrubs usually survive dry seasons and can contribute significantly to animal feed by providing nutritional foliage, in places where grassland productivity depends largely on the rainfall.

### 7.8.4 BNF and N Transfer

Positive interactions among trees or shrub species in SPS may maximize above- and belowground resource utilization for growth (Cadisch et al. 2002). Intercropping legumes and non-legumes increases the opportunity for complementary N use (van Kessel and Hartley 2000). Mixing different species may also improve resilience, by improving nutrient cycling and enhancing resistance to pests or diseases. Additionally, mixing species can exploit interactions in which one species enhances the biological performance of another (Cadisch et al. 2002; Gathumbi et al. 2003). Perhaps the most important feature is the direct transfer of N from a $N_2$-fixing plant to a non-fixing plant, which can be better exploited in a mixed system.

Evidence suggests that in addition to the positive advantages to agriculture from the leguminous species, they also play a major role in the growth of non-leguminous plants if they are planted in close contact with them (Sierra et al. 2007). Reports on mixtures of leguminous trees with non-leguminous trees or annual crops show that the N concentration tends to increase in the leaves of the non-leguminous species in comparison to that of monocrop stands of the non-fixing plants (Khanna 1997; Chirwa et al. 2003). New evidence suggests that non-leguminous crops benefit from the direct transfer of N fixed by the plants (Thilakarathna et al. 2016). The roots of nitrogen-fixing species have more nodules when they grow in close contact with roots of non-nitrogen-fixing plants (Van Noordwijk and Dommergues 1990; Sanchez 1995; Young 1997). This increased nodulation may lead to the direct transfer of nitrogen to the non-nodulating plant. Symbiotic activity in some intercropped legume species can be stimulated if the associated plants in the mixture exert intense competition for soil N, forcing the legume to rely more on symbiosis for its N nutrition (Eaglesham et al. 1981; Rerkasem et al. 1988). Since legumes usually do not have to compete with non-leguminous plants for soil N uptake, legumes grown together with non-leguminous plants usually derive a higher percentage of their N from symbiosis (Graham and Vance 2000).

$N_2$ fixation is made by N-fixing bacteria, which require a source of chemical energy from the plant. Therefore, this process has been identified as Rhizobium – leguminous symbiosis. This association contributes between 30 and 50% of biological $N_2$ fixation and is based on the exchange of carbon by nitrogen between both symbionts (symbiotically associated organisms), also helping to reverse the low fertility of the soil. Therein lays the importance of proper inoculation with Rhizobium, since it is possible to considerably increase the atmospheric fixation of $N_2$ in agricultural systems (Rodrigues et al. 2013).

The capacity of $N_2$ fixation is an important characteristic of the Rhizobium bacteria, which infects and colonizes the roots of the legumes causing the deformations that are known as nodules; it is in these where the transformation of $N_2$ to mineral nitrogen is carried out. The $N_2$ fixation process has a high energy cost, because the

**Table 7.1** Milk and meat production under ISPS and grass monoculture

| System | Milk production (kg animal$^{-1}$ día$^{-1}$) | References |
|---|---|---|
| ISPS + concentrated feed (1,5 kg) | 9.20 | Bacab and Solorio (2011) |
| Grass-based only + concentrated feed (8 kg) | 10.4 | |
| | Meat production (kg animal$^{-1}$ día$^{-1}$) | |
| ISPS | 0.770 | Mayo et al. (2013) |
| Grass-based only | 0.28-0.62 | Ku et al. (2012) |

triple bond linking the two nitrogen atoms is difficult to break. The great advantages that they obtain from this symbiosis are multiple; among the most important, we can cite the following ones: (a) the plant can self-supply of N, increasing considerably the content of protein in its tissues; (b) the legume transfers N to other associated crops (Kurppa et al. 2010); (c) the legume helps to prevent loss of soil fertility by incorporating and leaving available nitrogen in the soil for the next crop in the rotation; and (d) there is a greater efficiency of the use of atmospheric nitrogen by crops compared to the application of nitrogen fertilizers, because the applications of the latter are lost leaching fractions, which then become pollutants of soils, waters, animals and even humans (Parsons 2004).

Several studies (Solorio 2005; Sierra et al. 2007; Burchill et al. 2014; Mitran et al. 2018) show that in agroforestry systems, crops associated with legumes can increase their N content, since in the roots of trees and shrubs when interacting closely with the roots of crops, the N fixed by the legumes is transferred directly and is used by the grasses and is expressed through the increase in the protein content of the grasses. In this sense, the importance of nitrogen transfer is oriented to the ability of legumes to transfer nitrogen directly to grazing animals including the pastures associated with it (Table 7.1). Therefore, they have the capacity to increase the protein content in forage, which is reflected in higher weight gain in grazing animals and higher milk yield compared to monoculture pastures (Table 7.1).

Recent studies with SPS established with *Leucaena*, which evaluated the fixation and transfer of atmospheric nitrogen, indicate the ability of *Leucaena* to fix high amounts of N up to 285 kg N ha$^{-1}$ (Solorio 2005; Sarabia 2013), having greater potential of fixation and transference when they have high densities of *Leucaena*. In recent studies with *Leucaena* established at high densities, N transference to pastures greater than 50% has been found (Sarabia 2013). Despite the great importance of atmospheric N fixation for agricultural systems, there are few studies to evaluate the effect of climatic and edaphic conditions on the influence of legumes to incorporate nitrogen into the associated crops (Dreyfus et al. 1988).

## 7.9 Challenges

Despite the research evidence showing the advantages of SPS and the large interest in soil fertility maintenance in tropical agroecosystems, few reliable estimates and data of N cycling in SPS are available. Most of the research involving woody plants has been focused on grassland monoculture, while other research on more than one species relates only to their growth in a rotation system. Strategies for improving N use in livestock systems are becoming critically important, and today one of the most promising strategies for the GHG mitigation relies on the improvement of the fodder quality. Tropical soils have not enough N soil to crops, improving the overall use efficiency of available N, through intercropping systems where non-$N_2$-fixing plants are grown in close contact with $N_2$-fixing legumes (van Kessel and Hartley 2000; Graham and Vance 2003; Gogoi et al. 2018).

Silvopastoral systems adoption by farmers are still limited; livestock producers are reluctant to integrate the SPS into their farming systems. Research should be addressed to identify specific barrier (Dagang and Nair 2003). Long-term investment return and high costs for initial implementation could be another barrier for their adoption. Inefficient support of state policies, a national programme should be orientated to the implementation and to give a long-term support for the livestock producers.

## 7.10 Conclusions

Extensive livestock systems have resulted in degradation of natural resources and loss of productivity. In order to counter this effect, farmer uses the acquisition of inputs which results in a vicious circle of low productivity and environmental pollution. Silvopastoral systems have been increasingly adopted in different eco-regions of Latin America. They represent a viable alternative to contribute to providing excellent quality forage for tropical livestock. Additionally, they have several advantages over traditional monoculture-based livestock systems. The inclusion of legumes into animal production systems can incorporate significant amounts of N into the soil and transfer much of it directly or indirectly to the animal, soil and the associated grasses.

Silvopastoral systems can play also a major role in the rehabilitation of fragmented ecosystems, contributing towards mitigating the impacts of climate change on livestock and reducing GHG emissions, mainly $N_2O$, $CO_2$ and $CH_4$.Grass pastures can be restored by improving the carbon sequestration and increasing atmospheric N fixation in the soil through the integrations of leguminous shrubs and trees. Livestock feeding with leguminous species rich in N, tannins and saponins would be one of the best strategies for methane mitigation and reduce gaseous N losses from manure.

**Acknowledgements** The authors would like to thank the research funding agencies CAPES for the scholarships granted to the post-graduate students participating in the study. To the University of Yucatan for financially supporting the project 'Development and Consolidation of the Ruminant Nutrition and Tropical Environment Research Group' (PADECCA).

## References

Alvarez S, Rufino MC, Vayssières J, Salgado P, Tittonell P, Tillard E, Bocquier F (2014) Whole-farm nitrogen cycling and intensification of crop-livestock systems in the highlands of Madagascar: an application of network analysis. Agricul Syst 126:25–37

Ashoka P, Meena RS, Kumar S, Yadav GS, Layek J (2017) Green nanotechnology is a key for eco-friendly agriculture. J Clean Prod 142:4440–4441

Bacab PH, Solorio SF (2011) Oferta y consumo de forraje y producción de leche en ganado de doble propósito manejado en sistemas silvopastoriles en Tepalcatepec, Michoacán. Trop Subtropical Agroecosyst 13(3):271–278

Bacab H, Solorio F, Solorio B (2012) Efecto de la altura de poda en *Leucaena leucocephala* y su influencia en el rebrote y rendimiento de *Panicum maximum*. Avances en Investigación Agropecuaria 16(1):65–77

Binkley D, Bell R, Sollins P (1992) Comparison of methods for estimating soil nitrogen transformations in adjacent conifer and alder–conifer forests. Can J For Res 22(6):858–863

Boddey RM, de Carvalho IDN, Rezende CDP, Cantarutti RB, Pereira JM, Macedo R, Tarre R, Alves JRB, Urquiaga S (2015) The benefit and contribution of legumes and biological $N_2$ fixation to productivity and sustainability of mixed pastures. In: Proceedings of the 1st international conference on forages in warm climates, vol 1. Universidade Federal de Lavras, Lavras, pp 103–140

Broom D, Galindo A, Murgueitio E (2013) Sustainable, efficient livestock production with high biodiversity and good welfare for animal. Proc R Soc Biol Sci 280:2013–2025

Buragohain S, Sharma B, Nath JD, Gogaoi N, Meena RS, Lal R (2017) Impact of ten years of bio-fertilizer use on soil quality and rice yield on an inceptisol in Assam, India. Soil Res 56:49. https://doi.org/10.1071/SR17001

Burchill W, James E, Li D, Lanigan G, Williams M, Lannetta P, Humphreys J (2014) Comparisons of biological nitrogen fixation in association with white clover (*Trifolium repens L.*) under four fertilizer nitrogen inputs as measured using two 15N techniques. Plant Soil 385(1–2):287–302

Cadisch G, Gathumbi S, Ndufa JK, Giller KE (2002) Resource acquisition of mixed species fallows –competition or complementarity? In: Vanlauwe B, Sanginga N, Merckx R (eds) Integrated plant nutrient management in Sub-Saharan Africa. CABI Publishing, Wallingford, pp 143–154

Cairns MA, Dirzo R, Zadroza F (1995) Forest of Mexico: a diminishing resource. J For 93(7):21–23

Casanova Lugo F, Petit Aldana J, Solorio Sánchez FJ, Parsons D, RamírezAvilés L (2014) Forage yield and quality of *Leucaena leucocephala* and *Guazuma ulmifolia* in mixed and pure fodder banks systems in Yucatan, Mexico. Agrofor Syst 88(1):29–39

CGIAR (2004) Biological nitrogen fixation for increased crop productivity, enhanced human health and sustained soil fertility. www.cgiar.org/pdf/cpnitrogen.pdf

Chirwa PW, Black CR, Ong CK, Maghembe JA (2003) Tree and crop productivity in gliricidia/maize/pigeonpea cropping systems in southern Malawi. Agrofor Syst 59(3):265–277

Crews TE, Peoples MB (2004) Legume versus fertilizer sources of nitrogen: ecological tradeoffs and human needs. Agric Ecosyst Environ 102(3):279–297

Cuartas CC, Naranjo RJ, Tarazona MA, Murgueitio RE, Chará OJ, Ku VJ, Solorio SF, Flores EMX, Solorio SB, Barahona RR (2014) Contribution of intensive silvopastoral systems to animal performance and to adaptation and mitigation of climate change. Rev Colombiana Ciencia Pecuaria 27(2):76–94

Dadhich RK, Meena RS (2014) Performance of Indian mustard (*Brassica juncea* L.) in response to foliar spray of thiourea and thioglycollic acid under different irrigation levels. Indian J Ecol 41(2):376–378

Dadhich RK, Meena RS, Reager ML, Kansotia BC (2015) Response of bio-regulators to yield and quality of Indian mustard (*Brassica juncea* L. Czernj. and Cosson) under different irrigation environments. J Appl Natl Sci Found 7(1):52–57

Dagang AB, Nair PKR (2003) Silvopastoral research and adoption in Central America: recent findings and recommendations for future directions. Agrofor Syst 59(2):149–155

Dakora FD, Keya SO (1997) Contribution of legume nitrogen fixation to sustainable agriculture in Sub-Saharan Africa. Soil Biol Biochem 29(5–6):809–817

Dalzell SA, Shelton HM, Mullen BF, Larsen PH, Mclaughlin KG (2006) Leucaena: a guide to establishment and management. Meat & Livestock Australia Ltd, Sydney. 70 p

Datta R, Baraniya D, Wang YF, Kelkar A, Moulick A, Meena RS, Yadav GS, Ceccherini MT, Formanek P (2017a) Multi-function role as nutrient and scavenger off reeradical in soil. Sustain MDPI 9:402. https://doi.org/10.3390/su9081402

Datta R, Kelkar A, Baraniya D, Molaei A, Moulick A, Meena RS, Formanek P (2017b) Enzymatic degradation of lignin in soil: a review. Sustain MDPI 9:1163. https://doi.org/10.3390/su9071163. 1–18

Dhakal Y, Meena RS, De N, Verma SK, Singh A (2015) Growth, yield and nutrient content of mungbean (*Vigna radiata* L.) in response to INM in eastern Uttar Pradesh, India. Bangladesh J Bot 44(3):479–482

Dhakal Y, Meena RS, Kumar S (2016) Effect of INM on nodulation, yield, quality and available nutrient status in soil after harvest of green gram. Legum Res 39(4):590–594

Don GK (2012) Estimation of net gain of soil carbon in a nitrogen-fixing tree and crop intercropping system in sub-Saharan Africa: results from re-examining a study. Agrofor Syst 86(2):175–184

Dreyfus B, Diem H, Dommergues Y (1988) Future directions for biological nitrogen fixation research. Plant Soil 108(1):191–199

Dubeux JC Jr, Muir JP, Nair PR, Sollenberger LE, Silva HM, Mello AD (2015) The advantages and challenges of integrating tree legumes into pastoral systems. In: Proceedings of the 1st international conference on forages in warm climates. Universidade Federal de Lavras, Lavras, pp 141–164

Eaglesham ARJ, Ayanaba R, Rao RV, Eskew DL (1981) Improving the nitrogen nutrition of Maize by intercropping with cowpea. Soil Biol Biochem 13:169–171

Fagbola O, Osonubi O, Mulongoy K (1998) Contribution of arbuscular mycorrhizal (AM) fungi and hedgerow trees to the yield and nutrient uptake of cassava in an alley-cropping system. J Agric Sci 131(1):79–85

FAO (2008) Current world fertilizer trends and outlook to 2011/12. Rome, Italia

FAO (2012) Livestock: policy brief, cattle ranching and deforestation. http://www.fao.org/3/a-a0262e

FAO (2013) FishStatJ—Software for fishery statistical time series. Fisheries and Aquaculture Information and Statistics Service, Food and Agriculture Organization of the United Nations, Rome. www.fao.org/fishery/statistics/software/fishstatj/en. Release data March 2013

FAOSTAT (2014) Food and Organization of the United Nations, Statistics Division. http://faostat3.fao.org/home

Forrester D, Bauhus J, Khanna PK (2004) Growth dynamics in a mixed-species plantation of *Eucalyptus globulus* and *Acacia mearnsii*. For Ecol Manag 193:81–95

Franco AA, De Faria SM (1997) The contribution of N2-fixing tree legumes to land reclamation and sustainability in the tropics. Soil Biol Biochem 29(5–6):897–903

Gathumbi SM, Cadisch G, Buresh RJ, Giller KE (2003) Subsoil nitrogen capture in mixed legume stands as assessed by deep nitrogen-15 placement. Soil Sci Soc Am J 67(2):573–582

Giller K (2001) Nitrogen fixation in tropical cropping systems, 2nd edn. CABI Publishing, Wallinford, pp 222–250

Giller KE, Cadisch G (1995) Future benefits from biological nitrogen fixation: an ecological approach to agriculture. In: Management of biological nitrogen fixation for the development of more productive and sustainable agricultural systems. Springer, Dordrecht, pp 255–277

Giller K, Beare M, Lavelle P, Izac A, Swift M (1997) Agricultural intensification, soil biodiversity and agroecosystem function. Appl Soil Ecol 6(1):3–16

Gogoi N, Baruah KK, Meena RS (2018) Grain legumes: impact on soil health and agroecosystem. In: Meena RS et al (eds) Legumes for soil health and sustainable management. Springer, Singapore. https://doi.org/10.1007/978-981-13-0253-4_16
Goh K, Mansur I, Mead D, Sweet G (1996) Biological nitrogen fixing capacity and biomass production of different understorey pastures in a Pinusradiata-pasture agroforestry system in New Zealand. Agrofor Syst 34(1):33–49
González Torres A, Figueroa Viramontes U, Delgado J, Núñez Hernández G, Cueto Wong J, Preciado Rangel P, Palomo Gil A (2009) Calibración del spad-502 para evaluar requerimientos de nitrógeno en maíz forrajero. Terra Latinoamericana 27(4):303–309
Graham PH, Vance CP (2000) Nitrogen fixation in perspective: an overview of research and extension needs. Field Crop Res 65(2–3):93–106
Graham PH, Vance CP (2003) Legumes: importance and constraints to greater use. Plant Physiol 131(3):872–877
Grande D, de Leon F, Nahed J, Perez Gil F (2010) Importance and function of scattered trees in pastures in the Sierra region of Tabasco, Mexico. Res J Biol Sci 5(1):75–87
Greenland DJ (1975) Bringing the green revolution to the shifting cultivator. Science 190(4217):841–844
Herrero M, Petr H, Hugo V, An Notenbaert MCR, Philip K, Thornton MB, Franz W, Delia G, Michael O (2013) Biomass use, production, feed efficiencies, and greenhouse gas emissions from global livestock systems. Proc Natl Acad Sci 110(52):20888–20893
IFA, IFDC, IPI, PPI, FAO (2002) Fertilizer use by crop. Food and Agriculture Organization, Rome
Intergovernmental Panel on Climatic Change (IPCC) (2007) Mitigation of climate change. Contribution of working group III to the fourth assessment report of the intergovernmental panel on climatic change, Technical Summary. WMO-UNEP. Cambridge University Press, Cambridge
Jiménez TJ, Ibrahim M, Pezo D, Guevara HF, Gómez CH, NahedToral J, Pinto R (2011) Comparison of animal productivity and profitability between a silvopastoral system (*Brachiariabrizanta* associated with *Leucaena leucocephala*) and a conventional system (*B. brizantha*+chicken manure). Res J Biol Sci 6(2):75–81
Kakraliya SK, Singh U, Bohra A, Choudhary KK, Kumar S, Meena RS, Jat ML (2018) Nitrogen and legumes: a meta-analysis. In: Meena RS et al (eds) Legumes for soil health and sustainable management. Springer, Singapore. https://doi.org/10.1007/978-981-13-0253-4_9
Kamara AY, Akobundu IO, Sanginga N, Jutzi SC (2000) Effect of mulch from selected multipurpose trees (MPTs) on growth, nitrogen nutrition and yield of maize (Zeamays L.). J Agron Crop Sci 184(2):73–80
Khanna PK (1997) Comparison of growth and nutrition of young monocultures and mixed stands of *Eucalyptus globulus* and *Acacia mearnsii*. For Ecol Manag 94:105–113
Ku VJC, Aguilar PC, Ayala BA, Solorio SF, Ramírez AL, Briceño PE (2012) Alternativas nutricionales para incrementar la productividad y la calidad de la carne y de la leche de los rumiantes en el trópico. En, XXXVI Congreso Nacional de Buiatria, Mérida
Ku VJ, Briceño E, Ruiz A, Mayo R, Ayala A, Aguilar C, Solorio FJ, Ramírez L (2014) Manipulation of the energy metabolism of ruminants in the tropics: options for improving meat and milk production and quality. Cuba J Agric Sci 48(1):43–53
Kumar S, Meena RS, Pandey A, Seema (2017a) Soil acidity management and an economics response of lime and sulfur on sesame in an alley cropping system. Int J Curr Microbiol Appl Sci 6(3):2566–2573
Kumar S, Meena RS, Yadav GS, Pandey A (2017b) Response of sesame (*Sesamum indicum* L.) to sulphur and lime application under soil acidity. Int J Plant Soil Sci 14(4):1–9
Kumar S, Meena RS, Bohra JS (2018a) Interactive effect of sowing dates and nutrient sources on dry matter accumulation of Indian mustard (*Brassica juncea* L.). J Oilseed Brassica 9(1):72–76
Kumar S, Meena RS, Lal R, Yadav GS, Mitran T, Meena BL, Dotaniya ML, EL-Sabagh A (2018b) Role of legumes in soil carbon sequestration. In: Meena RS et al (eds) Legumes for soil health and sustainable management. Springer, Singapore. https://doi.org/10.1007/978-981-13-0253-4_4

Kurppa M, Leblanc H, Nygren P (2010) Detection of nitrogen transfer from $N_2$-fixing shade trees to cacao saplings in 15N labelled soil: ecological and experimental considerations. Agrofor Syst 80(2):223–239
Layek J, Das A, Mitran T, Nath C, Meena RS, Singh GS, Shivakumar BG, Kumar S, Lal R (2018) Cereal+legume intercropping: an option for improving productivity. In: Meena RS et al (eds) Legumes for soil health and sustainable management. Springer, Singapore. https://doi.org/10.1007/978-981-13-0253-4_11
Lenné JM, Thomas D (2006) Integrating Crop—Livestock Research and Development in Sub-Saharan Africa: option, imperative or impossible? Outlook Agric 35(3):167–175
Lessa ACR, Madari BE, Paredes DS, Boddey RM, Urquiaga S, Jantalia CP, Alves BJ (2014) Bovine urine and dung deposited on Brazilian savannah pastures contribute differently to direct and indirect soil nitrous oxide emissions. Agric Ecosyst Environ 190:104–111
Liu Y, Pan X, Li J (2015) A 1961–2010 record of fertilizer use, pesticide application and cereal yields: a review. Agron Sustain Dev 35(1):83–93
Mafongoya P, Giller K, Odee D, Gathumbi S, Ndufa S, Sitompul S (2004) Benefiting from $N_2$-fixation and managing rhizobia. In: Noordwijk V, Cadisch G, Ong CK (eds) Below-ground interactions in tropical agroecosystems. Concepts and models with multiple plant components. CABI, Wallingford, pp 227–242
Mayo ER, Utrilla M, Aguilar PC, Solorio SF, Ayala BA, Briceño PE, Ramírez AL, Ku VJ (2013) Productive performance of growing cattle grazing a silvopastoral system with *Leucaena leucocephala*. In: 22nd International Grasslands Congress. Sydney, Australia, 15–19, September 2013
McAllister CH, Beatty PH, Good AG (2012) Engineering nitrogen use efficient crop plants: the current status. Plant Biotechnol J 10(9):1011–1025
Meena RS, Lal R (2018a) Legumes and sustainable use of soils. In: Meena RS et al (eds) Legumes for soil health and sustainable management. Springer, Singapore. https://doi.org/10.1007/978-981-13-0253-4_1
Meena RS, Lal R (2018b) Legumes and sustainable use of soils. In: Legumes for soil health and sustainable management. Springer, Singapore, pp 1–31
Meena H, Meena RS (2017) Assessment of sowing environments and bio-regulators as adaptation choice for clusterbean productivity in response to current climatic scenario. Bangladesh J Bot 46(1):241–244
Meena RS, Yadav RS (2014) Phonological performance of groundnut varieties under sowing environments in hyper arid zone of Rajasthan, India. J Appl Nat Sci 6(2):344–348
Meena RS, Yadav RS (2015) Yield and profitability of groundnut *(Arachis hypogaea* L) as influenced by sowing dates and nutrient levels with different varieties. Legum Res 38(6):791–797
Meena RS, Yadav RS, Meena VS (2014) Response of groundnut (*Arachis hypogaea* L.) varieties to sowing dates and NP fertilizers under Western Dry Zone of India. Bangladesh J Bot 43(2):169–173
Meena RS, Dhakal Y, Bohra JS, Singh SP, Singh MK, Sanodiya P (2015a) Influence of bioinorganic combinations on yield, quality and economics of Mungbean. Am J Exp Agric 8(3):159–166
Meena RS, Meena VS, Meena SK, Verma JP (2015b) The needs of healthy soils for a healthy world. J Clean Prod 102:560–561
Meena RS, Meena VS, Meena SK, Verma JP (2015c) Towards the plant stress mitigate the agricultural productivity: a book review. J Clean Prod 102:552–553
Meena RS, Yadav RS, Meena H, Kumar S, Meena YK, Singh A (2015d) Towards the current need to enhance legume productivity and soil sustainability worldwide: a book review. J Clean Prod 104:513–515
Meena RS, Yadav RS, Reager ML, De N, Meena VS, Verma JP, Verma SK, Kansotia BC (2015e) Temperature use efficiency and yield of groundnut varieties in response to sowing dates and fertility levels in Western Dry Zone of India. Am J Exp Agric 7(3):170–177
Meena H, Meena RS, Singh B, Kumar S (2016a) Response of bio-regulators to morphology and yield of clusterbean [*Cyamopsis tetragonoloba* (L.) Taub.] under different sowing environments. J Appl Nat Sci 8(2):715–718

Meena RS, Bohra JS, Singh SP, Meena VS, Verma JP, Verma SK, Shiiag SK (2016b) Towards the prime response of manure to enhance nutrient use efficiency and soil sustainability a current need: a book review. J Clean Prod 112:1258–1260

Meena RS, Gogaoi N, Kumar S (2017a) Alarming issues on agricultural crop production and environmental stresses. J Clean Prod 142:3357–3359

Meena RS, Kumar S, Pandey A (2017b) Response of sulfur and lime levels on productivity, nutrient content and uptake of sesame under guava (*Psidium guajava* L.) based agri-horti system in an acidic soil of eastern Uttar Pradesh, India. J Crop Weed 13(2):222–227

Meena RS, Meena PD, Yadav GS, Yadav SS (2017c) Phosphate solubilizing microorganisms, principles and application of microphos technology. J Clean Prod 145:157–158

Meena H, Meena RS, Lal R, Singh GS, Mitran T, Layek J, Patil SB, Kumar S, Verma T (2018a) Response of sowing dates and bio regulators on yield of clusterbean under current climate in alley cropping system in eastern U.P., India. Legum Res 41(4):563–571

Meena RS, Kumar V, Yadav GS, Mitran T (2018b) Response and interaction of *Bradyrhizobium japonicum* and *Arbuscular mycorrhizal* fungi in the soybean rhizosphere: a review. Plant Growth Regul 84:207–223

Meena BL, Fagodiya RK, Prajapat K, Dotaniya ML, Kaledhonkar MJ, Sharma PC, Meena RS, Mitran T, Kumar S (2018c) Legume Green manuring: an option for soil sustainability. In: Meena RS et al (eds) Legumes for soil health and sustainable management. Springer, Singapore. https://doi.org/10.1007/978-981-13-0253-4_12

Mekonnen MM, Hoekstra AY (2012) A global assessment of the water footprint of farm animal products. Ecosystems 15(3):401–415

Mercado JAR, Van NM, Cadisch G (2011) Positive nitrogen balance of Acacia mangium woodlots as fallows in the Philippines based on 15N natural abundance data of N2 fixation. Agrofor Syst 81(3):221–233

Mitran T, Meena RS, Lal R, Layek J, Kumar S, Datta R (2018) Role of soil phosphorus on legume production. In: Meena RS et al (eds) Legumes for soil health and sustainable management. Springer, Singapore. https://doi.org/10.1007/978-981-13-0253-4_15

Mthembu BE, Everson CS, Everson TM (2018) Tree legumes-temperate grass agroforestry system effects on inorganic soil nitrogen as ecosystem services provision for smallholder farming systems in South Africa. J Crop Improv 32(2):141–155

Murgueitio E, Calle Z, Uribe F, Calle A, Solorio B (2011) Native trees and shrubs for the productive rehabilitation of tropical cattle ranching lands. For Ecol Manag 261(10):1654–1663

Murgueitio R, Chará J, Solarte A, Uribe F, Zapata C, Rivera J (2013) Agroforestería Pecuaria y Sistemas Silvopastoriles Intensivos (SSPi) para la adaptación ganadera al cambio climático con sostenibilidad. Revista Colombiana de Ciencias Pecuarias 26:313–316

Murgueitio E, Barahona R, Chará J, Flores M, Mauricio R, Moli J (2015) The intensive silvopastoral systems in Latin America sustainable alternative to face climatic change in animal husbandry. Los Sistemas silvopastoriles intensivos en América Latina alternative sostenible Para enfrentar el cambio climáticoen la ganadería. Cuba J Agric Sci 49(4):541

Parrota JA (1999) Productivity, nutrient cycling, and succession in single- and mixed-species plantations of *Casuarina equisetifolia*, *Eucalyptus robusta*, and *Leucaena leucocephala* in Puerto Rico. For Ecol Manag 124:45–77

Parsons R (2004) Plant – microbe metabolism. Revista Científica UDO Agrícola 4(1):1–20

Pelletier N, Tyedmers P (2010) Forecasting potential global environmental costs of livestock production 2000–2050. PNAS 107(43):18371–18374

Peoples M, Herridge D (1999) Qualification of biological nitrogen fixation in agricultural system. In: Pedrosa FO, Hungria M, Yates G, Newton W (eds) Nitrogen fixation: from molecular to crop productivity. Kluwer Academic Publishers, Dordrecht, pp 519–524

Peoples M, Brockwell J, Herridge D, Rochester I, Alves B, Urquiaga S, Boddey R, Dakora F, Bhattarai S, Maskey S, Sampet C, Rerkasem B, Khan D, Hauggaard Nielsen H, Jensen E (2009) The contributions of nitrogen-fixing crop legumes to the productivity of agricultural systems. Symbiosis 48:1–17

Ram K, Meena RS (2014) Evaluation of pearl millet and mungbean intercropping systems in Arid region of Rajasthan (India). Bangladesh J Bot 43(3):367–370

Rerkasem B, Rerkasem K, Peoples MB, Herridge DF, Bergersen FJ (1988) Measurement of nitrogen in maize (*Zea mays* L.) –ricebean (*Vigna umbellata* (Thunb) Ohui and Ohashi) intercrop. Plant Soil 1(08):125–135

Rodrigues A, Silveira J, Bonifacio A, Figueiredo M (2013) Metabolism of nitrogen and carbon: optimization of biological nitrogen fixation and cowpea development. Soil Biol Biochem 67:226–234

Rothe A, Binkley D (2001) Nutritional interactions in mixed species forests: a synthesis. Can J For Res 31:1855–1870

Sanchez PA (1995) Science in agroforestry. Agrofor Syst 30:5–55

Sarabia SL (2013) Efecto de la frecuencia de poda en *Leucaena leucocephala* y *Panicum maximum* sobre la fijación y transferencia de nitrógeno atmosférico en sistemas silvopastoriles intensivos. [tesis maestría]. Campus de Ciencias Biológicas y Agropecuarias, Universidad Autónoma de Yucatán, México

Schroth G, Sinclair FL (2003) Impact of trees on the fertility of agricultural soils. In: Trees, crops and soil fertility: concepts and research methods. CABI Publishing, Wallingford

Schroth G, Lehmann J, Rodrigues MRL, Barros E, Macêdo JLV (2001) Plant-soil interactions in multistrata agroforestry in the humid tropics. Agrofor Syst 53:85–102

Sierra J, Daudin D, Domenach A, Nygren P (2007) Desfontaines nitrogen transfer form signature of tree to the associated grass estimated by the isotopic signature of tree root exudates: a comparison of the $^{15}N$ leaf feeding and natural $^{15}N$ abundance methods. Eur J Agron 27(2):178–186

Sihag SK, Singh MK, Meena RS, Naga S, Bahadur SR, Gaurav YRS (2015) Influences of spacing on growth and yield potential of dry direct seeded rice (*Oryza sativa* L) cultivars. Ecoscan 9(1–2):517–519

Silveira ML, Liu K, Sollenberger Follet LERF and Vendramini JMB (2013) Short-term effects of grazing intensity and nitrogen fertilization on soil organic carbon pools under perennial grass pastures in the southeastern USA. Soil Biol Biochem 58:42–49. https://doi.org/10.1016/j.soilbio.2012.11.003

Sofi PA, Baba ZA, Hamid B, Meena RS (2018) Harnessing soil Rhizobacteria for improving drought resilience in legumes. In: Meena RS et al (eds) Legumes for soil health and sustainable management. Springer, Singapore. https://doi.org/10.1007/978-981-13-0253-4_8

Solorio SF (2005) Soil fertility and nutrient cycling in pure and mixed fodder bank systems using leguminous and non-leguminous shrubs. [Doctoral thesis]. Institute of Atmospheric and Environmental Science. University of Edinburgh

Solorio SF, Bacab PH, Castillo CJ, Ramírez AL, Casanova LF (2009) Potencial de los Sistemas Silvopastoriles en México. II Congreso sobre Sistemas Silvopastoriles Intensivos. pp 21–30

Solorio SF, Solorio SB., Casanova LF, Ramírez AL, Ayala BA, Ku VJ, Aguilar PC (2012) Situación actual global de la investigación y desarrollo tecnológico en el establecimiento, manejo y aprovechamiento de los sistemas silvopastoriles intensivos. En Solorio SFJ, Sánchez BC, Ku VJ (eds) Memorias del IV Congreso Internacional sobre Sistemas Silvopastoriles Intensivos, Fundación Produce Michoacán, Universidad Autónoma de Yucatán, Morelia

Solorio S, Basu S, Sarabia S, Ayala B, Ramírez A, Aguilar P, Erales V, Ku V, Wright J (2016) The potential of silvopastoral systems for milk and meat organic production in the tropics. In: Nandwani D (ed) Organic farming for sustainable agriculture. Springer, Cham, pp 169–183

Steinfeld H, Gerber P, Wassenaar TD, Castel V, Rosales M, Rosales M, de Haan C (2006) Livestock's long shadow: environmental issues and options. Food and Agriculture Organization, Rome

Stockdale E, Lampkin N, Hovi M, Keatinge R, Lennartsson E, Macdonald D, Padel S, Tattersall E, Wolfe M, Watson C (2001) Agronomic and environmental implications of organic farming systems. Adv Agron 70:261–327

Stocking MA (2003) Tropical soils and food security: the next 50 years. Science 302:1356–1359

Sturludóttir E (2011) Forage quality and yield in grass-legume mixtures in Northern Europe and Canada. Master's thesis, Faculty of Physical Sciences, University of Iceland

Tessema Z, Baars R (2006) Chemical composition, dry matter production and yield dynamics of tropical grasses mixed with perennial forage legumes. Trop Grasslands 40(3):150–156
Thilakarathna MS, McElroy MS, Chapagain T, Papadopoulos YA, Raizada MN (2016) Belowground nitrogen transfer from legumes to non-legumes under managed herbaceous cropping systems. A review. Agron Sustain Dev 36(4):58
Thornton PK, van de Steeg J, Notenbaert A, Herrero M (2009) The impacts of climate change on livestock and livestock systems in developing countries: a review of what we know and what we need to know. Agric Syst 101(3):113–127
Unkovich M, Herridge D, Peoples M, Cadisch G, Boddey B, Giller K, Chalk P (2008) Measuring plant-associated nitrogen fixation in agricultural systems. Australian Centre for International Agricultural Research (ACIAR), Canberra
van Kessel C, Hartley C (2000) Agricultural management of grain legumes: has it led to an increase in nitrogen fixation? Field Crop Res 65:165–181
Van Noordwijk M, Dommergues YR (1990) Root nodulation: the twelfth hypothesis. Agrofor Today 3(3):9–10
Varma D, Meena RS, Kumar S (2017a) Response of mungbean to fertility and lime levels under soil acidity in an alley cropping system in Vindhyan Region, India. Int J Chem Stu 5(2):384–389
Varma D, Meena RS, Kumar S, Kumar E (2017b) Response of mungbean to NPK and lime under the conditions of Vindhyan Region of Uttar Pradesh. Legum Res 40(3):542–545
Vendramini J, Silveira M, Dubeux JC, Sollenberger LE (2007) Environmental impacts and nutrient recycling on pastures grazed by cattle. Rev Bras Zootec 36:139–149
Verma JP, Jaiswal DK, Meena VS, Meena RS (2015a) Current need of organic farming for enhancing sustainable agriculture. J Clean Prod 102:545–547
Verma JP, Meena VS, Kumar A, Meena RS (2015b) Issues and challenges about sustainable agriculture production for management of natural resources to sustain soil fertility and health: a book review. J Clean Prod 107:793–794
Verma SK, Singh SB, Prasad SK, Meena RN, Meena RS (2015c) Influence of irrigation regimes and weed management practices on water use and nutrient uptake in wheat (*Triticum aestivum* L. Emend. Fiori and Paol.). Bangladesh J Bot 44(3):437–442
Waring RH, Schlesinger WH (1985) Forest ecosystems: concepts and management. Academic, New York
Williams DR, Alvarado F, Green RE, Manica A, Phalan B, Balmford A (2017) Land-use strategies to balance livestock production, biodiversity conservation and carbon storage in Yucatán, Mexico. Glob Chang Biol 23(12):5260–5272
Yadav GS, Babu S, Meena RS, Debnath C, Saha P, Debbaram C, Datta M (2017a) Effects of godawariphosgold and single supper phosphate on groundnut (*Arachis hypogaea*) productivity, phosphorus uptake, phosphorus use efficiency and economics. Indian J Agric Sci 87(9):1165–1169
Yadav GS, Lal R, Meena RS, Babu S, Das A, Bhomik SN, Datta M, Layak J, Saha P (2017b) Conservation tillage and nutrient management effects on productivity and soil carbon sequestration under double cropping of rice in North Eastern Region of India. Ecol Indic. http://www.sciencedirect.com/science/article/pii/S1470160X17305617
Yadav GS, Lal R, Meena RS, Datta M, Babu S, Das LJ, Saha P (2017c) Energy budgeting for designing sustainable and environmentally clean/safer cropping systems for rainfed rice fallow lands in India. J Clean Prod 158:29–37
Yadav GS, Das A, Lal R, Babu S, Meena RS, Saha P, Singh R, Datta M (2018a) Energy budget and carbon footprint in a no-till and mulch based rice–mustard cropping system. J Clean Prod 191:144–157
Yadav GS, Das A, Lal R, Babu S, Meena RS, Patil SB, Saha P, Datta M (2018b) Conservation tillage and mulching effects on the adaptive capacity of direct-seeded upland rice (*Oryza sativa* L.) to alleviate weed and moisture stresses in the North Eastern Himalayan Region of India. Arch Agron Soil Sci 64:1254. https://doi.org/10.1080/03650340.2018.1423555
Young A (1997) Agroforestry for soil management, 2nd edn. CAB International, Wallingford

# Enhanced Phosphorus Fertilizer Use Efficiency with Microorganisms

8

Hassan Etesami

**Abstract**

Phosphorus (P) is an essential nutrient in plant development and growth, and its deficiency is one of the major factors limiting crop yields worldwide. Although soils generally possess a large amount of total P (400–1000 mg $kg^{-1}$), only a small ratio (1.00–2.50%) is immediately available for uptake of plants since 75–90% of added P is precipitated by metal–cation (calcium, iron, and aluminum) complexes and quickly becomes fixed in soils. The nature of calcareous soils in the arid and semiarid regions of the world has made P use efficiency (PUE) low (10–25%) in this land. For this reason, farmers have added a significant amount of these chemical fertilizers to the cultivated land to achieve the desired result every year. Low-use efficiency of the P fertilizers and their continuous long-term use have led to environmental pollution. The use of chemical P fertilizers cannot be omitted at this time without intensely diminishing food production. However, it is known that the compound use of phosphate-solubilizing microorganisms (PSMs) and chemical P fertilizers can reduce the negative impacts of overuse of these fertilizers and improve PUE in an efficient and environmentally prudent manner. Among the PSMs, it can be mentioned arbuscular mycorrhizal fungi (AMF) and plant growth-promoting rhizobacteria (PGPR). AMF increase the growth, yield, and absorption of nutrients in the plant mostly by increasing the effective absorptive area of the roots by formation of an extensive extraradical hyphal network, and PGPR also contribute directly to increasing the solubilization of insoluble P compounds in the soil and thereby plant growth through mechanisms like producing organic and inorganic acids, increasing root surface area, and improving beneficial symbiosis with host plants at different stages of plant growth. In addition, it is known that the plants inoculated with a combination

H. Etesami (✉)
Agriculture & Natural resources Campus, Faculty of Agricultural Engineering & Technology, Department of Soil Science, University of Tehran, Tehran, Iran
e-mail: hassanetesami@ut.ac.ir



R. S. Meena (ed.), *Nutrient Dynamics for Sustainable Crop Production*,
https://doi.org/10.1007/978-981-13-8660-2_8

of PGPR and AMF can express synergistic effect to augment plant growth indices while maintaining safe natural resources such as P stocks. This chapter is a critical summary of the efforts in using phosphate-solubilizing bacteria (PSB) and phosphate- solubilizing-AMF for augmenting the use efficiency of P fertilizers.

**Keywords**
Arbuscular mycorrhizal fungi · Bio-fertilizer · Compound use · Fertilizer use efficiency · PGPR · Sustainable agriculture

## Abbreviations

| | |
|---|---|
| ACC | 1-Aminocyclopropane-1-carboxylate |
| AMF | Arbuscular mycorrhizal fungi |
| CK | Cytokinins |
| GAs | Gibberellins |
| IAA | Indole-3-acetic acid |
| NFR | $N_2$-fixing rhizobia |
| N | Nitrogen |
| P | Phosphorus |
| PGP | Plant growth promoting |
| PGPEB | Plant growth-promoting endophytic bacteria |
| PGPR | Plant growth-promoting rhizobacteria |
| PSB | Phosphate-solubilizing bacteria |
| PSF | Phosphorus-solubilizing fungi |
| PSMs | Phosphate-solubilizing microorganisms |
| PUE | P use efficiency |
| RP | Rock phosphate |
| TCP | Tricalcium phosphate |

## 8.1 Introduction

Soil is one of the largest vital systems in the planet Earth and will face major challenges in the coming decades. By 2050, the human population is expected to exceed 9.7 billion (Nations 2015), which will result in up to 50% increase in food and fuel demand. Given this issue, supplying food will face a huge challenge for many parts of the world. In general, with the growing world population, more food is needed to be produced via intensive agriculture, which requires large quantities of fertilizer (Fallah Nosratabad et al. 2017; Vitousek et al. 1997; Ashoka et al. 2017). Chemical fertilizers are essential components of modern agriculture because they provide necessary plant macro- and micronutrients (Adesemoye and Kloepper 2009; Martinez 2010; Kumar et al. 2017b). Phosphorus (P) fertilizers are one of these fertilizers. Phosphorus is a necessary nutrient and the most limiting element after nitrogen (N) for plants. Phosphorus plays several key roles in the plant, including participation in energy transfer reactions, root expansion, root and stem strength, flower and seed formation, photosynthesis, molecular nitrogen ($N_2$) fixation in legumes, product quality, resistance to plant diseases, deformation of sugar into starch, and transfer of genetic traits in plants (Cockefair 1931). In addition, P is part of the protein of cells and plays a special role as part of the nucleus protein, cell membrane, and nucleic acids (Cockefair 1931; Theodorou and Plaxton 1993). Accordingly, sufficient P nutrition is necessary for proper growth and yield of all plants (Cockefair 1931).

There are large amounts of P in the form of apatite minerals, which are the original source of all P, complexes of iron(III) phosphate ($FePO_4$), aluminum phosphate ($AlPO_4$) and calcium phosphate ($Ca_3(PO_4)_2$), and P adsorbed on clay particles in the soil. In addition, organic P, which originates from organic sources like microbial residues, manures, and plant tissues (as inositol phosphatases (phytate), phosphoesters, phospholipids and nucleic acids (phosphodiesters), and phosphotriesters), accounts for up to 30–65% of total P in soils (Islam and Hossain 2012; Rodríguez and Fraga 1999). Despite the high amount of P in the soil (400–1000 mg $kg^{-1}$), only a very low concentration of P (1.00–2.50%) is available to plants (Chen et al. 2008; Meena and Meena 2017). Since mineral and organic P are immobilized and mostly unavailable, many soils are actually P-deficient (Adesemoye and Kloepper 2009; Dey 1988; Meena and Lal 2018). Phosphorus is absorbed by the roots of plants from the soil solution mainly as orthophosphate ions ($H_2PO_4^-$ and $HPO_4^{2-}$). Phosphorus in pH 5.5–7 is relatively available to plant, respectively, at soil pH less than 5.5 and more than 7, due to the high reactivity of P with some metal complexes such as Fe, Al, and Ca resulting in the precipitation or adsorption of between 74 and 90% of P in the soil (Gyaneshwar et al. 2002; Leytem and Mikkelsen 2005; Yadav et al. 2018a). In general, the solubility of these P compounds (Ca–P, Fe–P, and Al–P), as well as organic P, is extremely low, and only very small amounts of soil P are in solution at any one time. In addition to soil pH, the amount of plant available P in the soil is controlled by other factors such as Ca ion concentration, soil organic matter content, type and amount of clay, soil moisture, soil texture, secretion, and density of root (Al-Rohily et al. 2013; Meena and Yadav 2015).

In order to compensate for the shortage of P, large quantities of phosphate fertilizers are added to the soil by farmers annually. The majority of P fertilizers are absorbed by solid particles and stored in a solid phase of soil (Fallah Nosratabad et al. 2017; Leytem and Mikkelsen 2005; Buragohain et al. 2017; Kumar et al. 2017a). In calcareous soils, such as Iran's soils, which have evolved in dry and semiarid climates, high pH, high calcium carbonate content, the low amounts of organic matter, and low soil moisture (drought stress) have caused the amount of plant available P to be less than the amount of P needed to provide the optimal growth of most agricultural products (Al-Rohily et al. 2013; Leytem and Mikkelsen 2005; Meena et al. 2015d). The use of chemical fertilizers containing this nutrient, especially superphosphates, which are one of the common ways of compensating for the deficiency of this nutrient in soil, is not very effective in calcareous and alkaline soils because most of the P in the fertilizer, after entering the soil, gradually turns into insoluble form and is stored in a plant nonavailable form in the soil (Leytem and Mikkelsen 2005; Yadav et al. 2017c; Dadhich and Meena 2014).

Although the application of chemical fertilizers including P fertilizers, as the best means to resolve P deficiency in crop plants (Etesami and Maheshwari 2018; Meena and Yadav 2014); Dhakal et al. 2015), initially has had an impact on the increase in yield, the excessive use of these inputs has led to a reduction in soil fertility, environmental degradation, and unexpected harmful environmental effects such as surface runoff of P, changes in the food web, eutrophication of aquatic ecosystems, and reduction in biodiversity (Adesemoye and Kloepper 2009; Verma et al. 2015c). In addition, the use efficiency of chemical fertilizers is now theoretically up to its highest level, which means that more use of chemical fertilizers can hardly increase yields. It has been known that the use efficiency of P fertilizers in calcareous and alkaline soils does not exceed 20%. Sometimes up to 90% of applied P in soil is precipitated by metal complexes in the soil (Gyaneshwar et al. 2002; Meena et al. 2015e; Yadav et al. 2018b).

Although most plants (P-efficient plants) have evolved diverse array of strategies to uptake sufficient P under P- restricting conditions and cope with P-stressed conditions (i.e., carbon metabolism, modifications to root morphology, exudation of organic and inorganic acids, protons, and enzymes of acid phosphatase, membrane structure, etc.) (Islam and Hossain 2012; Karthikeyan et al. 2002; Lambers et al. 2006, 2015; Mudge et al. 2002), studies have shown that this strategy cannot meet the plant's need for this nutrient (Etesami and Beattie 2017; Meena et al. 2016a).

The phosphorus mobility in soil is very low and cannot respond to the rapid absorption of P by plant. This leads to the emergence and development of phosphate-depleted areas adjacent to the contact surface of roots with soil. Therefore, the plant needs an auxiliary system that can easily go beyond these P-depleted areas and, by developing a wide network around the root system, receive P from a larger volume of adjacent soil (Etesami and Beattie 2017; Lambers et al. 2006; Varma et al. 2017a; Datta et al. 2017a).

Biological fertilizers are considered to be the most effective plant assistants for the supply of P at the optimal level, which are prepared based on the selection of a variety of useful soil microorganisms (Etesami and Maheshwari 2018; Meena et al. 2018b). Today, biological fertilizers are considered as a supplement for chemical fertilizers aimed at increasing soil fertility and producing agricultural products in sustainable agriculture (Adesemoye and Kloepper 2009; Sihag et al. 2015; Kumar et al. 2018a). Biological fertilizers have significant advantages in comparison with chemical substances, including that they do not produce toxic substances in the food cycle, have self-replicating properties, and cause soil physical and chemical properties to be improved (Al Abboud et al. 2014; Singh et al. 2016; Yadav et al. 2017b). Generally, the microorganisms used to produce biological fertilizers originate from soil and are active in most soils. However, in many cases, their quantity and quality are not optimal, and therefore the use of their inoculations is necessary. In these biological fertilizers, the cell population density is such that it can provide up to more than one million living cells for each inoculated plant, while naturally there is no such number of bacteria, especially in the plant's rhizosphere (Etesami and Maheshwari 2018; Nosratabad et al. 2017; Meena et al. 2015c).

The role of phosphate-solubilizing microorganisms (PSMs) is well-known in solubilizing insoluble phosphates and increasing its availability to plants (Sharma et al. 2013; Kumar et al. 2018b). Although there are several PSMs in the soil, usually this number of bacteria is not noticeable in comparison with other common bacteria and is located in the rhizosphere of different plants (Rodríguez and Fraga 1999). Less than 10% of all microorganisms in the soil are able to dissolve insoluble phosphates (Gupta et al. 1998). Therefore, the amount of P released by these bacteria is not usually sufficient to increase the growth of the plants. Therefore, inoculation of plants with specific bacteria with a much larger population than that found in the soil is necessary to benefit from the P-solubilizing properties of those bacteria in increasing plant growth and yield considerably (Etesami and Maheshwari 2018; Meena et al. 2016b).

PGPR (plant growth-promoting rhizobacteria), the bacteria that colonize plant roots and promote plant growth, and AMF (arbuscular mycorrhizal fungi), which are composed of a group of root obligate biotrophs (obligate symbiosis) that barter mutual benefits with about 80% of plants, augment the availability of micro- and macronutrients to growing plants by influencing solubility or uptake conditions (i.e., augmenting the solubility of P and Fe) (Berruti et al. 2016; Vessey 2003). Under conditions of nutritional deficiencies, these microorganisms augment the availability of micro- and macronutrients by different ways. Tolerance to nutrient deficiency stress can be explained by nutrient mobilization in the rhizosphere and via generation of phytohormones especially IAA (indole-3-acetic acid), ACC (1-aminocyclopropane-1-carboxylate) deaminase activity, siderophore production, and phosphate solubilization (Etesami and Beattie 2017; Glick 2014). Plant-associated beneficial microorganisms can be applied to make better the ability of crop plants to withstand and produce yield in nutrient-poor growth environments (Etesami 2018b; Etesami and Maheshwari 2018). As an example, PGPR- IAA-induced changes in root architecture might lead to an enlargement in total root surface area, consequently improving micro- and macronutrients and water uptake, which may have positive effects on plant growth in a general sense (Etesami et al. 2015a; Etesami and Alikhani 2016a; Etesami and Maheshwari 2018; Glick 2012; Somers et al. 2008).

There are several reports that show the potential of various bacterial strains to solubilize insoluble inorganic phosphates such as TCP (tricalcium phosphate), dicalcium phosphate ($CaHPO_4$), hydroxyapatite [$Ca_5(PO_4)_3(OH)$], and rock phosphate (phosphorite) and mineralize organic phosphates (Islam and Hossain 2012; Khan et al. 2007; Sharma et al. 2013). In addition to enhancing P availability, PSB can, through other mechanisms such as fixing atmospheric nitrogen, producing plant hormones, e.g., such as GAs (gibberellins), CK (cytokinins), and auxins (i.e., IAA), and synthesizing the enzyme ACC deaminase, which lessens plant levels of ethylene, thereby diminishing environmental stresses (abiotic and biotic stresses) on plants, sequestering Fe for plants by production of siderophores, and antifungal activity, improve plant growth (Bianco and Defez 2010; Chabot et al. 1996; Duarah et al. 2011; Hamdali et al. 2008; Islam and Hossain 2012; Naik et al. 2008; Thakuria et al. 2004; Yildirim et al. 2011; Zaidi et al. 2006; Varma et al. 2017b).

The availability of macronutrient elements including P can be a major restriction to plant growth in many agricultural environments of the world, especially the tropics where soils are highly low in macro- and micronutrients (Etesami and Maheshwari 2018). PGPR and AMF take part in the geochemical cycling of micro- and macronutrients and determine their availability for plants and soil microbial community by different action mechanisms (Adesemoye et al. 2008; Desai et al. 2016; Sofi et al. 2018). In this chapter, a summary of the efforts in using PSB and phosphate-solubilizing-AMF for increasing the use efficiency of P fertilizers is discussed. Review of the literature shows that the PSMs as both co-inoculation and single inoculation can take advantage of plant uptake of P and thereby augment the use efficiency of applied chemical P fertilizers.

## 8.2 Plant–Microbe Interactions

Long ago, the study of the interactions between plant and their associated microorganisms (whether beneficial microorganisms or harmful microorganisms) has been very interesting for microbiologists and botanists. The knowledge gained from this research could lead to the development of novel agricultural applications. Plant communities affect soil microorganisms via interactions inside the rhizosphere, the region of soil where microbial communities are directly influenced by plant root systems and exudates (Fig. 8.1) (Berg and Smalla 2009; Buée et al. 2009).

The rhizosphere is a rich niche for diverse microorganisms compared to surrounding bulk soils (Bais et al. 2006). Microbial root colonization often initiates with the recognition of specific compounds in the root exudates by microorganisms. These compounds probably have also major roles in belowground community interactions (Compant et al. 2010). Plant roots secrete a wide range of organic compounds between 6% and 21% of the carbon fixed including sugars (i.e., glucose, xylose, fructose, maltose, sucrose, and ribose), organic acids (i.e., citric, malic, lactic, succinic, oxalic, and pyruvic acids), putrescine, amino acids, fatty acids, nucleotides, and vitamins (Etesami and Maheshwari 2018; Meena et al. 2015a), which can be used as nutrients or signals by microbial populations. These signal molecules can also be used for cross talk between the plant and microbes (Lugtenberg 2015). From the other point of view, plant-associated microorganisms release some metabolites like phytohormones, small molecules, or volatile compounds, which may operate directly or indirectly to actuate plant immunity or adjust plant growth and morphogenesis (Ortíz-Castro et al. 2009).

Recent advance in plant–microorganism interaction research revealed that plants are able to shape their rhizosphere and endorhiza microbiome (Berendsen et al. 2012). Under stress conditions, stressed plants can require the presence of associated microorganisms for their growth and establishment in different ecosystems

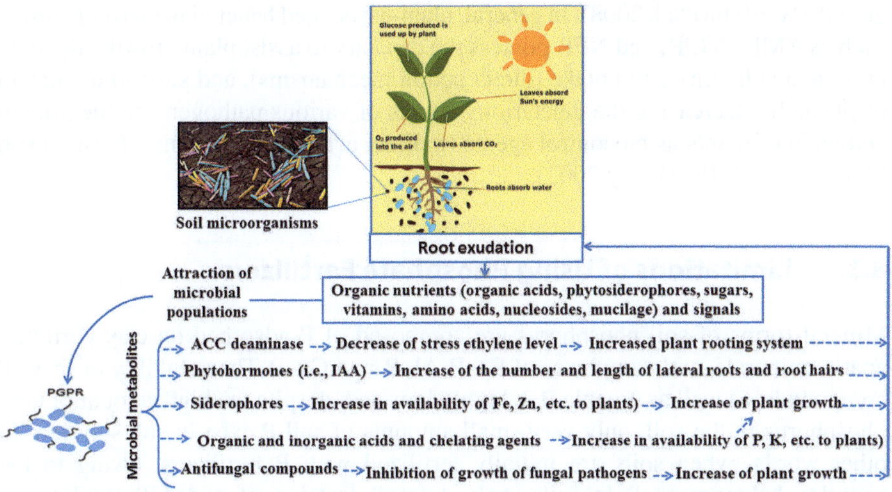


**Fig. 8.1** A diagrammatic representation of how interactions occur between plants and their associated bacteria. Up to 40% of photosynthetically fixed carbon is secreted into the rhizosphere by plants. These carbon materials attract microbial populations, especially those able to metabolize plant-exuded compounds and proliferate in this microbial habitat. Microorganisms can use these compounds as substrates, resulting in an increased microbial biomass and activity around the roots, the so-called rhizosphere effect. Plants can influence bacterial gene expression, especially genes encoding plant-beneficial traits, by releasing these root exudates. Root exudation-mediated plant-associated PGPR can modulate root development and growth through the production of phytohormones (indole-3-acetic acid, IAA), 1-aminocyclopropane-1-carboxylate (ACC) deaminase, siderophores, organic and inorganic acids, etc. The PGPR result in a reduction of the growth rate of primary root and an increase of the number and length of lateral roots and root hairs. PGPR also modify root physiology by changing gene transcription and metabolite biosynthesis in plant cells and thereby increase root exudations, resulting in the microbial activity, and this process continues in a cycle

(Hardoim et al. 2008; Kakraliya et al. 2018). Symbiotic microorganisms exist in all plants living in the natural ecosystems. This relationship may be the key factor involved in plants' stress tolerance ability. Indeed, local adjustment of plants to their growth environment is driven by genetic differentiation in plant closely associated microorganisms (Etesami and Beattie 2018; Rodriguez and Redman 2008). It has been proven that transplanting different plant species in the absence of microorganisms is notoriously difficult, which hints at a role of microorganisms in plant growth under stressful conditions (Leifert et al. 1989).

In general, the microorganisms may affect plant growth in one of three ways. The interaction may be (1) beneficial, such as interaction between plant and AMF, PGPR, plant growth-promoting endophytic bacteria (PGPEB), and the N-fixing rhizobia (NFR), (2) harmful such as the interaction between plant and phytopathogenic microorganisms (plant disease-causing soil microorganisms), and (3) neutral for the plant, and sometimes the impact of a microorganism may vary as the soil conditions alter (Cheng et al. 2010; Yadav et al. 2017a). Majority of plants harbor a diverse community of microorganisms that can positively affect host plant growth (Hardoim et al. 2008). In general, plant-associated beneficial microorganisms such as AMF, PGPR, and NFR possess the capacity to assist plant growth, augment nutrient availability and uptake (direct action mechanisms), and support the health of plants by decreasing the deleterious effects of various pathogens on the growth and yield of plants as biocontrol agents (indirect action mechanisms) (Etesami and Maheshwari 2018; Vessey 2003).

## 8.3 Limitations of Using Phosphate Fertilizers

Mineral forms of soil phosphorus are composed of P adsorbed on clay particles, apatite minerals, and complexes of Fe–P, Al–P, and Ca–P. The solubility of these P compounds, as well as organic P, is highly low, and, despite the high amount of total phosphorus in the soil, only very small amounts of soil P exist in soil solution. In other words, when soils are initially fertilized with P fertilizers, owing to the complex behavior of P in soils, only a small fraction of added P fertilizer to agricultural soils is taken up by plants. In addition to worldwide concern about the energy and costs connected with mining the phosphorite (rock phosphate) and its conveyance from manufacturing sites to farm crop fields (Sharma et al. 2013), the use efficiency of P fertilizers (recovery of fertilizer P) in calcareous and alkaline soils does not exceed 20%. Sometimes up to 90% of applied P in soil is precipitated by metal complexes in the soil (Gyaneshwar et al. 2002). The alteration of plant-available phosphorus to less available forms in soil is the reason for the low initial efficiency of P fertilizers. Soil P availability is affected by soil pH. Phosphorus in pH 6.5 is more available to plants. In other words, pH-dependent chemical fixation determines the quantity of available P. In highly calcareous soils, soluble P readily forms insoluble minerals with calcium ($Ca_3(PO_4)_2$), which is indeed a problem (Bertrand et al. 2003). Soils with high clay content, particularly those dominated by Al- and Fe oxide minerals ($Fe_2O_3$ and $Al_2O_3$), retain P most forcefully. When organic matter is accruing in soil, retention of P in organic matter (P immobilization) is also only an inefficiency process (Williams and Donald 1957).

In the near future, most countries face the energy crisis and environmental hazards due to pollutants so that the process of producing chemical fertilizers may not be easily possible. Although the available P sources in the world are so high that the risk of a critical shortage is not very serious at least until the next century, it is very likely that the excess costs of preparing and producing P fertilizers in the near future are very probable. Therefore, it is needed to reduce application of P fertilizers in agricultural land by PSMs. Mycorrhizal fungi and PSB cannot replace all P fertilizers. However, these microorganisms can reduce the plant's need for chemical P fertilizers by increasing the plant's ability to absorb more P and other mineral elements from the soil and enhancing the efficiency of P fertilizer use. As an example, research has shown that mycorrhizal fungi and PSB could supply the plant with P from an RP source in calcareous soils (Ghorchiani et al. 2018; Dhakal et al. 2016).

## 8.4 Strategies for Increasing Fertilizer Phosphorus Use Efficiency

Some strategies that can improve fertilizer PUE include (McLaughlin 2012) (1) altering timing of fertilizer application. In soils with high P-retention capacities like calcareous soil and acidic soil, P fertilizers must not be applied too long before planting because increase in the time of contact with soil diminishes P availability rapidly in those soils (McLaughlin 2012). It is known that the best time to apply P fertilizers in soils with high fixation capacity is at sowing. Applying small amounts and splitting applications at sowing and topdressing later in the crop growth cycle are managements that must be considered about this element in highly sandy soils, (2) altering rate of fertilizer application. The best and only way to determine the correct rate of P fertilizers to apply is based on soil testing. Adding P fertilizers to soils that contain sufficient amounts of plant-available P is wasteful and could lead to P losses to water bodies (McLaughlin 2012), (3) altering placement in the soil. Since adsorption of P is high in soils with high P retention, band placement of P is the best management practice for soluble P fertilizers because this method of fertilizer application decreases the amount of soil fertilizer contact and limits strong adsorption (McLaughlin 2012). On the other hand, the best method of fertilizer application for sparingly soluble fertilizers like reactive RP is broadcast application because this method of fertilizer application promotes dissolution in the soil, (4) picking out crop species or varieties efficient at scavenging P from soils (McLaughlin 2012). Since P is diffusion-limited in most soils, it is known that genotypes/species with efficient and extensive root systems (to access a greater soil volume) and with effective associations with mycorrhizal fungi are more efficient in taking up P from soils and thereby increase P use efficiency (Lynch 2007) and (5) different fertilizer formulations. Use of acidifying fertilizers in alkaline soils or the compound use of PSM and chemical P fertilizers improves P use efficiency. Among the strategies mentioned above, it seems that microbial mediated P management is an eco-friendly and cost-effective way for sustainable development of agricultural crop (Etesami and Maheshwari 2018; Meena et al. 2014).

## 8.5 Biodiversity of Phosphate-Solubilizing Microorganisms

Both fungi and bacteria play a central role in the natural P cycle and convert insoluble forms of P to available forms to plants. Diverse genera of PSMs inhabit in soil and plant rhizosphere. These microorganisms occur in both fertile and P-deficient soils (Etesami and Maheshwari 2018; Oehl et al. 2001) and were isolated from diverse environment including rhizoplane, rhizosphere, and endorhiza (endosphere) of different plants (Etesami and Alikhani 2016b; Etesami et al. 2014a, b; Islam et al. 2007; Islam and Hossain 2012; Kumar et al. 2001; Oliveira et al. 2009; Panhwar et al. 2011a, b; Pei-Xiang et al. 2012; Sharma et al. 2013) and from salinity and heavy metal-stressed environments (Etesami 2018a; Etesami and Beattie 2018; Etesami and Maheshwari 2018; Zhu et al. 2011; Layek et al. 2018). Soil PSMs include bacterial genera viz. *Micrococcus, Bacillus, Flavobacterium, Enterobacter, Klebsiella, Azotobacter, Vibrio, Chryseobacterium, Xanthobacter, Erwinia, Acinetobacter, Pantoea, Burkholderia, Arthrobacter, Achromobacter, Agrobacterium, Pseudomonas*, fungi of *Aspergillus, Trichoderma, Rhizoctonia solani, Glomus manihotis, Fusarium, Helminthosporium, Alternaria*, and *Penicillium* and some *Actinomyces* such as *Streptomyces* and *Micromonospora*, as well as some cyanobacteria (i.e., *Anabaena* sp., *Calothrix braunii, Nostoc* sp., and *Scytonema* sp.) (Behera et al. 2014; Sharma et al. 2013). It has also been reported that some rhizobial strains can also dissolve organic and inorganic phosphates. Of the bacteria mentioned above, the bacteria with multiple plant growth-promoting (PGP) traits such as *Pseudomonas, Bacillus, Burkholderia, Streptomyces*, and *Pantoea* have been reported among the most efficient PSB as well as important bioinoculants (Islam and Hossain 2012; Rodríguez and Fraga 1999). Bacteria are more effective at solubilizing phosphorus than fungi (Alam et al. 2002; Sharma et al. 2013; Ram and Meena 2014) and play a remarkable role in mediating the transformation of complex form of essential micro- and macronutrient elements into more available form for swift acquisition by the plants (Sharma et al. 2013). PSB comprise 1–50% of the soil microbial population, while phosphorus-solubilizing fungi (PSF) comprise only 0.1–0.5% (Chen et al. 2006; Kucey 1983). In general, P solubilization by microorganisms depends on many factors including nutritional, physiological, and growth condition of the culture (Behera et al. 2014).

## 8.6 Mechanisms of PSB in Increasing P Availability

PSB directly and indirectly contribute to increasing the available P to the plant. In the direct method, the presence of microorganisms is necessary, for example, when the microorganisms increase the P available to the plant by releasing organic and inorganic acids. In an indirect way, the presence of these microorganisms during the increase of plant phosphorus is not necessary. In this case, microorganisms secrete enzymes capable of mineralizing organic P. According to previous findings, there have been some potential mechanisms by which PSB could increase the availability of P (P release from insoluble phosphates) and thereby promote plant growth (Sharma et al. 2013). One of the first mechanisms suggested in the literature is the production of low-molecular-weight organic acids (Goldstein 1986; Kim et al. 1997a). By chelating the cations bound to phosphate through their hydroxyl and carboxyl groups, the released organic acids convert insoluble P forms into soluble forms (Kpomblekou-a and Tabatabai 1994). Different organic acids (in terms of both amount and type) such as oxalic acid, malic acid, succinic acid, propionic acid, 2-ketogluconic acid, 2-hydroxyglutaric acid, formic acid, citric acid, and lactic acid (Chen et al. 2006; Rodríguez and Fraga 1999) have been produced by PSB (i.e., *Acinetobacter* sp., *Sinorhizobium meliloti, Bacillus* spp., *B. megaterium, Burkholderia* sp., *Enterobacter* sp., *E. agglomerans, Microbacterium* sp., *Pseudomonas* sp., *P. fluorescens, P. trivialis, P. poea, Serratia* sp., *Ralstonia* sp., *Pantoea* sp., and *Klebsiella* sp.), but gluconic acid has been reported to be as the principal organic acid produced by these bacteria (Bianco and Defez 2010; Castagno et al. 2011; Islam and Hossain 2012; Ogut et al. 2010; Panhwar et al. 2011a, b; Perez-Lopez et al. 2007; Sharma et al. 2013; Vyas and Gulati 2009).

In some studies, there is a direct relationship between the amount of produced organic acid and the amount of solubilized P, but in some studies, such a relationship has not been observed (Vyas and Gulati 2009). Some researchers reported the solubilization of insoluble phosphates by PSB without producing organic acids (Chen et al. 2006; Illmer and Schinner 1992). These findings suggests that organic acids cannot be the only mechanism for solubilizing phosphorus by bacteria, but mechanisms such as the production of inorganic acids and the release of proton ($H^+$) as a result of absorption of cations such as $NH_4^+$ are also involved in this work (Illmer and Schinner 1992).

An important part of soil P is as organic P, which is in fact not available to the plant. Therefore, these organic P compounds need to be converted into mineral form by enzymes. Secretion of hydrolytic enzymes (e.g., phosphatases and phytases) is another mechanism of PSB such as *Bacillus megaterium* and *S. meliloti* to increase P availability to the plant (Bianco and Defez 2010; Dey et al. 2004; Rodríguez et al. 2006; Sharma et al. 2013; Verma et al. 2015a). Since P mobility in soils is low, it is necessary that the roots move themselves to the sites where P is accumulated. The rooting system is the main channel for water absorption and mineral elements in all plants. One of the known mechanism by which IAA-producing PSB affect P uptake is by increasing development and growth of plant roots, causing root systems with larger surface area and enhanced number of root hairs, which are then able to access more P (Etesami and Beattie 2017; Etesami and Maheshwari 2018).

Application of PSB such as *Bacillus* spp., *Acinetobacter* sp., *B. megaterium*, *Pseudomonas* sp., *P. trivialis*, and *P. poea* alone or in combination with low rate of P fertilizers or with varying doses of P fertilizers has been shown to remarkably augment P availability in soils as well as high P uptake by major crops (Duarah et al. 2011; Gyaneshwar et al. 2002; Ogut et al. 2010; Oliveira et al. 2009; Panhwar et al. 2011a, b; Sapsirisopa et al. 2009; Sharma et al. 2007; Toro et al. 1997; Vyas and Gulati 2009; Yildirim et al. 2011; Meena et al. 2017a), augment the efficiency of P fertilizer, and diminish about 25–50% of the required P to crop plants (Adesemoye et al. 2010; Attia et al. 2009; Duarah et al. 2011; Güneş et al. 2009; Gyaneshwar et al. 2002; Kennedy et al. 2004; Kumar et al. 2010; Yildirim et al. 2011).

Since most of the PSB are heterotrophic and dependent on carbon and energy sources (Nahas 2007), to ensure their growth, organic acid production, and hence solubilization of insoluble phosphate compounds, metabolizable carbon compounds must be applied as an energy source to the PSB (Vassilev and Vassileva 2003; Meena et al. 2015b), especially in soils of arid and semiarid regions. Previous studies have also shown that use of PSB along with organic amendments could be a promising management strategy to increase PUE of insoluble P resources (i.e., RP) for crop production (Abbasi et al. 2013; Adnan et al. 2017; Fallah Nosratabad et al. 2017; Dadhich et al. 2015). Some examples of PSB that have been able to increase P availability from P sources with low P solubility in the presence or absence of an organic amendment are shown in Table 8.1.

**Table 8.1** Role of PSB (phosphate-solubilizing bacteria) in increasing the solubility and availability of P from P sources with low P solubility in the presence or absence of an organic amendment

| PSB | P sources | Effect | References |
|---|---|---|---|
| *Pseudomonas*, *Pantoea*, *Mycobacterium*, *Bacillus*, *Rhizobia*, *Burkholderia*, *Arthrobacter*, and *Enterobacter* | Rock phosphate (RP), single super phosphate (SSP), farmyard manure (FYM), and poultry manure (PM) | PSB could increase Olsen-extractable P in all P sources compared to the control, but this increase was higher in organic sources (PM and FYM) than mineral P sources (SSP and RP) | Adnan et al. (2017) |
| *Bacillus*, *Rhodococcus*, *Arthrobacter*, *Serratia*, *Chryseobacterium*, *Delftia*, *Gordonia*, and *Phyllobacterium* | Tricalcium phosphate (TCP) | These PSB could solubilize considerable amount of TCP in the medium by secreting organic acids | Chen et al. (2006) |
| *Agrobacterium tumefaciens* | Poultry manure (PM) and rock phosphate (RP) | The combined use of phosphate-solubilizing bacterium and PM with RP increased Olsen-extractable P (25 mg P $kg^{-1}$) that was maintained at high levels without any loss | Abbasi et al. (2015) |
| *Pantoea cypripedii* and *P. plecoglossicida* | Rock phosphate (RP) | The combined use of PSB and RP increased the growth indices and total P uptake in maize and wheat crops compared to control | Kaur and Reddy (2015) |
| *Bacillus* spp. | Rock phosphate (RP) | PSB solubilized significantly high amounts of P (20.05–24.08 mg $kg^{-1}$) compared to control (19–23.10 mg $kg^{-1}$) treatments. | Panhwar et al. (2011b) |
| *Bacillus* sp. | Rock phosphate (RP) | The application of PSB enhanced soluble P in the soil solution | Panhwar et al. (2013) |
| *Bacillus* sp. | Rock phosphate (RP) and compost | PSB along with compost indicated an increase of 12.9% and 4.3% in P contents in straw and grains of chickpea, respectively, compared to control | Ditta et al. (2018) |
| *Pseudomonas*, *Azospirillum*, and *Agrobacterium* | Poultry manure (PM), rock phosphate (RP), and compost | PSB along with PM and compost resulted in more increase in wheat plant yield, P uptake, and P utilization efficiency (PUE) compared to control | Abbasi et al. (2013) |

(continued)

**Table 8.1** (continued)

| PSB | P sources | Effect | References |
|---|---|---|---|
| *P. fluorescens* | Tricalcium phosphate (TCP) | PSB reduced the transformation of Olsen-P to $Ca_{10}$-P, thus increasing P availability in soil solution | Shi et al. (2017) |
| *B. megaterium* | Rock phosphate (RP) and organic manure | PSB along with organic fertilizers were effective at solubilizing RP | Alzoubi and Gaibore (2012) |

In addition to PSB, IAA, ACC deaminase, and siderophore-producing bacteria can also indirectly provide P for the plant (Etesami et al. 2015b, c; Etesami and Beattie 2017). The mechanisms by which these bacteria lead to an increase in P availability are shown in Fig. 8.2.

## 8.7 Mechanisms of Phosphate-Solubilizing-AMF in Increasing P Availability

Mycorrhiza is a symbiotic relationship between the roots of plants and fungi. AMF belong to phylum Glomeromycota, which form symbiotic associations. This mycorrhizal symbiosis is one of the oldest types of symbioses known between mycorrhizal fungi and a wide variety of plants. More than 80% of the plants on Earth are benefiting from this mycorrhizal symbiosis. In other words, AMF are widely distributed and can be found on all Earth ecosystems where plants can grow (Redecker et al. 2013; Datta et al. 2017b). Arbuscular mycorrhizal fungi can also colonize and establish symbiotic, reciprocally advantageous associations with the roots of most agricultural crop plants (Munyanziza et al. 1997) and augment the effective absorptive area of the roots by forming an extensive extraradical hyphal network, which boosts the efficiency of the absorption of micro- and macronutrients. By a high-affinity P-uptake mechanism and scavenging the available P through their hyphae, which are important in the absorption of P and P transfer from the AMF to plants and act as a bridge between the soil and plant roots (Bianciotto and Bonfante 2002; Harrison and van Buuren 1995; Liu et al. 2000), AMF influence P content and enhance P nutrition in plants as has been widely reported over the years (Barea et al. 2002; Giovannetti et al. 2006).

The root system is known as the main channel for water absorption and mineral elements in all plants. One of the scientific solutions proposed to increase the growth and efficiency of the root system of plants is the use of symbiotic microorganisms, such as mycorrhizal fungi, along with appropriate chemical and organic inputs in the vicinity of the root system of the plants. Mycorrhizal symbiosis is one of the most well-known and, at the same time, the most extensive and most important symbiosis on the planet Earth. The most important effect of the mycorrhizal symbiosis association is the increase in the absorption of mineral elements and especially P in host plants. This effect is more evident especially in areas where

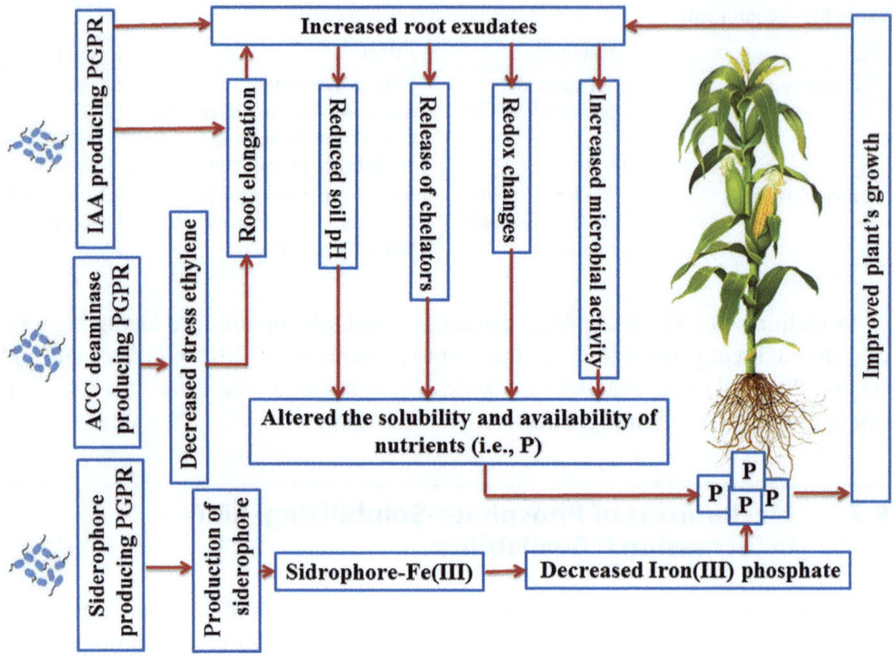


**Fig. 8.2** Schematic representation of mechanisms by which IAA, ACC deaminase, and siderophore-producing PGPR may affect P availability in the rhizosphere. 1-aminocyclopropane-1-carboxylate deaminase producing PGPR hydrolyzes the ethylene precursor 1-aminocyclopropane-1-carboxylate (ACC) to ammonia and α-ketobutyrate (α-KB) and thus prevents the production of stress ethylene. The result of decreased ethylene production is the increase in root length and subsequently increased root exudates. One of the most common roles of IAA in the plant is the increase in root length and root exudation (role in loosening plant cell walls). The root exudation subsequently provides additional nutrients to support the growth of plants and PGPR in rhizosphere. Root exudates also contain different chemical molecules such as chelating agents that mobilize the availability of P in P-deficient soils. Siderophore prevents iron phosphate ($FePO_4$) precipitation through chelating iron ($Fe^{3+}$), and in the absence of free iron, P can be absorbed by the plant. Phosphorus uptake by the plant increases plant growth, and increase in the plant growth inevitably leads to an increase in root exudates, and the exudates also lead to an increase in the number of PGPR and increased P availability in the rhizosphere, and this process continues in a cycle

plant-available P in soils is low or due to drought, the diffusion coefficient of P has been significantly reduced. The absorption of nutrients, e.g., P, which is carried out through diffusion process and moving toward the root, depends on rate of their diffusion in the soil and on the distances that must be traversed to reach the absorbing surfaces the root. It is known that the mycorrhizal roots have higher density than non-mycorrhizal roots. The presence of extraradical hyphal network, which penetrates up to 24 cm away from the root surface, results in decreasing the distance that P must go through to reach the absorbing surfaces of the root, and thereby the rate of absorption of the element increases (Smith and Smith 2011; Meena et al. 2017b).

The rate of development of extraradical hyphae is, on average, 800 times the rate of development of the root system of the plant. Therefore, the P-depleted region around the hyphae of mycorrhizal fungi is more restricted than that of around root hairs, which is why more P is absorbed in the mycorrhizal symbiosis (Smith and Smith 2011; Meena et al. 2018a). On the other hand, the thickness of hyphae of mycorrhizal fungi is one-tenth of that of the root hairs, so these fungal hyphae penetrate into the pores of the soil where the roots cannot penetrate them and thereby absorb more P. In mycorrhizal plants, P is absorbed through fungal hyphal network and transmitted through the cytoplasmic channel of the fungal network to the plant, in which the transfer rate of P to the plant is much higher than its transfer rate in the soil (Smith and Smith 2011). Arbuscular mycorrhizal fungi increase the absorption area of the root zone by 10–100%, thereby improving the ability of plants to use more soil resources. The roots of mycorrhizal plants can explore more soil volume due to their extramatrical hyphae that make easy them for taking up and translocating more P than by plants non-inoculated with mycorrhizal fungi (Guo et al. 2010; Gogoi et al. 2018).

There is evidence of activity of acid phosphatase and alkaline phosphatase enzymes in mycorrhizal fungi, which indicates the ability of these fungi to use phosphorus existing in organic compounds (Antibus et al. 1992). On the other hand, these fungi, by secretion of organic acids such as oxalic acid/oxalates, which have a higher affinity to combine with Ca, Fe, and Al ions in comparison to P, release P from insoluble metal compounds and absorb the released P (Miyasaka and Habte 2001). The secreted oxalates are eventually degraded by actinomycetes and converted to $CO_2$. The carbon dioxide released, by lowering pH in alkaline soils, releases more P from insoluble P compounds and makes it available to the plant (Miransari 2010; Smith and Read 2010; Meena et al. 2017c). It has been estimated that about 80% of the P taken up by a mycorrhizal plant is supplied by AM fungus (Marschner and Dell 1994). In general, it is believed that mycorrhizal fungi can be a good alternative to a part of the chemical fertilizers used, especially phosphate fertilizers, in different systems (Ghorchiani et al. 2018).

## 8.8 Synergistic Effects of PSB and Phosphate-Solubilizing-AMF in Increasing P Availability

One of the ways to boost the efficiency of microorganisms is co-inoculation of microorganisms (Etesami et al. 2015c; Nadeem et al. 2014) that through various mechanisms leads to stimulating plant growth (Bashan et al. 2004). There is an accruing and synergistic effect of PSB combined with AMF (Table 8.2). They (dual inoculation of PSB and phosphate-solubilizing-AMF) have disclosed better performance in terms of sustainable plant growth on nutrient-poor environments (Lee et al. 2015; Mohamed et al. 2014; Nadeem et al. 2014; Xun et al. 2015; Zarei et al. 2006; Verma et al. 2015b). Increased yields of crop plants (Mäder et al. 2011), augmented fruit quality (Bona et al. 2016; Ordookhani et al. 2010), improved nutrient use efficiency of chemicals fertilizers, enhanced phytoremediation

**Table 8.2** The synergistic effects of AMF (arbuscular mycorrhizal fungi) and PSB (phosphate-solubilizing bacteria) on the plants grown in soil with low available P

| PSB | AM fungi | Experimental plant | Effect | References |
|---|---|---|---|---|
| *Pseudomonas fluorescens* | *Funneliformis mosseae* | Maize (*Zea mays* L.) | Under the conditions of fertilization with phosphate rock, dual inoculation of maize plants with *AM fungus* and *phosphate-solubilizing bacterium* led to a significant increase in colonization of root, the grain yield of maize, plant vegetative and reproductive traits, and P and N content in plant tissue in comparison with non-inoculated controls and the plants inoculated with these microorganisms alone | Ghorchiani et al. (2018) |
| *Coccus* sp., *Streptococcus* sp., and *Bacillus* sp. | unknown | Maize (*Zea mays* L.) | Compared to control and single inoculation, co-inoculation of maize with AM *fungus* and PSB significantly promoted mineralization of phosphate rock in soil and improved all growth parameters including shoot (56%), height (41%), root yield (52%), and N (80%) and P (91%) uptake by the maize plants | Wahid et al. (2016) |
| *Bacillus polymyxa* | *Rhizophagus fasciculatus* | *Terminalia paniculata* and *T. tomentosa* | The combined inoculation of phosphate-solubilizing bacterium and AM fungus brought marked increase in plant growth, dry matter, and P uptake when, compared to individual inoculants or non-inoculated plants. The increase in growth was attributed to the increase in P uptake in shoots of the seedlings | Jang et al. (2016) |

(continued)

**Table 8.2** (continued)

| PSB | AM fungi | Experimental plant | Effect | References |
|---|---|---|---|---|
| *B. subtilis* | *Claroideoglomus etunicatum*, *Funneliformis mosseae*, and *Rhizophagus intraradices* | *Acacia gerrardii* | The combined inoculation of *A. gerrardii* with *B. subtilis* and AM fungi resulted in a significant increase in shoot and root dry weight, nodule number, and leghemoglobin content in comparison with non-inoculated controls and the plants inoculated with these microorganisms alone under salinity stress. Co-inoculation and single inoculation of *A. gerrardii* increased the N, P, K, Mg, and Ca contents and phosphatase activities in salt-stressed *A. gerrardii* tissues and diminished concentration of Na and Cl | Hashem et al. (2016) |
| *Bacillus polymyxa* | *Glomus mosseae* | Onion (*Allium cepa* L.) | Co-inoculation of onion with AM fungus and phosphate-solubilizing bacterium significantly augmented shoot fresh and dry weights, plant height, root fresh and dry weights, average bulb diameter, and total yield in comparison with non-inoculated controls and the plants inoculated with these microorganisms alone | Mohamed (2015) |

(continued)

**Table 8.2** (continued)

| PSB | AM fungi | Experimental plant | Effect | References |
|---|---|---|---|---|
| *Burkholderia cepacia* | *G. etunicatum* | Wheat | Co-inoculation of wheat with *B. cepacia* and *G. etunicatum* increased all growth and yield parameters in comparison with non-inoculated controls and the wheat inoculated with these microorganisms alone. Co-inoculation also increased crop yield and N concentration more than 50% and 90%, respectively | Saxena and Jha (2014) |
| *Pseudomonas fluorescens* | *G. mosseae* and *G. intraradices* | Wheat | Combined application of *P. fluorescens*, *G. mosseae*, and *G. intraradices* augmented shoot dry matter yield, seed grain spike number, and grain yield by 52%, 19%, and 26%, respectively, in comparison with non-inoculated controls and the plants inoculated with these microorganisms alone | Yousefi et al. (2011) |
| *Mortierella* sp. | *G.s aggregatum* and *G. mosseae* | *Kostelelzkya virginica* | Compared to single inoculation, co-inoculation of *G.s aggregatum*, *G. mosseae*, and *Mortierella* sp. augmented the AMF colonization (%) and *bacterial* populations under salinity stress (i.e., 100, 200, and 300 mM NaCl). Co- inoculation of salinity-stressed *K. virginica with bacterium* and AMF had significant effects on electrical conductivities of rhizosphere and bulk soils, pH values, and shoot and root dry weights and concentration of available P. | Zhang et al. (2011) |

(continued)

**Table 8.2** (continued)

| PSB | AM fungi | Experimental plant | Effect | References |
|---|---|---|---|---|
| *Pseudomonas striata* | *G. intraradices and G. mosseae* | Maize (*Zea-mays* L.) | Both single inoculation and co-inoculation of maize with *P. striata*, *G. intraradices*, and *G. mosseae significantly* increased crop productivity, grain protein content, mycorrhizal root colonization, and inorganic P, thus showing a synergistic interaction between AMF and phosphate-solubilizing bacterium. Single inoculation of maize with AMF or co- inoculation of maize with AMF + *P. striata* + 75% $P_2O_5$ remained at par with single application of 100% $P_2O_5$ dose with regard to productivity, soil fertility status, and nutrient uptake (particularly P) | Suri et al. (2011) |
| *Enterobacter* sp. and *B. subtilis* | *G. intraradices* | Onion (*Allium cepa* L.) | Co-inoculation of onion with AM fungus and phosphate-solubilizing bacterium significantly augmented onion biomass and accumulation of N and P in onion tissues | Toro et al. (1997) |

(Xun et al. 2015), and reduced application of chemical fertilizers (Adesemoye et al. 2009) are some of the most combined applications of PGPR/PSB and AMF/ phosphate-solubilizing-AMF used so far. Among co-inoculation, the interactions between AM fungi and PSB have been the subject of great interest. There is much speculation that PSB and AMF work together to provide the benefits to plant (Ordoñez et al. 2016). It has been well-proven that AMF and PSB acted synergistically and increased the growth of different plants as compared to that of the plants inoculated with each of them alone (Bona et al. 2017; Bouhraoua et al. 2015; Gamalero et al. 2004; Jangandi et al. 2016; Kalavathi et al. 2000; Kim et al. 1997a, b, 2010; Kothamasi et al. 2006; Mäder et al. 2011; Marulanda et al. 2009; Nadagouda and Lakshman 2010; Ordoñez et al. 2016; Ordookhani et al. 2010; Sabannavar and Lakshman 2009; Sandhya et al. 2013; Saxena et al. 2013, 2015; Souchie et al. 2006; Toro et al. 1997; Wahid et al. 2016; Zhang et al. 2014; Zhang et al. 2016; Meena

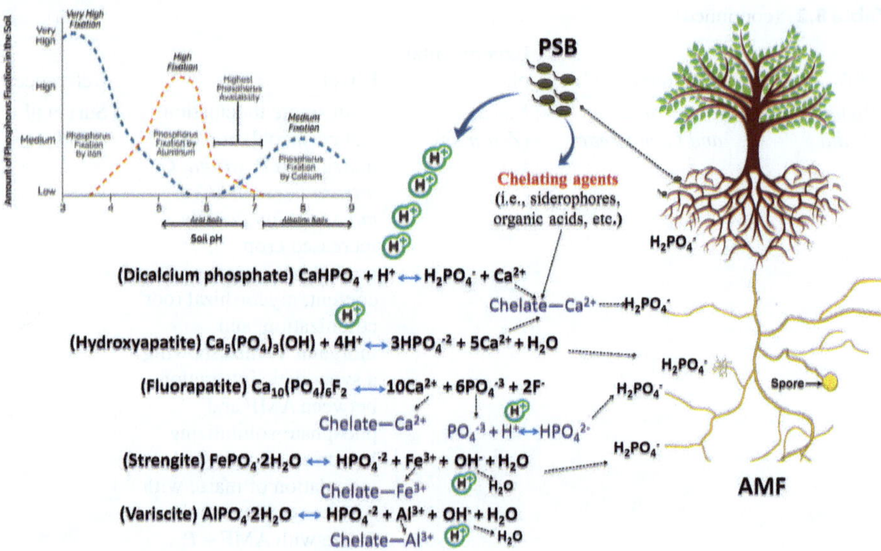


**Fig. 8.3** The synergistic effects of arbuscular mycorrhizal fungi (AMF) and phosphate-solubilizing bacteria (PSB) in increasing availability of P to plant. Phosphate-solubilizing microorganisms enhance the capacity of plants to acquire P from soil through alteration of sorption equilibria that may result in increased net transfer of orthophosphate ions ($H_2PO_4^-$ and $HPO_4^{2-}$) into soil solution. Organic anions and protons are particularly effective at solubilizing precipitated forms of P (e.g., Ca phosphates under alkaline conditions and Fe and Al phosphates under acidic conditions), chelating metal ions that are commonly associated with complexed forms of soil P. According to Le Chatelier's principle, by increasing the concentration of any substance, the balance moves to the consumption of that material, and, by lowering the concentration of each substance, the balance proceeds to produce that material. Chelating agents, such as siderophore and organic anions, by reaction with $Fe^{3+}$, $Al^{3+}$, and $Ca^{2+}$, remove these ions from the reaction, causing the balance to be moved to the right and thereby producing more $H_2PO_4^-$ and $HPO_4^{2-}$. The addition of $H^+$ ion also causes the balance to be adjusted to the right in order to reduce the $H^+$ ion, thereby producing more $H_2PO_4^-$ and $HPO_4^{2-}$. Arbuscular mycorrhizal fungi can take up and transfer orthophosphate ions to plant roots by their effective mycorrhizal mycelium, reaching microhabitats where orthophosphate is made available by P-mobilizing bacteria (PSB) and preventing quickly its immobilization by microbial biomass

et al. 2018c). In general, in this synergistic effect, AMF can only exploit soluble P sources. However, a large amount of P in the soil is in an unsolvable form, in which PSB can potentially make these insoluble forms available for uptake by AMF hyphae and plants. PSB probably augment the availability of P, which subsequently can be efficiently absorbed by AMF hyphae (Fig. 8.3) (Nazir et al. 2010; Toro et al. 1997; Mitran et al. 2018).

## 8.9 Conclusions and Future Prospects

Nonnormative and nonscientific use of phosphorus fertilizers is nothing but waste of money on one side and, on the other hand, the degradation and pollution of basic resources, namely, soil and water. It is known that PSB and PS-mycorrhizal fungi, if used as seed inoculation, can provide between 25% and 50% of the P requirement of the plant in soils with high total P and low plant available P. Therefore, it is recommended that inoculants of these microorganisms along with 50% of the chemical P fertilizer recommended by the soil test be used. It is known that different species of PSB and PS-mycorrhizal fungi have different abilities to dissolve low-soluble P compounds, and usually it has been observed that the use of inoculants including several PSM has a much better effect in increasing the availability of P than the use of only one type of these microorganisms. Since most of PSM are heterotrophic and, as a result, dependent on organic matter in terms of carbon source supply, adding organic matter to the soil when using PSM usually results in increasing their efficiency. As a very good feature of PSM, it is possible to use them simultaneously with mycorrhizal fungi and other PGPR such as IAA, siderophores, and ACC deaminase producers. In this way, in addition to supplying the host plant with P, other plant nutrients will also be supplied, and at the same time, the plant will better grow as a result of the production of growth-promoting hormones by IAA-producing bacteria. When the plant is exposed to environmental stresses, especially drought and salinity, and there is a limit to the use of chemical fertilizers due to their effect on increasing the osmotic pressure of the soil solution and reducing the plant's ability to absorb water, the use of AM fungi can be a very suitable option. By increasing the level of root absorption, AM fungi not only increase the ability of the host plant to absorb water and mineral elements, by modifying the physical structure of the soil, but also create a more favorable environment for the growth of the host plant roots and ultimately reduce consumption of chemical fertilizers, especially phosphorus fertilizers. The possibility of using PSM with rock phosphate, sulfur, organic matter, and *Thiobacillus* bacteria is another potential of microorganisms. In general, broader research is needed on the efficacy of these microorganisms, along with various sources of organic and inorganic materials in different soils and climates and in the presence of the indigenous microflora under field conditions.

**Acknowledgment** I wish to thank the University of Tehran for providing the necessary facilities for this study.

## References

Abbasi MK, Mansha S, Rahim N, Ali A (2013) Agronomic effectiveness and phosphorus utilization efficiency of rock phosphate applied to winter wheat. Agron J 105:1606–1612

Abbasi MK, Musa N, Manzoor M (2015) Mineralization of soluble P fertilizers and insoluble rock phosphate in response to phosphate-solubilizing bacteria and poultry manure and their effect on the growth and P utilization efficiency of chilli (*Capsicum annuum* L.). Biogeosciences 12:4607–4619

Adesemoye AO, Kloepper JW (2009) Plant–microbes interactions in enhanced fertilizer-use efficiency. Appl Microbiol Biotechnol 85:1–12
Adesemoye AO, Torbert HA, Kloepper JW (2008) Enhanced plant nutrient use efficiency with PGPR and AMF in an integrated nutrient management system. Can J Microbiol 54:876–886
Adesemoye AO, Torbert HA, Kloepper JW (2009) Plant growth-promoting rhizobacteria allow reduced application rates of chemical fertilizers. Microb Ecol 58:921–929
Adesemoye AO, Torbert HA, Kloepper JW (2010) Increased plant uptake of nitrogen from 15N-depleted fertilizer using plant growth-promoting rhizobacteria. Appl Soil Ecol 46:54–58
Adnan M et al (2017) Phosphate-solubilizing bacteria nullify the antagonistic effect of soil calcification on bioavailability of phosphorus in alkaline soils. Sci Rep 7:16131
Al Abboud MA, Ghany TMA, Alawlaqi MM (2014) Role of biofertilizers in agriculture: a brief review. Mycopathologia 11:95–101
Alam S, Khalil S, Ayub N, Rashid M (2002) In vitro solubilization of inorganic phosphate by phosphate solubilizing microorganisms (PSM) from maize rhizosphere. Int J Agric Biol 4:454–458
Al-Rohily KM, Ghoneim AM, Modaihsh AS, Mahjoub MO (2013) Phosphorus availability in calcareous soil amend with chemical phosphorus fertilizer, cattle manure compost and sludge manure. Int J Soil Sci 8:17–24
Alzoubi MM, Gaibore M (2012) The effect of phosphate solubilizing bacteria and organic fertilization on availability of Syrian rock phosphate and increase of triple superphosphate efficiency. World J Agric Sci 8:473–478
Antibus RK, Sinsabaugh RL, Linkins AE (1992) Phosphatase activities and phosphorus uptake from inositol phosphate by ectomycorrhizal fungi. Can J Bot 70:794–801
Ashoka P, Meena RS, Kumar S, Yadav GS, Layek J (2017) Green nanotechnology is a key for eco-friendly agriculture. J Clean Prod 142:4440–4441
Attia M, Ahmed MA, El-Sonbaty MR (2009) Use of biotechnologies to increase growth, productivity and fruit quality of Maghrabi banana under different rates of phosphorus. World J Agric Sci 5:211–220
Bais HP, Weir TL, Perry LG, Gilroy S, Vivanco JM (2006) The role of root exudates in rhizosphere interactions with plants and other organisms. Annu Rev Plant Biol 57:233–266. https://doi.org/10.1146/annurev.arplant.57.032905.105159
Barea JM, Toro M, Orozco MO, Campos E, Azcón R (2002) The application of isotopic (32 P and 15 N) dilution techniques to evaluate the interactive effect of phosphate-solubilizing rhizobacteria, mycorrhizal fungi and *Rhizobium* to improve the agronomic efficiency of rock phosphate for legume crops. Nutr Cycl Agroecosyst 63:35–42
Bashan Y, Holguin G, De-Bashan LE (2004) Azospirillum-plant relationships: physiological, molecular, agricultural, and environmental advances (1997–2003). Can J Microbiol 50:521–577
Behera BC, Singdevsachan SK, Mishra RR, Dutta SK, Thatoi HN (2014) Diversity, mechanism and biotechnology of phosphate solubilising microorganism in mangrove—a review. Biocatal Agric Biotechnol 3:97–110
Berendsen RL, Pieterse CMJ, Bakker PAHM (2012) The rhizosphere microbiome and plant health. Trends Plant Sci 17:478–486
Berg G, Smalla K (2009) Plant species and soil type cooperatively shape the structure and function of microbial communities in the rhizosphere. FEMS Microbiol Ecol 68:1–13. https://doi.org/10.1111/j.1574-6941.2009.00654.x
Berruti A, Lumini E, Balestrini R, Bianciotto V (2016) Arbuscular mycorrhizal fungi as natural biofertilizers: let's benefit from past successes. Front Microbiol 6:1559
Bertrand I, Holloway RE, Armstrong RD, McLaughlin MJ (2003) Chemical characteristics of phosphorus in alkaline soils from southern Australia. Soil Res 41:61–76
Bianciotto V, Bonfante P (2002) Arbuscular mycorrhizal fungi: a specialised niche for rhizospheric and endocellular bacteria. Antonie Van Leeuwenhoek 81:365–371
Bianco C, Defez R (2010) Improvement of phosphate solubilization and Medicago plant yield by an indole-3-acetic acid-overproducing strain of *Sinorhizobium meliloti*. Appl Environ Microbiol 76:4626–4632

Bona E et al (2016) Arbuscular mycorrhizal fungi and plant growth-promoting pseudomonads improve yield, quality and nutritional value of tomato: a field study. Mycorrhiza:1–11
Bona E et al (2017) Arbuscular mycorrhizal fungi and plant growth-promoting *pseudomonads* improve yield, quality and nutritional value of tomato: a field study. Mycorrhiza 27:1–11
Bouhraoua D, Aarab S, Laglaoui A, Bakkali M, Arakrak A (2015) Phosphate solubilizing bacteria efficiency on mycorrhization and growth of peanut in the northwest of Morocco. Am J Microbiol Res 3:176–180
Buée M, Boer W, Martin F, Overbeek L, Jurkevitch E (2009) The rhizosphere zoo: An overview of plant-associated communities of microorganisms, including phages, bacteria, archaea, and fungi, and of some of their structuring factors. Plant Soil 321:189–212. https://doi.org/10.1007/s11104-009-9991-3
Buragohain S, Sharma B, Nath JD, Gogaoi N, Meena RS, Lal R (2017) Impact of ten years of biofertilizer use on soil quality and riceyield on an inceptisol in Assam, India. Soil Res. https://doi.org/10.1071/SR17001
Castagno LN, Estrella MJ, Sannazzaro AI, Grassano AE, Ruiz OA (2011) Phosphate-solubilization mechanism and in vitro plant growth promotion activity mediated by Pantoea eucalypti isolated from Lotus tenuis rhizosphere in the Salado River Basin (*Argentina*). J Appl Microbiol 110:1151–1165
Chabot R, Antoun H, Cescas MP (1996) Growth promotion of maize and lettuce by phosphate-solubilizing *Rhizobium leguminosarum* biovar. *phaseoli*. Plant Soil 184:311–321
Chen YP, Rekha PD, Arun AB, Shen FT, Lai WA, Young CC (2006) Phosphate solubilizing bacteria from subtropical soil and their tricalcium phosphate solubilizing abilities. Appl Soil Ecol 34:33–41
Chen Z, Ma S, Liu LL (2008) Studies on phosphorus solubilizing activity of a strain of phosphobacteria isolated from chestnut type soil in China. Bioresour Technol 99:6702–6707
Cheng Z, McConkey BJ, Glick BR (2010) Proteomic studies of plant–bacterial interactions. Soil Biol Biochem 42:1673–1684. https://doi.org/10.1016/j.soilbio.2010.05.033
Cockefair EA (1931) The role of phosphorus in the metabolism of plants. Am J Bot 18:582–597
Compant S, Clément C, Sessitsch A (2010) Plant growth-promoting bacteria in the rhizo- and endosphere of plants: Their role, colonization, mechanisms involved and prospects for utilization. Soil Biol Biochem 42:669–678. https://doi.org/10.1016/j.soilbio.2009.11.024
Dadhich RK, Meena RS (2014) Performance of Indian mustard (*Brassica juncea* L.) in Response to foliar spray of thiourea and thioglycollic acid under different irrigation levels. Indian J Ecol 41(2):376–378
Dadhich RK, Meena RS, Reager ML, Kansotia BC (2015) Response of bio-regulators to yield and quality of Indian mustard (*Brassica juncea* L. Czernj. and Cosson) under different irrigation environments. J App Nat Sci 7(1):52–57
Datta R, Baraniya D, Wang YF, Kelkar A, Moulick A, Meena RS, Yadav GS, Ceccherini MT, Formanek P (2017a) Multi-function role as nutrient and scavenger off reeradical in soil. Sustain MDPI 9:402. https://doi.org/10.3390/su9081402
Datta R, Kelkar A, Baraniya D, Molaei A, Moulick A, Meena RS, Formanek P (2017b) Enzymatic degradation of lignin in soil: a review. Sustain MDPI 1163(9):1–18. https://doi.org/10.3390/su9071163
Desai S, Kumar GP, Amalraj LD, Bagyaraj DJ, Ashwin R (2016) Exploiting PGPR and AMF biodiversity for plant health management. In: Microbial inoculants in sustainable agricultural productivity. Springer, New Delhi, pp 145–160
Dey KB (1988) Phosphate solubilizing organisms in improving fertility status. In, 1988, pp 237–248
Dey R, Pal KK, Bhatt DM, Chauhan SM (2004) Growth promotion and yield enhancement of peanut (Arachis hypogaea L.) by application of plant growth-promoting rhizobacteria. Microbiol Res 159:371–394
Dhakal Y, Meena RS, De N, Verma SK, Singh A (2015) Growth, yield and nutrient content of mungbean (*Vigna radiata* L.) in response to INM in eastern Uttar Pradesh, India. Bangladesh J Bot 44(3):479–482

Dhakal Y, Meena RS, Kumar S (2016) Effect of INM on nodulation, yield, quality and available nutrient status in soil after harvest of green gram. Legum Res 39(4):590–594
Ditta A, Muhammad J, Imtiaz M, Mehmood S, Qian Z, Tu S (2018) Application of rock phosphate enriched composts increases nodulation, growth and yield of chickpea. Int J Recycl Organic Waste Agric 7:33–40
Duarah I, Deka M, Saikia N, Boruah HPD (2011) Phosphate solubilizers enhance NPK fertilizer use efficiency in rice and legume cultivation. 3 Biotech 1:227–238
Etesami H (2018a) Bacterial mediated alleviation of heavy metal stress and decreased accumulation of metals in plant tissues: Mechanisms and future prospects. Ecotoxicol Environ Saf 147:175–191. https://doi.org/10.1016/j.ecoenv.2017.08.032
Etesami H (2018b) Can interaction between silicon and plant growth promoting rhizobacteria benefit in alleviating abiotic and biotic stresses in crop plants? Agric Ecosyst Environ 253:98–112. https://doi.org/10.1016/j.agee.2017.11.007
Etesami H, Alikhani HA (2016a) Co-inoculation with endophytic and rhizosphere bacteria allows reduced application rates of N-fertilizer for rice plant. Rhizosphere 2:5–12. https://doi.org/10.1016/j.rhisph.2016.09.003
Etesami H, Alikhani HA (2016b) Rhizosphere and endorhiza of oilseed rape (Brassica napus L.) plant harbor bacteria with multifaceted beneficial effects. Biol Control 94:11–24
Etesami H, Beattie GA (2017) Plant-microbe interactions in adaptation of agricultural crops to abiotic stress conditions. In: Probiotics and plant health. Springer, Singapore, pp 163–200
Etesami H, Beattie G (2018) Mining halophytes for plant growth-promoting halotolerant bacteria to enhance the salinity tolerance of non-halophytic crops. Front Microbiol 9:148. https://doi.org/10.3389/fmicb.2018.00148
Etesami H, Maheshwari DK (2018) Use of plant growth promoting rhizobacteria (PGPRs) with multiple plant growth promoting traits in stress agriculture: action mechanisms and future prospects. Ecotoxicol Environ Saf 156:225–246. https://doi.org/10.1016/j.ecoenv.2018.03.013
Etesami H, Hosseini HM, Alikhani HA (2014a) Bacterial biosynthesis of 1-aminocyclopropane-1-caboxylate (ACC) deaminase, a useful trait to elongation and endophytic colonization of the roots of rice under constant flooded conditions. Physiol Mol Biol Plants 20:425–434
Etesami H, Hosseini HM, Alikhani HA, Mohammadi L (2014b) Bacterial biosynthesis of 1-aminocyclopropane-1-carboxylate (ACC) deaminase and indole-3-acetic acid (IAA) as endophytic preferential selection traits by rice plant seedlings. J Plant Growth Regul 33:654–670
Etesami H, Alikhani H, Mirseyed Hosseini H (2015a) Indole-3-Acetic Acid and 1-Aminocyclopropane-1-Carboxylate deaminase: bacterial traits required in rhizosphere, rhizoplane and/or endophytic competence by beneficial bacteria. In: Maheshwari DK (ed) Bacterial metabolites in sustainable agroecosystem, vol 12. Sustainable development and biodiversity. Springer International Publishing, Cham, pp 183–258. https://doi.org/10.1007/978-3-319-24654-3_8
Etesami H, Alikhani HA, Hosseini HM (2015b) Indole-3-acetic acid (IAA) production trait, a useful screening to select endophytic and rhizosphere competent bacteria for rice growth promoting agents. MethodsX 2:72–78
Etesami H, Alikhani HA, Hosseini HM (2015c) Indole-3-acetic acid and 1-aminocyclopropane-1-carboxylate deaminase: Bacterial traits required in rhizosphere, rhizoplane and/or endophytic competence by beneficial bacteria. In: Bacterial metabolites in sustainable agroecosystem. Springer, Cham, pp 183–258
Fallah Nosratabad AR, Etesami H, Shariati S (2017) Integrated use of organic fertilizer and bacterial inoculant improves phosphorus use efficiency in wheat (Triticum aestivum L.) fertilized with triple superphosphate. Rhizosphere 3:109–111. https://doi.org/10.1016/j.rhisph.2017.03.001
Gamalero E, Trotta A, Massa N, Copetta A, Martinotti MG, Berta G (2004) Impact of two fluorescent pseudomonads and an arbuscular mycorrhizal fungus on tomato plant growth, root architecture and P acquisition. Mycorrhiza 14:185–192
Ghorchiani M, Etesami H, Alikhani HA (2018) Improvement of growth and yield of maize under water stress by co-inoculating an arbuscular mycorrhizal fungus and a plant growth promoting rhizobacterium together with phosphate fertilizers. Agric Ecosyst Environ 258:59–70

Giovannetti M, Avio L, Fortuna P, Pellegrino E, Sbrana C, Strani P (2006) At the root of the wood wide web: self recognition and nonself incompatibility in mycorrhizal networks. Plant Signal Behav 1:1–5

Glick BR (2012) Plant growth-promoting bacteria: mechanisms and applications. Scientifica 2012:1–15

Glick BR (2014) Bacteria with ACC deaminase can promote plant growth and help to feed the world. Microbiol Res 169:30–39

Gogoi N, Baruah KK, Meena RS (2018) Grain legumes: impact on soil health and agroecosystem. In: Meena RS et al (eds) Legumes for soil health and sustainable management. Springer, Singapore. https://doi.org/10.1007/978-981-13-0253-4_16

Goldstein AH (1986) Bacterial solubilization of mineral phosphates: historical perspective and future prospects. Am J Altern Agric 1:51–57

Güneş A, Ataoğlu N, Turan M, Eşitken A, Ketterings QM (2009) Effects of phosphate-solubilizing microorganisms on strawberry yield and nutrient concentrations. J Plant Nutr Soil Sci 172:385–392

Guo Y, Ni Y, Huang J (2010) Effects of rhizobium, arbuscular mycorrhiza and lime on nodulation, growth and nutrient uptake of lucerne in acid purplish soil in China. Trop Grassl 44:109–114

Gupta RP, Vyas MK, Pandher MS (1998) Role of phosphorus solubilizing microorganisms in P-economy and crop yield Soil–Plant–Microbe Interaction in Relation to Nutrient Management, pp 95-101

Gyaneshwar P, Kumar GN, Parekh LJ, Poole PS (2002) Role of soil microorganisms in improving P nutrition of plants. In: Food security in nutrient-stressed environments: exploiting plants' genetic capabilities. Springer, Dordrecht, pp 133–143

Hamdali H, Hafidi M, Virolle MJ, Ouhdouch Y (2008) Rock phosphate-solubilizing Actinomycetes: screening for plant growth-promoting activities. World J Microbiol Biotechnol 24:2565–2575

Hardoim PR, van Overbeek LS, JDv E (2008) Properties of bacterial endophytes and their proposed role in plant growth. Trends Microbiol 16:463–471. https://doi.org/10.1016/j.tim.2008.07.008

Harrison MJ, van Buuren ML (1995) A phosphate transporter from the mycorrhizal fungus Glomus versiforme. Nature 378:626

Hashem A, Abd_Allah EF, Alqarawi AA, Al-Huqail AA, Wirth S, Egamberdieva D (2016) The interaction between arbuscular mycorrhizal fungi and endophytic bacteria enhances plant growth of Acacia gerrardii under salt stress. Front Microbiol 7:1089

Illmer P, Schinner F (1992) Solubilization of inorganic phosphates by microorganisms isolated from forest soils. Soil Biol Biochem 24:389–395

Islam MT, Hossain MM (2012) Plant probiotics in phosphorus nutrition in crops, with special reference to rice. In: Bacteria in agrobiology: plant probiotics. Springer, Berlin, pp 325–363

Islam MT, Deora A, Hashidoko Y, Rahman A, Ito T, Tahara S (2007) Isolation and identification of potential phosphate solubilizing bacteria from the rhizoplane of Oryza sativa L. cv. BR29 of Bangladesh. Z Naturforsch C 62:103–110

Jang S, Negalur CB, Narayan M, Lakshman HC (2016) Effect of phosphate solubilizing bacteria and arbuscular mycorrhizal fungi with and without rock phosphate on four forest tree seedlings. Int J Bioassays 6:5204–5207

Jangandi S, Negalur CB, Narayan M, Lakshman HC (2016) Effect of phosphate solubilizing bacteria and arbuscular mycorrhizal fungi with and without rock phosphate on four forest tree seedlings. Int J Bioassays 6:5204–5207

Kakraliya SK, Singh U, Bohra A, Choudhary KK, Kumar S, Meena RS, Jat ML (2018) Nitrogen and legumes: a meta-analysis. In: Meena RS et al (eds) Legumes for soil health and sustainable management. Springer, Singapore. https://doi.org/10.1007/978-981-13-0253-4_9

Kalavathi BP, Santhanakrishnan P, Divya MP (2000) Effect of VA-mycorrhizal fungus and phosphorus solubilising bacterium in neem. Indian Forester 126:67–70

Karthikeyan AS, Varadarajan DK, Mukatira UT, D'Urzo MP, Damsz B, Raghothama KG (2002) Regulated expression of Arabidopsis phosphate transporters. Plant Physiol 130:221–233

Kaur G, Reddy MS (2015) Effects of phosphate-solubilizing bacteria, rock phosphate and chemical fertilizers on maize-wheat cropping cycle and economics. Pedosphere 25:428–437

Kennedy IR, Choudhury A, Kecskés ML (2004) Non-symbiotic bacterial diazotrophs in crop-farming systems: can their potential for plant growth promotion be better exploited? Soil Biol Biochem 36:1229–1244

Khan MS, Zaidi A, Wani PA (2007) Role of phosphate-solubilizing microorganisms in sustainable agriculture—a review. Agron Sustain Dev 27:29–43

Kim KY, Jordan D, Krishnan HB (1997a) Rahnella aquatilis, a bacterium isolated from soybean rhizosphere, can solubilize hydroxyapatite. FEMS Microbiol Lett 153:273–277

Kim KY, Jordan D, McDonald GA (1997b) Effect of phosphate-solubilizing bacteria and vesicular-arbuscular mycorrhizae on tomato growth and soil microbial activity. Biol Fertil Soils 26:79–87

Kim K et al (2010) Synergistic effects of inoculating arbuscular mycorrhizal fungi and Methylobacterium oryzae strains on growth and nutrient uptake of red pepper (*Capsicum annuum* L.). Plant Soil 327:429–440

Kothamasi D, Kothamasi S, Bhattacharyya A, Kuhad RC, Babu CR (2006) Arbuscular mycorrhizae and phosphate solubilising bacteria of the rhizosphere of the mangrove ecosystem of Great Nicobar island, India. Biol Fertil Soils 42:358–361

Kpomblekou-a K, Tabatabai MA (1994) Effect of organic acids on release of phosphorus from phosphate rocks1. Soil Sci 158:442–453

Kucey RMN (1983) Phosphate-solubilizing bacteria and fungi in various cultivated and virgin Alberta soils. Can J Soil Sci 63:671–678

Kumar V, Behl RK, Narula N (2001) Establishment of phosphate-solubilizing strains of Azotobacter chroococcum in the rhizosphere and their effect on wheat cultivars under green house conditions. Microbiol Res 156:87–93

Kumar H, Bajpai VK, Dubey RC, Maheshwari DK, Kang SC (2010) Wilt disease management and enhancement of growth and yield of Cajanus cajan (L) var. Manak by bacterial combinations amended with chemical fertilizer. Crop Prot 29:591–598

Kumar S, Meena RS, Pandey A, Seema (2017a) Soil acidity management and an economics response of lime and sulfur on sesame in an alley cropping system. Int J Curr Microb App Sci 6(3):2566–2573

Kumar S, Meena RS, Yadav GS, Pandey A (2017b) Response of sesame (*Sesamum indicum* L.) to sulphur and lime application under soil acidity. Int J Plant Soil Sci 14(4):1–9

Kumar S, Meena RS, Lal R, Yadav GS, Mitran T, Meena BL, Dotaniya ML, EL-Sabagh A (2018a) Role of legumes in soil carbon sequestration. In: Meena RS et al (eds) Legumes for soil health and sustainable management. Springer, Singapore. https://doi.org/10.1007/978-981-13-0253-4_4

Kumar S, Meena RS, Bohra JS (2018b) Interactive effect of sowing dates and nutrient sources on dry matter accumulation of Indian mustard (*Brassica juncea* L.). J Oilseed Brassica 9(1):72–76

Lambers H, Shane MW, Cramer MD, Pearse SJ, Veneklaas EJ (2006) Root structure and functioning for efficient acquisition of phosphorus: matching morphological and physiological traits. Ann Bot 98:693–713

Lambers H, Martinoia E, Renton M (2015) Plant adaptations to severely phosphorus-impoverished soils. Curr Opin Plant Biol 25:23–31

Layek J, Das A, Mitran T, Nath C, Meena RS, Singh GS, Shivakumar BG, Kumar S, Lal R (2018) Cereal+Legume intercropping: an option for improving productivity. In: Meena RS et al (eds) Legumes for soil health and sustainable management. Springer, Singapore. https://doi.org/10.1007/978-981-13-0253-4_11

Lee Y, Krishnamoorthy R, Selvakumar G, Kim K, Sa T (2015) Alleviation of salt stress in maize plant by co-inoculation of arbuscular mycorrhizal fungi and Methylobacterium oryzae CBMB20. J Korean Soc Appl Biol Chem 58:533–540. https://doi.org/10.1007/s13765-015-0072-4

Leifert C, Waites WM, Nicholas JR (1989) Bacterial contaminants of micropropagated plant cultures. J Appl Bacteriol 67:353–361

Leytem AB, Mikkelsen RL (2005) The nature of phosphorus in calcareous soils Better Crops, vol 89, pp 11–13

Liu A, Hamel C, Hamilton RI, Ma BL, Smith DL (2000) Acquisition of Cu, Zn, Mn and Fe by mycorrhizal maize (*Zea mays* L.) grown in soil at different P and micronutrient levels. Mycorrhiza 9:331–336

Lugtenberg B (2015) Life of Microbes in the Rhizosphere. In: Lugtenberg B (ed) Principles of plant-microbe interactions: microbes for sustainable agriculture. Springer International Publishing, Cham, pp 7–15. https://doi.org/10.1007/978-3-319-08575-3_3
Lynch JP (2007) Roots of the second green revolution. Aust J Bot 55:493–512
Mäder P et al (2011) Inoculation of root microorganisms for sustainable wheat–rice and wheat–black gram rotations in India. Soil Biol Biochem 43:609–619
Marschner H, Dell B (1994) Nutrient uptake in mycorrhizal symbiosis. Plant Soil 159:89–102
Martinez C (2010) Sustainable development and biodiversity. Environ Policy Law 40:273
Marulanda A, Barea J-M, Azcón R (2009) Stimulation of plant growth and drought tolerance by native microorganisms (AM fungi and bacteria) from dry environments: mechanisms related to bacterial effectiveness. J Plant Growth Regul 28:115–124
McLaughlin M (2012) Technical Bulletin: phosphorus fertilizer use efficiency in soils. Fertiliser Technology Research Centre, pp 1–4
Meena RS, Lal R (2018) Legumes and sustainable use of soils. In: Meena RS et al (eds) Legumes for soil health and sustainable management. Springer, Singapore. https://doi.org/10.1007/978-981-13-0253-4_1
Meena H, Meena RS (2017) Assessment of sowing environments and bio-regulators as adaptation choice for clusterbean productivity in response to current climatic scenario. Bangladesh J Botany 46(1):241–244
Meena RS, Yadav RS (2014) Phonological performance of groundnut varieties under sowing environments in hyper arid zone of Rajasthan, India. J App Nat Sci 6(2):344–348
Meena RS, Yadav RS (2015) Yield and profitability of groundnut (*Arachis hypogaea* L.) as influenced by sowing dates and nutrient levels with different varieties. Legum Res 38(6):791–797
Meena RS, Yadav RS, Meena VS (2014) Response of groundnut (*Arachis hypogaea* L.) varieties to sowing dates and NP fertilizers under Western Dry Zone of India. Bangladesh J Bot 43(2):169–173
Meena RS, Dhakal Y, Bohra JS, Singh SP, Singh MK, Sanodiya P (2015a) Influence of bioinorganic combinations on yield, quality and economics of Mungbean. Am J Exp Agri 8(3):159–166
Meena RS, Meena VS, Meena SK, Verma JP (2015b) The needs of healthy soils for a healthy world. J Clean Prod 102:560–561
Meena RS, Meena VS, Meena SK, Verma JP (2015c) Towards the plant stress mitigate the agricultural productivity: a book review. J Clean Prod 102:552–553
Meena RS, Yadav RS, Meena H, Kumar S, Meena YK, Singh A (2015d) Towards the current need to enhance legume productivity and soil sustainability worldwide: a book review. J Clean Prod 104:513–515
Meena RS, Yadav RS, Reager ML, De N, Meena VS, Verma JP, Verma SK, Kansotia BC (2015e) Temperature use efficiency and yield of groundnut varieties in response to sowing dates and fertility levels in Western Dry Zone of India. Am J Exp Agri 7(3):170–177
Meena H, Meena RS, Singh B, Kumar S (2016a) Response of bio-regulators to morphology and yield of clusterbean [*Cyamopsis tetragonoloba* (L.) Taub.] under different sowing environments. J App Nat Sci 8(2):715–718
Meena RS, Bohra JS, Singh SP, Meena VS, Verma JP, Verma SK, Shiiag SK (2016b) Towards the prime response of manure to enhance nutrient use efficiency and soil sustainability a current need: a book review. J Clean Prod 112:1258–1260
Meena RS, Gogaoi N, Kumar S (2017a) Alarming issues on agricultural crop production and environmental stresses. J Clean Prod 142:3357–3359
Meena RS, Kumar S, Pandey A (2017b) Response of sulfur and lime levels on productivity, nutrient content and uptake of sesame under guava (*Psidium guajava* L.) based agri-horti system in an acidic soil of eastern Uttar Pradesh, India. The J Crop and Weed 13(2):222–227
Meena RS, Meena PD, Yadav GS, Yadav SS (2017c) Phosphate solubilizing microorganisms, principles and application of microphos technology. J Clean Prod 145:157–158
Meena RS, Kumar V, Yadav GS, Mitran T (2018a) Response and interaction of *Bradyrhizobium japonicum* and *Arbuscular mycorrhizal* fungi in the soybean rhizosphere: a review. Plant Growth Regul 84:207–223

Meena BL, Fagodiya RK, Prajapat K, Dotaniya ML, Kaledhonkar MJ, Sharma PC, Meena RS, Mitran T, Kumar S (2018b) Legume green manuring: an option for soil sustainability. In: Meena RS et al (eds) Legumes for soil health and sustainable management. Springer, Singapore. https://doi.org/10.1007/978-981-13-0253-4_12

Meena H, Meena RS, Lal R, Singh GS, Mitran T, Layek J, Patil SB, Kumar S, Verma T (2018c) Response of sowing dates and bio regulators on yield of clusterbean under current climate in alley cropping system in eastern U.P. Indian Leg Res 41(4):563–571

Miransari M (2010) Contribution of arbuscular mycorrhizal symbiosis to plant growth under different types of soil stress. Plant Biol 12:563–569

Mitran T, Meena RS, Lal R, Layek J, Kumar S, Datta R (2018) Role of soil phosphorus on legume production. In: Meena RS et al (eds) Legumes for soil health and sustainable management. Springer, Singapore. https://doi.org/10.1007/978-981-13-0253-4_15

Miyasaka SC, Habte M (2001) Plant mechanisms and mycorrhizal symbioses to increase phosphorus uptake efficiency. Commun Soil Sci Plant Anal 32:1101–1147

Mohamed HM (2015) Impact of inoculation with arbuscular mycorrhizal, phosphate solubilizing bacteria and soil yeast on growth, yield and phosphorous content of onion plants. Int J Soil Sci 10:93

Mohamed AA, Eweda WEE, Heggo AM, Hassan EA (2014) Effect of dual inoculation with arbuscular mycorrhizal fungi and sulphur-oxidising bacteria on onion (*Allium cepa* L.) and maize (*Zea mays* L.) grown in sandy soil under green house conditions. Ann Agric Sci 59:109–118. https://doi.org/10.1016/j.aoas.2014.06.015

Mudge SR, Rae AL, Diatloff E, Smith FW (2002) Expression analysis suggests novel roles for members of the Pht1 family of phosphate transporters in *Arabidopsis*. Plant J 31:341–353

Munyanziza E, Kehri HK, Bagyaraj DJ (1997) Agricultural intensification, soil biodiversity and agro-ecosystem function in the tropics: the role of mycorrhiza in crops and trees. Appl Soil Ecol 6:77–85

Nadagouda MG, Lakshman HC (2010) Microbial solubilization of P and Arbuscular mycorrhizal fungi use for yield and phosphate uptake in improvement of nodulation and yield of [Vicia faba L.]. Int J Agric Sci 6:319–321

Nadeem SM, Ahmad M, Zahir ZA, Javaid A, Ashraf M (2014) The role of mycorrhizae and plant growth promoting rhizobacteria (PGPR) in improving crop productivity under stressful environments. Biotechnol Adv 32:429–448

Nahas E (2007) Phosphate solubilizing microorganisms: effect of carbon, nitrogen, and phosphorus sources. In: Velázquez E, Rodríguez-Barrueco C (eds) First international meeting on microbial phosphate solubilization. Springer Netherlands, Dordrecht, pp 111–115. https://doi.org/10.1007/978-1-4020-5765-6_15

Naik PR, Raman G, Narayanan KB, Sakthivel N (2008) Assessment of genetic and functional diversity of phosphate solubilizing fluorescent pseudomonads isolated from rhizospheric soil. BMC Microbiol 8:230

Nations U (2015) World population prospects: the 2015 revision United Nations. Econ Soc Aff 33:1–66

Nazir R, Warmink JA, Boersma H, Van Elsas JD (2010) Mechanisms that promote bacterial fitness in fungal-affected soil microhabitats. FEMS Microbiol Ecol 71:169–185

Nosratabad ARF, Etesami H, Shariati S (2017) Integrated use of organic fertilizer and bacterial inoculant improves phosphorus use efficiency in wheat (*Triticum aestivum* L.) fertilized with triple superphosphate. Rhizosphere 3:109–111

Oehl F, Oberson A, Probst M, Fliessbach A, Roth H-R, Frossard E (2001) Kinetics of microbial phosphorus uptake in cultivated soils. Biol Fertil Soils 34:31–41

Ogut M, Er F, Kandemir N (2010) Phosphate solubilization potentials of soil Acinetobacter strains. Biol Fertil Soils 46:707–715

Oliveira CA et al (2009) Phosphate solubilizing microorganisms isolated from rhizosphere of maize cultivated in an oxisol of the Brazilian Cerrado Biome. Soil Biol Biochem 41:1782–1787

Ordoñez YM, Fernandez BR, Lara LS, Rodriguez A, Uribe-Vélez D, Sanders IR (2016) Bacteria with phosphate solubilizing capacity alter mycorrhizal fungal growth both inside and outside the root and in the presence of native microbial communities. PLoS One 11:e0154438

Ordookhani K, Khavazi K, Moezzi A, Rejali F (2010) Influence of PGPR and AMF on antioxidant activity, lycopene and potassium contents in tomato. Afr J Agric Res 5:1108–1116
Ortíz-Castro R, Contreras-Cornejo HA, Macías-Rodríguez L, López-Bucio J (2009) The role of microbial signals in plant growth and development. Plant Signal Behav 4:701–712
Panhwar QA, Radziah O, Rahman AZ, Sariah M, Razi IM, Naher UA (2011a) Contribution of phosphate-solubilizing bacteria in phosphorus bioavailability and growth enhancement of aerobic rice. Span J Agric Res 9:810–820
Panhwar QA, Radziah O, Zaharah AR, Sariah M, Razi IM (2011b) Role of phosphate solubilizing bacteria on rock phosphate solubility and growth of aerobic rice
Panhwar QA, Jusop S, Naher UA, Othman R, Razi MI (2013) Application of potential phosphate-solubilizing bacteria and organic acids on phosphate solubilization from phosphate rock in aerobic rice. Sci World J 2013:1–10
Pei-Xiang Y et al (2012) Phosphate solubilizing ability and phylogenetic diversity of bacteria from P-rich soils around Dianchi Lake drainage area of China. Pedosphere 22:707–716
Perez-Lopez R, Alvarez-Valero AM, Nieto JM (2007) Changes in mobility of toxic elements during the production of phosphoric acid in the fertilizer industry of Huelva (SW Spain) and environmental impact of phosphogypsum wastes. J Hazard Mater 148:745–750
Ram K, Meena RS (2014) Evaluation of pearl millet and mungbean intercropping systems in Arid Region of Rajasthan (India). Bangladesh J Bot 43(3):367–370
Redecker D, Schüßler A, Stockinger H, Stürmer SL, Morton JB, Walker C (2013) An evidence-based consensus for the classification of arbuscular mycorrhizal fungi (*Glomeromycota*). Mycorrhiza 23:515–531
Rodríguez H, Fraga R (1999) Phosphate solubilizing bacteria and their role in plant growth promotion. Biotechnol Adv 17:319–339
Rodriguez R, Redman R (2008) More than 400 million years of evolution and some plants still can't make it on their own: plant stress tolerance via fungal symbiosis. J Exp Bot 59:1109–1114
Rodríguez H, Fraga R, Gonzalez T, Bashan Y (2006) Genetics of phosphate solubilization and its potential applications for improving plant growth-promoting bacteria. Plant Soil 287:15–21
Sabannavar SJ, Lakshman HC (2009) Effect of rock phosphate solubilization using mycorrhizal fungi and phosphobacteria on two high yielding varieties of *Sesamum indicum* L. World J Agric Sci 5:470–479
Sandhya A, Vijaya T, Sridevi A, Narasimha G (2013) Influence of vesicular arbuscular mycorrhiza (VAM) and phosphate solubilizing bacteria (PSB) on growth and biochemical constituents of Marsdenia volubilis. Afr J Biotechnol 12:5648–5654
Sapsirisopa S, Chookietwattana K, Maneewan K, Khaengkhan P (2009) Effect of salt-tolerant Bacillus inoculum on rice KDML 105 cultivated in saline soil. Asian J Food Ag-Ind 2:S69–S74
Saxena J, Jha A (2014) Impact of a phosphate solubilizing bacterium and an arbuscular mycorrhizal fungus (Glomus etunicatum) on growth, yield and P concentration in wheat plants. Clean Soil Air Water 42:1248–1252
Saxena J, Chandra S, Nain L (2013) Synergistic effect of phosphate solubilizing rhizobacteria and arbuscular mycorrhiza on growth and yield of wheat plants. J Soil Sci Plant Nutr 13:511–525
Saxena J, Saini A, Ravi I, Chandra S, Garg V (2015) Consortium of Phosphate-solubilizing Bacteria and Fungi for Promotion of Growth and Yield of Chickpea (*Cicer arietinum*). J Crop Improv 29:353–369
Sharma K, Dak G, Agrawal A, Bhatnagar M, Sharma R (2007) Effect of phosphate solubilizing bacteria on the germination of Cicer arietinum seeds and seedling growth. J Herb Med Toxicol 1:61–63
Sharma SB, Sayyed RZ, Trivedi MH, Gobi TA (2013) Phosphate solubilizing microbes: sustainable approach for managing phosphorus deficiency in agricultural soils. Springerplus 2:587
Shi X-K, Ma J-J, Liu L-J (2017) Effects of phosphate-solubilizing bacteria application on soil phosphorus availability in coal mining subsidence area in Shanxi. J Plant Interact 12:137–142
Sihag SK, Singh MK, Meena RS, Naga S, Bahadur SR, Gaurav YRS (2015) Influences of spacing on growth and yield potential of dry direct seeded rice (*Oryza sativa* L.) cultivars. Ecoscan 9(1-2):517–519

Singh M, Dotaniya ML, Mishra A, Dotaniya CK, Regar KL, Lata M (2016) Role of biofertilizers in conservation agriculture. In: Conservation agriculture. Springer, Singapore, pp 113–134
Smith SE, Read DJ (2010) Mycorrhizal symbiosis. Academic, Amsterdam
Smith SE, Smith FA (2011) Roles of arbuscular mycorrhizas in plant nutrition and growth: new paradigms from cellular to ecosystem scales. Annu Rev Plant Biol 62:227–250
Sofi PA, Baba ZA, Hamid B, Meena RS (2018) Harnessing soil Rhizobacteria for improving drought resilience in legumes. In: Meena RS et al (eds) Legumes for soil health and sustainable management. Springer, Singapore. https://doi.org/10.1007/978-981-13-0253-4_8
Somers E, Vanderleyden J, Srinivasan M (2008) Rhizosphere bacterial signalling: a love parade beneath our feet. Crit Rev Microbiol
Souchie EL, Saggin-Júnior OJ, Silva EMR, Campello EFC, Azcón R, Barea JM (2006) Communities of P-solubilizing bacteria, fungi and arbuscular mycorrhizal fungi in grass pasture and secondary forest of Paraty, RJ-Brazil. An Acad Bras Cienc 78:183–193
Suri VK, Choudhary AK, Chander G, Verma TS, Gupta MK, Dutt N (2011) Improving phosphorus use through co-inoculation of vesicular arbuscular mycorrhizal fungi and phosphate-solubilizing bacteria in maize in an acidic Alfisol. Commun Soil Sci Plant Anal 42:2265–2273
Thakuria D, Talukdar NC, Goswami C, Hazarika S, Boro RC, Khan MR (2004) Characterization and screening of bacteria from rhizosphere of rice grown in acidic soils of Assam. Curr Sci:978–985
Theodorou ME, Plaxton WC (1993) Metabolic adaptations of plant respiration to nutritional phosphate deprivation. Plant Physiol 101:339–344
Toro M, Azcon R, Barea J (1997) Improvement of arbuscular mycorrhiza development by inoculation of soil with phosphate-solubilizing Rhizobacteria to improve rock phosphate bioavailability ((sup32) P) and nutrient cycling. Appl Environ Microbiol 63:4408–4412
Varma D, Meena RS, Kumar S (2017a) Response of mungbean to fertility and lime levels under soil acidity in an alley cropping system in Vindhyan Region, India. Int J Chem Stu 5(2):384–389
Varma D, Meena RS, Kumar S, Kumar E (2017b) Response of mungbean to NPK and lime under the conditions of Vindhyan Region of Uttar Pradesh. Legum Res 40(3):542–545
Vassilev N, Vassileva M (2003) Biotechnological solubilization of rock phosphate on media containing agro-industrial wastes. Appl Microbiol Biotechnol 61:435–440
Verma JP, Jaiswal DK, Meena VS, Meena RS (2015a) Current need of organic farming for enhancing sustainable agriculture. J Clean Prod 102:545–547
Verma JP, Meena VS, Kumar A, Meena RS (2015b) Issues and challenges about sustainable agriculture production for management of natural resources to sustain soil fertility and health: a book review. J Clean Prod 107:793–794
Verma SK, Singh SB, Prasad SK, Meena RN, Meena RS (2015c) Influence of irrigation regimes and weed management practices on water use and nutrient uptake in wheat (*Triticum aestivum* L. Emend. Fiori and Paol.). Bangladesh J Bot 44(3):437–442
Vessey JK (2003) Plant growth promoting rhizobacteria as biofertilizers. Plant Soil 255:571–586
Vitousek PM et al (1997) Human alteration of the global nitrogen cycle: sources and consequences. Ecol Appl 7:737–750
Vyas P, Gulati A (2009) Organic acid production in vitro and plant growth promotion in maize under controlled environment by phosphate-solubilizing *fluorescent Pseudomonas*. BMC Microbiol 9:174
Wahid F, Sharif M, Steinkellner S, Khan MA, Marwat K, Khan S (2016) Inoculation of arbuscular mycorrhizal fungi and phosphate solubilizing bacteria in the presence of rock phosphate improves phosphorus uptake and growth of maize. Pak J Bot 48:739–747
Williams CH, Donald CM (1957) Changes in organic matter and pH in a podzolic soil as influenced by subterranean clover and superphosphate. Aust J Agric Res 8:179–189
Xun F, Xie B, Liu S, Guo C (2015) Effect of plant growth-promoting bacteria (PGPR) and arbuscular mycorrhizal fungi (AMF) inoculation on oats in saline-alkali soil contaminated by petroleum to enhance phytoremediation. Environ Sci Pollut Res 22:598–608

Yadav GS, Babu S, Meena RS, Debnath C, Saha P, Debbaram C, Datta M (2017a) Effects of godawariphosgold and single supper phosphate on groundnut (*Arachis hypogaea*) productivity, phosphorus uptake, phosphorus use efficiency and economics. Indian J Agric Sci 87(9):1165–1169

Yadav GS, Lal R, Meena RS, Babu S, Das A, Bhomik SN, Datta M, Layak J, Saha P (2017b) Conservation tillage and nutrient management effects on productivity and soil carbon sequestration under double cropping of rice in North Eastern Region of India. Ecol Indic. http://www.sciencedirect.com/science/article/pii/S1470160X17305617

Yadav GS, Lal R, Meena RS, Datta M, Babu S, Das LJ, Saha P (2017c) Energy budgeting for designing sustainable and environmentally clean/safer cropping systems for rainfed rice fallow lands in India. J Clean Prod 158:29

Yadav GS, Das A, Lal R, Babu S, Meena RS, Saha P, Singh R, Datta M (2018a) Energy budget and carbon footprint in a no-till and mulch based rice–mustard cropping system. J Clean Prod 191:144–157

Yadav GS, Das A, Lal R, Babu S, Meena RS, Patil SB, Saha P, Datta M (2018b) Conservation tillage and mulching effects on the adaptive capacity of direct-seeded upland rice (*Oryza sativa* L.) to alleviate weed and moisture stresses in the North Eastern Himalayan Region of India. Arch Agron Soil Sc. https://doi.org/10.1080/03650340.2018.1423555

Yildirim E, Karlidag H, Turan M, Dursun A, Goktepe F (2011) Growth, nutrient uptake, and yield promotion of broccoli by plant growth promoting rhizobacteria with manure. HortScience 46:932–936

Yousefi AA, Khavazi K, Moezi AA, Rejali F, Nadian HA (2011) Phosphate solubilizing bacteria and arbuscular mycorrhizal fungi impacts on inorganic phosphorus fractions and wheat growth. World Appl Sci J 15:1310–1318

Zaidi S, Usmani S, Singh BR, Musarrat J (2006) Significance of Bacillus subtilis strain SJ-101 as a bioinoculant for concurrent plant growth promotion and nickel accumulation in *Brassica juncea*. Chemosphere 64:991–997

Zarei M, Saleh-Rastin N, Alikhani HA, Aliasgharzadeh N (2006) Responses of lentil to co-inoculation with phosphate-solubilizing rhizobial strains and arbuscular mycorrhizal fungi. J Plant Nutr 29:1509–1522

Zhang H, Wu X, Li G, Qin P (2011) Interactions between arbuscular mycorrhizal fungi and phosphate-solubilizing fungus (*Mortierella* sp.) and their effects on Kostelelzkya virginica growth and enzyme activities of rhizosphere and bulk soils at different salinities. Biol Fertil Soils 47:543

Zhang L, Fan J, Ding X, He X, Zhang F, Feng G (2014) Hyphosphere interactions between an arbuscular mycorrhizal fungus and a phosphate solubilizing bacterium promote phytate mineralization in soil. Soil Biol Biochem 74:177–183

Zhang L, Xu M, Liu Y, Zhang F, Hodge A, Feng G (2016) Carbon and phosphorus exchange may enable cooperation between an arbuscular mycorrhizal fungus and a phosphate-solubilizing bacterium. New Phytol 210:1022–1032

Zhu F, Qu L, Hong X, Sun X (2011) Isolation and characterization of a phosphate-solubilizing halophilic bacterium *Kushneria* sp. YCWA18 from Daqiao Saltern on the coast of Yellow Sea of China Evidence-Based Complementary and Alternative Medicine 2011

# 9 Use of Organic and Biological Fertilizers as Strategies to Improve Crop Biomass, Yields and Physicochemical Parameters of Soil

Abdelilah Meddich, Khalid Oufdou, Abderrahim Boutasknit, Anas Raklami, Abdelilah Tahiri, Raja Ben-Laouane, Mohamed Ait-El-Mokhtar, Mohamed Anli, Toshiaki Mitsui, Said Wahbi, and Marouane Baslam

**Abstract**

Interest in the sustainability of soil resources has been stimulated by increasing concerns that soil is one of the most critical components of the earth's biosphere,

A. Meddich (✉) · A. Boutasknit · R. Ben-Laouane · M. Ait-El-Mokhtar · M. Anli · S. Wahbi
Department of Biology, Laboratory of Biotechnology and Plant Physiology, Faculty of Sciences Semlalia, Cadi Ayyad University, Marrakesh, Morocco
e-mail: a.meddich@uca.ma

K. Oufdou
Laboratory of Biology and Biotechnology of Microorganisms, Faculty of Sciences Semlalia, Cadi Ayyad University, Marrakesh, Morocco
e-mail: oufdou@uca.ac.ma

A. Raklami · A. Tahiri
Department of Biology, Laboratory of Biotechnology and Plant Physiology, Faculty of Sciences Semlalia, Cadi Ayyad University, Marrakesh, Morocco

Laboratory of Biology and Biotechnology of Microorganisms, Faculty of Sciences Semlalia, Cadi Ayyad University, Marrakesh, Morocco

T. Mitsui
Laboratory of Biochemistry, Faculty of Agriculture, Niigata University, Niigata, Japan
University, Niigata, Japan

Department of Life and Food Sciences, Graduate School of Science and Technology, Niigata University, Niigata, Japan

M. Baslam
Laboratory of Biochemistry, Faculty of Agriculture, Niigata University, Niigata, Japan
University, Niigata, Japan
e-mail: mbaslam@gs.niigata-u.ac.jp



R. S. Meena (ed.), *Nutrient Dynamics for Sustainable Crop Production*,
https://doi.org/10.1007/978-981-13-8660-2_9

participating in food production and management maintenance of environmental quality. In arid and semiarid regions, oases appear to be a major threat model in the soil component. The palm plantations contribute to the formation of oasis ecosystems by promoting the creation of a suitable microclimate for the development of underlying crops and offsetting the effects of drought. These ecosystems play key roles in multiple socioeconomic and environmental issues. Nevertheless, they remain fragile and undergo impacts of human and/or natural origins permanently such as extreme temperatures, soil salinity, drought, erosion, and low contents in organic matter and native fertility. In order to ensure good yields, farmers use an intensive amount of chemical fertilizer, but it can have detrimental effects on soil. In this chapter, we will focus on the improvement of the biomass and yield of different agricultural crops – i.e., cereals (wheat, corn), vegetable crops (lettuce, tomato, leek), leguminous (alfalfa), and trees (date palm) – in field via the enrichment of soil by setting up an efficient biological protocol integrating arbuscular mycorrhizal fungi (AMF), PGPR, and/or organic soil conditioners resulting from green waste, phosphogypsum, phosphate wash sludge, and agro-industrial poultry waste manure. Our results confirmed the advantages of various biological and organic fertilizers in improving the biomass and yields for different crops. The combination of AMF and compost green waste appeared to be interesting for the improvement of the growth, mineral nutrition, and physiological and water parameters of date palm (*Phoenix dactylifera* L.). Furthermore, the combination of low dose (5%, 10%) and indigenous AMF is clearly beneficial for the growth of alfalfa and tomato under a greenhouse. Concerning the experiments carried out in the field, it confirmed the advantages of biological and organic fertilizers in improving the yield for leguminous (alfalfa), vegetable crops (lettuce, tomato, and leek), and cereals (wheat). Application of the tripartite combination AMF-PGPR compost was more efficient in increasing the yield of the tested plants. Indeed, biological treatments had an important effect on the physicochemical properties of the soil. Finally, we have elucidated the positive impacts of biofertilizers used and the interest of adopting the innovative practices improving soil fertility, preserving water resources, respecting the environment, and ensuring the development of sustainable organic agriculture.

**Keywords**

Climate change · Compost · Arbuscular mycorrhizal fungi · PGPR · Symbiosis · Soil degradation · Date palm · Yield · Soil management · Sustainable agriculture · Underlying crops

## Abbreviations

AMF Arbuscular mycorrhizal fungi
F Mycorrhizal frequency

| | |
|---|---|
| Foa | *Fusarium oxysporum* f. sp. *albedinis* |
| MPN | Most probable number |
| PGPR | Plant growth-promoting rhizobacteria |
| GWDL | Grass waste and dead leaves |
| GWSP | Grass waste and sludge of phosphate |
| OCOMWWG | Olive cake, olive oil mill wastewater, and garbage |
| PGGW | Phosphogypsum and green waste |
| PMGW | Poultry manure and green waste |
| R | Stomatal resistance |
| RWC | Relative water content |
| SOC | Soil organic carbon |
| SOM | Soil organic matter |
| TOC | Total organic carbon |
| TKN | Total Kjeldahl nitrogen |

## 9.1 Introduction

Interest in the sustainability of soil resources has been stimulated by increasing concerns that soil is one of the most critical components of the earth's biosphere, involved in producing food and maintaining different scales of environmental quality. At the same time, the function of supporting world food and agriculture is a key for the preservation and advancement of all life on planet Earth. However, the decline in the soil quality around the world is a challenging task and will likely remain an important global issue in the years to come. International attention to protecting the environment has been increasing in recent decades focusing on protection from global warmth and desertification, which has affected the ecosystems worldwide. In arid and semiarid regions, oases appear to be a major threat model in the soil component. Therefore, it is important to understand the spatial distribution characteristics of the soil and the management practices to provide a wide range of ecosystem services and specific sustainability benefits associated with improving soil health practices. Oasis environment is considered as a model ecosystem where agriculture is possible by the microclimate determined by the date palm (*Phoenix dactylifera* L.) for the development of underlying crops (arboriculture, cereals, and horticultural crops) and offsetting the effects of environmental stresses (Ehsine et al. 2014). This ecosystem plays key roles in integrating economic, social, and environmental issues. Nevertheless, this agroecosystem remains fragile and often vulnerable to human and/or natural impacts such as urbanization, abiotic (i.e., heat, drought, salinity, desertification, depletion of soil organic matter, and nutrients) and biotic stresses (i.e., bayoud palm caused by *Fusarium oxysporum* f. sp. *albedinis* (Foa)), genetic erosion, and aging (Oihabi 1991; Saaidi 1992; Ziouti 1998; Botes and Zaid 2002; Awad 2006; Jaiti et al. 2008; Meddich et al. 2015a; Meddich and Boumezzough 2017; Ashoka et al. 2017; Kumar et al. 2017b).

In the oasis, the date palm is the oldest and most widely cultivated tree that is commercially the most important tree in the life of its people and their heritage. The importance of the date palm occurs because of its great contribution to the creation, maintenance, and development of the economy in the oases. The economic utility of these palms is multifold, including staple food, beverages, ornamentals, building wood, and industrial materials (Balick and Beck 1990; Meena et al. 2015d; Gogoi et al. 2018). In addition to its commercial and nutritional value, the date palm tree has a minimum water demand, tolerates harsh weather, and tolerates high levels of salinity. Morocco, one of the important countries for date palm cultivation, has 41% of the world's date palm trees (14% of the date production), and nearly 340 of 2000 varieties recorded around the world are grown here. Morocco occupied the 3rd largest producer countries (Oihabi 1991) with $15 \times 10^6$ date palm trees at the end of the nineteenth century. However, this number currently decreased dramatically to $6.6 \times 10^6$ palm trees spread on 51,000 ha (FAOSTAT 2018). Yet it's important to note that this shrinking owing to bayoud disease is caused by Foa, tree aging, lack of maintenance, and the environmental conditions affecting negatively the development of date palm and its underlying cultures (FAO 2012). In addition, the low contents in organic matter and native fertility remain one of the major constraints to date palm production. As a result of these circumstances and to ensure good yields, many farmers increase the frequency of watering and the use of an intensive amount of chemical fertilizer to increase the level of nutrients found in soil. Inorganic fertilizers, which are highly absorbed by the ground, enhance plants' growth and yields of fruits and vegetables in a relatively short period of time leaving the rest of the chemicals to leach. As a result of leaching, the chemical fertilizer adversely affects soil chemical properties, water irrigation, and the amount, activity, and diversity of microorganisms beneficial to plant and soil health. The need to respond to these situations by adopting appropriate and sustainable strategies to ensure the protection and restoration of our oasis is a time-demanding task. Organic and biologic fertilizers such as arbuscular mycorrhizal fungi (AMF), plant growth-promoting rhizobacteria (PGPR), and compost have emerged as safe and effective alternatives to the chemical fertilizers in order to improve the sustainability of agroecosystems as well as to increase soil quality and crop production per unit area of arable land.

AMF are a key integral component of the soil rhizosphere and are essential for the stability, sustainability, and functioning of the ecosystems. The external mycelium of AMF, considered as an extension of host plant roots, acts as a direct link between roots and soil nutrient reserves. The abundance of extraradical hyphae is a major factor in soil structure as they promote soil aggregate formation, which is important in resistance of soil erosion. The roots of date palm are receptive to the AMF and are capable to grow in the arid area (Oihabi 1991; Al-Yahya'ei et al. 2011; Meddich et al. 2015b; Meena and Meena 2017; Yadav et al. 2017c). The positive effects of mycorrhizal symbiosis on the growth and health of date palm have been reported (Al-Karaki 2013; Meddich et al. 2015a; Buragohain et al. 2017). Previous reports have revealed that AMF (1) promoted the growth of date palm seedlings in nursery conditions (Shabbir et al. 2011) than the controls treated with chemical fertilizers (Symanczik et al. 2014), (2) increased nutrient availability in soil cultures

(Al-Karaki et al. 2007), and (3) improved the absorption of water and nutrients under salt and drought stress conditions (Bearden and Petersen 2000; Baslam et al. 2013). Additionally, AMF were involved to improve the salt and drought tolerance in other crops such as lettuce (Ruiz-lozano et al. 1995; Baslam and Goicoechea 2012; Vicente-Sánchez et al. 2014), sorghum (Augé et al. 1995), corn (Subramanian and Charest 1997), clover (Oihabi and Meddich 1996; Meddich et al. 2000), and barley (Meddich 2001; Tao et al. 2014). AMF can protect also their host plant against biotic stress factors such as soil-borne fungal pathogens causing root rot or wilting and aboveground pathogens such as *Alternaria solani* in tomato (Linderman 1994; Azcon-Aguilar and Barea 1996; Thygesen et al. 2004; Fritz et al. 2006; Jung et al. 2012; Xiao et al. 2014; Meddich et al. 2015a; Meena and Yadav 2015).

PGPR bacteria promote the growth of plants and represent a beneficial and heterogeneous group of rhizosphere microorganisms on the root surface. They are capable of improving plant growth and increasing tolerance against biotic and abiotic stresses (Dimkpa et al. 2009; Grover et al. 2011; Glick 2012; Oufdou et al. 2014; Dadhich and Meena 2014). PGPR stimulate plant growth through direct mechanisms such as biological nitrogen fixation, phosphate solubilization, stress reduction, and production of phytohormones and siderophores or by indirect mechanisms such as stimulation of mycorrhizal symbiosis, antagonism toward phytopathogens, or removal of phytotoxic substances (Glick 2005; Haas and Défago 2005; Blaha et al. 2006; Couillerot et al. 2009; Zamioudis and Pieterse 2012; Datta et al. 2017a). The mode of action of PGPR is influenced by a number of biotic factors (plant genotypes, plant developmental stages, plant defense mechanisms, other members of the microbial community) and abiotic factors (soil composition, soil management, and climatic conditions) (Vacheron et al. 2013; Meena et al. 2015e; Meena and Lal 2018). Although studies on boosting plant growth through PGPR are widely available, information between the potential uses of PGPR for sustainable development and their present applications remain to be clarified.

The combination of socioeconomic development and population growth in many countries was accompanied by the increase of large quantities of solid and liquid wastes generated mainly by households, green space maintenance services, and industry and farming livestock units (Laarousi et al. 2006; Yadav et al. 2018b; Verma et al. 2015c; Mitran et al. (2018). Despite the fact that organic wastes are full of enormous potential, their use is currently very limited. Morocco, one of the green waste producers, has launched several strategies falling within the framework of sustainable development, which are aimed to preserve the country's natural resources. Therefore, in Marrakesh city, the choice of these green wastes as organic waste is justified by their high abundance in the gardens (18,000 t/year) and their improving effect of the mixture structure by ensuring source carbon for microbial growth. Morocco holds more than 72% of all phosphate rock reserves in the world, being the natural phosphate residue especially phosphogypsum, a mixture of calcium phosphate in various forms and gypsum, never valorized locally. The estimated production of this residue in Morocco is $20 \times 10^6$ t/year with $4 \times 10^6$ t of $P_2O_5$ (El Cadi et al. 2014). Li et al. (2018) and Kammoun et al. (2017) reported that phosphogypsum – as the main by-product of phosphoric acid production – might be effective in reducing $NH_3$ and $CH_4$ emissions throughout the composting process with an increment of $SO_4^{2-}$ content of the compost. Similarly, phosphate sludge waste is generated at significant quantities estimated at $2 \times 10^6$ t/year.

Currently, animal organic manures like poultry manure are receiving more attention as fertilizers due to the high cost of inorganic fertilizers and its limited ability to improve soil quality for sustainable production systems (Arancon et al. 2008; Kumar et al. 2018b; Meena et al. 2016a). The increase in poultry production driven by the recent demand for low-cholesterol meat products conjunctly with high protein sources and the economic incentive has led to an expansion in the poultry industry worldwide (Sarangi et al. 2016; Dhakal et al. 2015). FAO projections suggest that global meat production and consumption will continue rising over the coming years. Manure from poultry units, containing organic and mineral substances, is often produced in large quantities and discharged into landfills without any exploitation. Other types of waste are generated in significant quantities by industrialists, in particular, liquid effluents (400,000 $m^3$/year) and olive cake (180,000 t/year) (CFC/COI 2008).

In this sense, composting these wastes can be a valuable economic and ecological solution allowing the return of organic matter to the soil as a stable humus-like product and its reintegration to the biogeochemical systems (Francou 2003). Compost is an effective way to increase healthy plant production, reduce costs and the use of chemical fertilizers, and conserve natural resources. Compost provides a stable organic matter that improves the physical, chemical, and biological properties of soils, thereby enhancing soil quality and crop production. Organic amendments are known to improve soil productivity by influencing soil organic matter (SOM) pool. The SOM is considered to be an important criterion of soil quality and therefore is a major determinant of sustainability of agricultural systems. Soil organic carbon (SOC) influences productivity via soil structure, water-holding capacity, soil-buffering capacity, and as a source of plant nutrients. Stable soil structure is in fact required for better soil physical environment. The compost enriches the organic matter in soil with organic molecules, diversified degradation products, and humus substances that improve the belowground structure by interaction with minerals and aggregation of clay particles allowing the production of microaggregates and hence soil stability (Stevenson 1994; Clapp et al. 2001; Seul et al. 2009). This organic matter also decreases the density and promotes the root growth and penetration by improving nutrition, photosynthesis, and plant biomass (Schnitzer and Poapest 1967; Rauthan and Schnitzer 1981; Nardi et al. 1996; Tejada et al. 2009; Varma et al. 2017a; Meena and Yadav 2014). Similarly, the compost increases the cation exchange capacity and soil water retention by ensuring a good flow of water and limiting leaching (Giusquiani et al. 1995; Takeda et al. 2009). It stimulates the activity of microorganisms and thus accelerates the cycle of elements and mineral alteration. The gradual decomposition requires large amounts of macro- and micronutrients necessary for plant nutrition (Clapp et al. 2001). Compost can inhibit the development of pathogenic microorganisms (Tautorus and Townsley 1983, Vassilev et al. 2009; Yadav et al. 2017b). Little information is available about the integrative potential of microorganisms with organic fertilizers and its effects on crops in the ability of agricultural systems to adapt to climatic and other global changes.

This chapter aims to address the importance of the functional rhizospheric microbiome as a sustainable and effective strategy in plant fitness and disease protection. It also highlights the beneficial interactions among plants and different AMF, PGPR, and compost in boosting agricultural productivity in food security.

Here we will focus on the improvement of the biomass and yield of different agricultural crops – cereals (wheat and corn), vegetable crops (lettuce, tomato, and leek), leguminous (alfalfa), and trees (date palm) – in field via the enrichment of soil by setting up an efficient biological protocol integrating by AMF, exotic and endogenous species; and/or rhizobia, autochthonous bacteria inoculum rhizobia strain; and/or organic soil conditioners resulting from green waste, phosphogypsum, phosphate wash sludge, and agro-industrial poultry waste. This integrated nutrient management approach would improve the fertility of soils and preserve the hydrous resources to reduce the harmful environmental effects while achieving high-quality and high-yielding crops. Also, integrating biofertilizers for the development of plants might be considered as an appropriate strategy to reverse the land degradation trend and encourage sustainable patterns for the development of oasis zones with a vision to promote durable and biological agriculture. Our findings represent the first-of-its-kind study examining the combined application of indigenous/exotic AMF, rhizobia, and organic amendments for the improvement of morphological and physiological parameters, water status, and yields of underlying crops such as leguminous, cereals, and vegetable crops in oasis system. These key species have been selected based on their economic value, their interests and protected status, and their potential for more widespread use by farmers. In addition, our work will illustrate the impact of the interaction effects of the rhizosphere cultivable microorganisms and compost on soil physicochemical properties.

## 9.2 Methodology

### 9.2.1 Study Site

We grew the selected plants in the farm field spread over a total area of 3 hectares and equipped with a drip tape irrigation system. The farm is located in the municipality of Tamesloht, Marrakesh, Morocco (N 31 54 176°; W 008 02 087°; elevation 531 m) (Fig. 9.1). The regional climate of the experimental site is typically Mediterranean, with an average temperature of 20.5 °C and 281 mm of annual rainfall. We didn't apply any herbicides nor chemical fertilizers in the previous growing seasons.

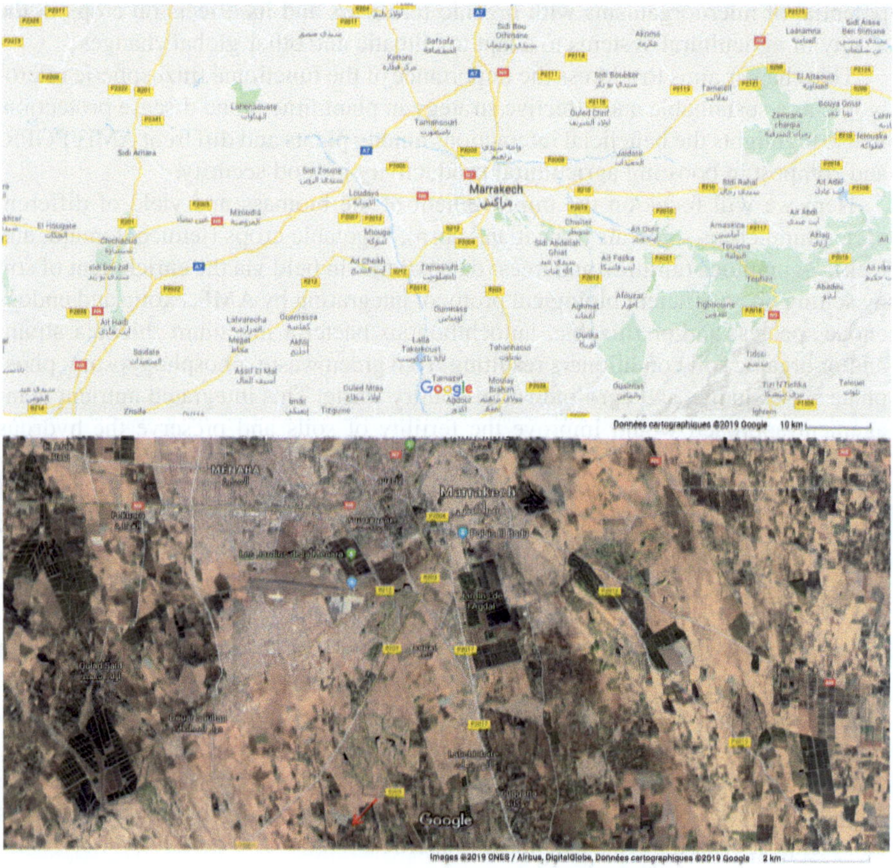


**Fig. 9.1** Geographical location of the study area. (Google Maps 2017)

## 9.2.2 Characterization of Arbuscular Mycorrhizal Fungi

The AMF used were (1) *Rhizophagus irregularis* (pure spores produced in vitro from *Glomus irregularis* (GI) isolate of DAOM 197198), kindly provided by M. Hijri Ph.D. (Research Institute of Plant Biology, University of Quebec, Montreal, Canada), and (2) consortium arbuscular mycorrhizal (CAM) isolated from Tafilalet palms located in Tafilalet, 500 Km southeast of Marrakesh, Morocco (Meddich et al. 2015a). The CAM contains a mixture of endogenous species, *Glomus* sp. (15 spores/gr soil), *Sclerocystis* sp. (9 spores/gr soil), and *Acaulospora* sp. (1 spore/gr soil). The endogenous (native) species were identified according to their color, size, attachment hyphae, and consistency (Koske and Tessier 1983; Morton and Benny 1990).

Fifty (50) of rhizosphere soil samples from 10 to 40 cm in depth – area of date palm tree roots rich with AMF – were collected from Tafilalet palm grove. Samples were taken at 1 m of palm stems and were spaced 80–100 m. Soil samples were mixed thoroughly to obtain a homogenous sample representative of the entire sampling interval.

The mycorrhizal potential of Tafilalet palm grove was determined by the most probable number (MPN) of propagules per unit of soil method (Plenchette et al. 1989) to reflect the ability of soil to initiate the formation of mycorrhizal associations from propagules, i.e., spores, mycelium, and roots of debris-carrying vesicles.

Corn (*Zea mays* L.) plants were used as a host plant to trap the native mycorrhizal complex naturally associated with date palm and for the multiplication of *G. irregularis*.

MPN propagules were calculated by the following formula: Log MPN = ($x$ log $a$) – K, where $x$ = average mycorrhizal pots, $a$ = dilution factor, and $y$ = s – x where s = dilution number and y is required for the determination of K in the table of Fisher and Yates (1970).

The rate of mycorrhizal root infection was microscopically estimated according to the method described by Trouvelot et al. (1986). The method calculates the parameters of infection as follows:

*F: Frequency of root mycorrhization (percentage of root segments infection)*

$$\text{F\%} = (N - n_0) / N \times 100$$

where $N$ = number of fragments observed and $n_0$ = number of fragments without a trace of mycorrhization.

*M: Intensity of cortical infection* (proportion of the cortical colonization in all the mycorrhizal root system)

$$\text{M\%} = \left[(95 \times n_5) + (70 \times n_4) + (30 \times n_3) + (5 \times n_2 + n_1)\right] / \text{N}$$

where $n_5$, $n_4$, ... $n_1$ = number of fragments noted 5,4,... and 1, respectively. Class 5, more than 91%; Class 4, from 51% to 90%; Class 3, from 11% to 50%; Class 2, less than 10 %; Class 1, trace; and Class 0, no mycorrhization.

### 9.2.3 Characterization of Rhizobacterial Strains

The autochthonous bacterial inoculum, kindly provided by Prof. Oufdou (Cadi Ayyad University, Marrakesh, Morocco), was isolated from the local bean (*Phaseolus vulgaris*) rhizospheric soil – attached to bean roots – and consists of two PGPR and two rhizobia strains. The bacteria selected have been described as plant growth promoters and nitrogen-fixing bacteria. Active culture of strains was prepared in Tryptic Soy Broth (TSB) medium (tryptone, 15 g/L; peptone of soya, 5 g/L;

and sodium chloride, 5 g/L) and agitated for 2–3 days at 28 °C to obtain an optical density (OD = 1) at 600 nm (equivalent to $10^9$ colony forming unit/ml).

PGPR strains were selected based on their ability for solubilization of complex insoluble phosphate (Raklami 2017). These microbial strains have the ability to solubilize K from K-bearing minerals. These PGPR can produce indole acetic acid (AIA) and exopolysaccharides at very low levels and are incapable of producing hydrogen cyanide (HCN).

### 9.2.4 Preparation of High-Grade Compost by an Enrichment Technique

In our study, the main raw materials used for composting are

- Grass waste (GW) or dandelion
- Dead leaves (DL)
- Waste from livestock units of poultry
- Phosphates washing sludge (GWSP)
- Phosphogypsum
- Pomace olive consisting of olive cake and olive oil mill wastewater (OCOMWWG)
- Household waste

All of the main raw materials were analyzed for physiochemical, nutrients, and heavy metals. Composting was carried out in a composting area consisting of a metal frame of 2400 m at the municipal nursery of Marrakesh. In this experiment, there are five treatments compost piles. The mixtures used for all piles were arranged:

- Grass waste and dead leaves (GWDL)
- Grass waste and phosphates washing sludge (GWSP)
- Pomace olive and household waste (OCOMWWG)
- Phosphogypsum and green waste
- Poultry manure and green waste

The moisture content was maintained at 50–60% by the addition of water throughout the active composting period by frequent checking. To maintain the moisture and prevent excessive loss of heat, drying windrows runoff, and leaching phenomena, the heaps of composting material were then deposited and covered using plastic sheets. The mixtures were turned at 3-day intervals to permit the ventilation, porosity, and high decomposition until the end of the composting. The temperature was measured daily with a thermometer at random depths. The maturity of composts is considered complete when the temperature inside the heap decreased to the surrounding temperature (around 90 days).

### 9.2.5 Inoculation Methods and Growth Under Greenhouse and Field Conditions

Three crops (lettuce, *Lactuca sativa*; tomato, *Solanum lycopersicum*; and leek, *Allium ampeloprasum*), two cereals (wheat, *Triticum aestivum*, and corn, *Zea mays*), one leguminous (alfalfa, *Medicago sativa*), and one tree (date palm, *Phoenix dactylifera*) species were tested for the microorganisms and compost effectiveness. The goal was to select plants of economic importance, compatible as date palm underlying culture, and widespread use and capable of producing high biomass under local climatic conditions. Seeds were sterilized in 2.5% sodium hypochlorite, incubated under the corresponding temperature of each plant in the dark condition. They were then placed and cultured on seedling nursery trays and cultured.

Seedlings were transplanted in 5 L pots filled with sterilized soils sampled directly from the research sites to replicate the in situ rhizosphere condition and hence give a better prediction of plant growth promotion effects of AMF and/or PGPR and/or compost for the field trials. Plants were grown under semi-controlled greenhouse conditions; the average temperature was 24.5 °C, average relative humidity was 70%, and light intensity was 330 $\mu mol\ m^{-2}\ s^{-1}$. The young palm trees and its underlying cultures were amended by the compost produced locally (Meddich et al. 2017) based on green waste (couch grass, dead leaves), agro-industrial waste (solid and liquid wastes olives), household waste, animal waste (poultry manure), phosphogypsum, and/or phosphate wash sludge. The combination of compost addition and/or AMF and/or PGPR has been evaluated for their growth promotion effect on plants. Furthermore, the physicochemical properties of different waste mixtures and composts were determined.

Under our field experiments, the agricultural soil properties used for plants growth were sandy loam texture (sand, 74.75%; silt, 13.55%; and clay, 11.69%); pH, 8.12; electrical conductivity, 138.3 μs/cm; organic matter, 0.87%; limestone content, 5.04%; phosphorus available, 57.42 ppm; and total nitrogen, 9.98 mg/g. Seedlings were treated by the various biological fertilizers (compost, PGPR, indigenous rhizobia, and/or native or exotic AMF strains).

At transplanting, half of the plants were inoculated (2.8 g) near the root system with the mycorrhizal and disinfected corn roots used as trap AMF (Strullu 1986). The inoculum was infective propagules (mycelium, spores, and roots). A filtrate was added to plants that did not receive the mycorrhizal inoculum (NM plants) in an attempt to restore other soil free-living microorganisms accompanying AMF.

Plants were inoculated two times in different days, 4 ml and 8 ml of each suspension with the symbiotic bacteria PGPR and the rhizobia strains to increase the level of these bacteria in soil and ensure the infection of newly formed roots. The liquid suspension of these strains at a concentration of $10^8$ cells/ml for each selected strain was inoculated.

According to our previous studies, we used the low doses of compost (5–10%) in this experiment.

Our field trial was conducted to test the effectiveness of the native biofertilizers as single or co-inoculations on crops biomass. The uninoculated (control) plants for

**Fig. 9.2** Field plot layout showing the field design equipped with drip irrigation system

each crop were grown under the same environmental conditions without any biological nor organic amendments. The experimental design was a randomized complete block subdivided into several basic blocks of 1.5 m × 0.8 m each. The plots and their blocks were equipped with a drop by drop system (Fig. 9.2). A device of 12 blocks repeated for the same treatment and the same culture was used to evaluate the impact of the various biological and organic treatments. The evaluation of the yield of the studied crops was determined by measuring the fresh weight, the biomass produced, and/or the number of fruits produced.

### 9.2.6 Studied Parameters

We evaluated the efficacy of AMF-PGPR and/or compost combination for crop yield production and their impact on soil quality and properties. The AMF infectivity, plant growth, water content, and the physiological parameters for amended and non-amended plants were measured. Nutritional analyses were conducted in treated and non-treated (control) plants.

#### 9.2.6.1 Physicochemical Properties of Composts and Soils

Samples were taken at 0–15 cm depth before and after the trial experiment in order to measure the soil physicochemical properties. Field texturing was determined by Robinson's method (Baize 1988). The total organic carbon content was determined according to Aubert (1978) by the oxidation method of organic material in cold condition with an excess of potassium dichromate $K_2Cr_2O_7$ in the presence of concentrated sulfuric acid. The total limestone was determined using a Bernard calcimeter. After each brushing waste for composting, the sampling was performed at ten different levels of windrows (deep, surface, side, and center), as described in the method of quartering (AFNOR 1999).

The different soil-size fractions of minerals were determined. The hydrogen peroxide ($H_2O_2$ to 20 V) was used to remove the organic carbon matter and the sodium hexametaphosphate (50 g/L) for clay dispersion. The portions of coarse sand and fine aggregate were recovered by passing 200 μm and 50 μm sieve, respectively. The sift-clay fraction was sorted according to Robinson's pipette method. Soil pH was measured by an electrometric procedure using a suspension of 10 g of fresh sample in 20 ml of distilled water. The measurement of bulk soil electrical conductivity (EC) was quantified by a probe. The compost temperature was measured continuously at depth of 30, 70, and 100 cm. Each temperature measurement is an average of six temperature readings taken at three equally spaced locations along the sides of the pile.

Ash content was determined by calcining the previously dried samples in a muffle furnace at 600 °C for 6 h. The increase in temperature has been achieved by heat bearing (105 °C [1 h], 200 °C [1 h], 600 °C [6 h]) to prevent from the sudden destruction of the organic matter.

The measurement of the total nitrogen was based on the transformation of organic nitrogen into ammonium nitrogen. After sample mineralization by concentrated sulfuric acid and in the presence of Kjeldahl catalyst, the formed ammonia was displaced by NaOH (40%). Then, the entrained ammonia by the water vapor was fixed by the boric acid and titrated with sulfuric acid. NKT content was determined by the distillation unit Velp-UDK132 according to the protocol described by Rodier (1984).

Ammonium levels were determined according to the Kjeldahl method (AFNOR 1975) from a fresh sample (2 g) using a distillation in an alkaline medium with 10 ml of sodium hydroxide 40%.

Nitrates were measured by passing the filtered sample through a column containing granulated copper-cadmium to reduce nitrate to nitrite. The nitrite (and reduced nitrate) was determined by diazotizing with sulfanilamide and coupling with N-(1-naphthyl)-ethylenediamine dihydrochloride to form a highly colored azo dye which was measured colorimetrically.

Total phosphorus was determined by a colorimetric assay as described by Olsen and Sommers (1982). The potassium content of the filtered extract was measured using a Jenway PFP7 flame spectrophotometer.

#### 9.2.6.2 Mycorrhization Parameters

Root samples were cleared and stained by trypan blue 0.01% in lactoglycerol (Phillips and Hayman 1970), and mycorrhizal colonization was determined by examining 1 cm root segments ($n$ = 20 per sample) under the microscope. Results were expressed as a percentage of infection (Hayman et al. 1976). The analysis of the state of the mycorrhizal root system was performed according to the method described by Trouvelot et al. (1986) to characterize the frequency and intensity of mycorrhization in the presence and absence of biofertilizers.

#### 9.2.6.3 Measurement of Plant Growth and Minerals Concentration

The response of control and treated plants by biofertilizers was evaluated by determining the shoots (SDM) and root (RDM) dry masses, a reliable indicator of biomass. SDM and RDM were measured after drying the fresh material into the oven at 80 °C until the weight was constant. Mineral determinations (P, $K^+$, $Ca^{2+}$, $Mg^{2+}$, and $Na^+$) were quantified by a wet digestion method (Pequerul et al. 1993). Dried, finely powered plant samples (0.5 g) were placed in the oven for 6 h at 550 °C. The obtained material was digested in 3 ml of 6N HCl, evaporated on a hot plate, and then recovered with hot distilled water. The solutions obtained were filtered, and the extracts were collected and subsequently stored. The digested solution was shaken gently and filtered through 0.2 μm filters (Whatman, England), and the solid fraction was discarded.

The content of phosphorus in the extract was determined according to Olsen and Sommers (1982). The $K^+$, $Ca^{2+}$, $Mg^{2+}$, and $Na^+$ elements were quantified by a flame photometer (AFP 100 flame photometer). The total content of nitrogen (N) in plants was carried out according to the method described by Kjeldahl.

#### 9.2.6.4 Physiological Parameters

The relative water content (RWC%) of plant leaves was determined according to Sade et al. (2009) as:

$$\%\mathrm{WC} = (\text{fresh weight} - \text{dry weight}) / (\text{turgid weight} - \text{dry weight}) \times 100$$

Turgid weight (TW) was calculated after fully hydrating fresh leaves in darkness at 4°C for 24 h. Results were expressed as percentages.

Leaf water potential (Wh) was measured using a pressure chamber (Scholander et al. 1965). The stomatal resistance was determined in fully expanded leaves of the same rank with an LI-1600 gas exchange system (LI-COR Inc., USA). Chlorophyll fluorescence was measured within plants leaves of the same row using a fluorimeter (OS-30p + OPTI-SCIENCES). The measured parameters correspond to the initial fluorescence ($F_0$), the maximum fluorescence ($F_m$), and the quantum efficiency noted Fv/Fm, where Fv is the variable fluorescence (Tardieu 2005).

### 9.2.7 Statistical Analysis

Values are presented as the mean ± standard deviation (SD). Means were tested by one-way analysis of variance (ANOVA) followed by Newman and Keuls test at $P < 0.05$ in the SPSS software (SPSS Inc., Chicago, IL, USA).

## 9.3 Results

### 9.3.1 AM Colonization Potential and Infectivity Parameters

Results of the mycorrhizal roots of corn plants grown on different dilutions of the studied soil are presented in Table 9.1. The symbol (+) represents the plants with at least one point of infection, while the symbol (−) shows the no-infection of roots. For the tested soil, the percentage inoculation with AMF remained 100%, independently of the soil dilution level achieved. The soil of the Tafilalet palm grove shows the most significant number of mycorrhizal propagules, estimated at 1,626.89.

Maize roots inoculated with AMF obtained from Tafilalet palm groves and *Glomus irregularis* (GI) showed higher mycorrhizal frequencies ranging from 98% to 100% (Fig. 9.3). Similarly, the colonization intensity of corn roots remained higher and exceeds 65% after 3 months of cultivation with AMF from palm grove and GI.

**Table 9.1** Mycorrhizal potential of Tafilalet palm grove

| Dilution | Site | Repetitions | | | | | Number of mycorrhizal plants |
|---|---|---|---|---|---|---|---|
| | | $R_1$ | $R_2$ | $R_3$ | $R_4$ | $R_5$ | |
| 1 | Palm grove of Tafilalet | + | + | + | + | + | 5 |
| 1/4 | | + | + | + | + | + | 5 |
| 1/16 | | + | + | + | + | + | 5 |
| 1/64 | | + | + | + | + | + | 5 |
| 1/256 | | + | + | + | + | + | 5 |
| 1/1024 | | + | + | + | + | + | 5 |

*R* Repetition

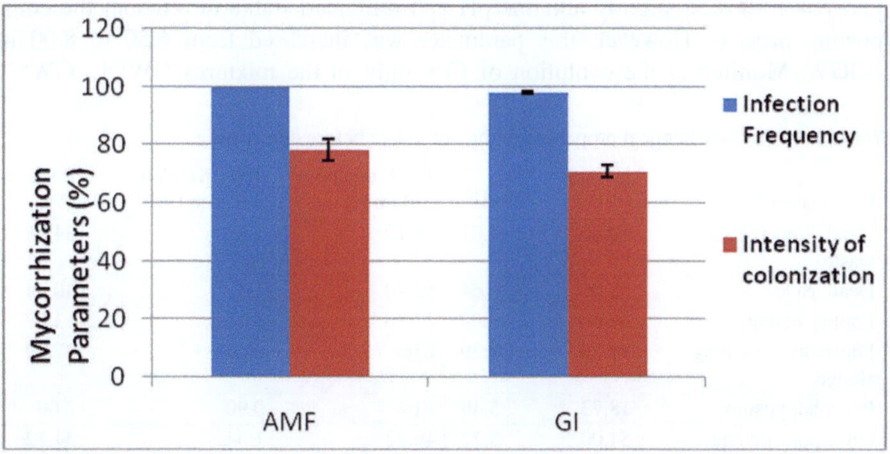


**Fig. 9.3** Frequency and intensity of mycorrhizal corn root after 3 months of culture. *AMF* mycorrhizal consortium obtained from Tafilalet palm grove, *GI* Glomus irregularis

### 9.3.2 Composting Agricultural, Animal, and Agro-industrial Wastes

#### 9.3.2.1 Physicochemical Characteristics of Waste Raw Materials

The physicochemical characteristics of the raw materials used in the composting process are presented in Table 9.2 and Fig. 9.4. We found that the pH of the waste varies between 5 and 8, favorable to microorganisms' growth. The dandelions, phosphate sludge, and poultry have an alkaline pH (~8). Poultry manure is rich in total nitrogen (>3%). The tested dandelions and household waste were rich in organic matter and contain relatively moderate amounts of total nitrogen. Most of the waste used in our study presents a C/N ratio between 14 and 40 enough for enhanced microbial activities and suitable for the composting process. Many studies report that pH and C/N ratio are considered important compost parameters owing to their effects on the quality and suitability of the final product for plant growth.

#### 9.3.2.2 Composting Mixture Process

In Fig. 9.5, we illustrate the different phases of composting. The biodegradation can be assessed by the temperature of the compost, which shows a gradual increase during the mesophilic phase and reaching maximum values during the thermophilic phase. For waste mixtures used in our study, we observed that the temperature increased rapidly in the first days of composting reaching maximum values between 56 and 69 °C. Then it gradually decreased to values ranging from 30 to 35 °C approaching the air temperature during the maturation phase.

We evaluated the pH values of the different combination of the main raw material. PMGW had the highest pH values ranged from 8.50 to 9.20, followed by GWDL with values between 7.23 and 8.40. While the lowest pH value was observed in OCOMWWG (5.49–6.06), the compost pH value ranging from 5.5 to 8.5 is considered acceptable (Table 9.3). The combination of grass waste with waste sludge phosphate showed slightly alkaline pH and remained stable throughout the composting process. However, this parameter was increased from 6.50 to 8.00 in PGGW. Monitoring the evolution of C/N ratio in the mixtures GWDL, GWSP,

**Table 9.2** Physicochemical properties of the raw waste before composting

| Raw materials | Humidity (%) | pH | Total organic carbon (%) | Total Kjeldahl nitrogen (%) | Ratio C/N |
|---|---|---|---|---|---|
| Dandelions (Grass waste) | 69.00 | 8.01 | 58.10 | 1.68 | 34,58 |
| Dead leaves | 58.00 | 6.64 | 55.40 | 1.37 | 40.44 |
| Poultry manure | 39.00 | 8.00 | 47.30 | 3.41 | 13.87 |
| Phosphates washing sludge | 59.00 | 8.29 | 2.00 | 0.073 | 27.39 |
| Phosphogypsum | 18.73 | 5.49 | 1.64 | 0.90 | 2.00 |
| Olive pomace and liquid effluents | 54.00 | 5.77 | 46.52 | 1.34 | 34.72 |
| Household waste (Garbage) | 84.40 | 5.20 | 60.80 | 2.41 | 25.23 |

**Fig. 9.4** Pictures of the raw material waste used for the composting process and the platform composting

OCOMWWG, PGGW, and PMGW showed a rapid decrease from 33 to 11, 64 to 12, 32 to 19, 47 to 18, and 17 to 12, respectively, after the third month (Table 9.3).

The ash content varies widely among the compost owing to the mineralization of organic matter and concentration of carbon, nitrogen, phosphorus, and other nutrients during composting. Notably, GWSP had the higher ash content relative to other mixtures.

As a result, the composts obtained from grass waste and dead leaves presented the lower levels of available phosphorus, which do not exceed 11 ppm, than those of other compost combinations. The values of this element remained high in the other composts, especially in poultry manure and green waste (1,800 ppm).

**Fig. 9.5** Pictures illustrating the different stages during composting

#### 9.3.2.3 Compost Characterization at the Maturity Stage

GWDL, GWSP, PGGW, and PMGW had alkaline pH 8.40, 7.77, 8.00, and 9.20, respectively (Table 9.3), but the OCOMWWG stayed neutral or slightly acid (pH = 6.06). The composts with a pH between 6 and 9 are compatible for most plants.

The results of total organic carbon (TOC) have demonstrated that the composts OCOMWWG (37.30%) and PMGW (36.05%) had significantly higher levels than the GWDL (18.40%), GWSP (7.64%), and PGGW (25.00%) (Table 9.3).

The TOC concentration declined for all the treatments between the initial day and the 3 months of the composting period.

One of the often-used criteria to assess the rate of decomposition in the composting process is the C/N ratio since it can reflect the maturity of the compost. The higher C/N ratio at the initial period of compost, C is not in an available form and drops significantly in all the combinations than the end of composting. After 3 months of composting, we found that all composts had lower C/N ratio values ranging from 11.2 (GWDL) to 18.8 (OCOMWWG). The C/N ratio which is less than 20 is indicative of an acceptable maturity and suitable for nursery plant production.

### 9.3.3 Impacts of Biofertilizers on Growth and Physiological Parameters of Date Palm (*P. dactylifera*) and the Underlying Crops

#### 9.3.3.1 The Case Study of Date Palm

##### 9.3.3.1.1 Effects of Biofertilizers on the AMF Infectivity, Growth, and Water Status of Date Palm

We assessed the effects of the single or combination application of compost (5%) and AMF (mycorrhizal consortium of Tafilalet) on *P. dactylifera* growth. As a result,

**Table 9.3** Physicochemical parameters during composting

| Composting stage | pH | TOC (%) | TKN (%) | C/N Ratio | Ashes (%) | P (mg/g) |
|---|---|---|---|---|---|---|
| GWDLi | 7.23 ± 0,13 | 41.60 ± 1.66 | 1.25 ± 0.24 | 33.28 ± 1.02 | 19.00 ± 1.46 | 0.009 ± 0.0010 |
| GWDL3 | 8.40 ± 0,08 | 18.40 ± 1.01 | 1.60 ± 0.32 | 11.20 ± 0.31 | 31.54 ± 2.43 | 0.011 ± 0.0012 |
| GWSPi | 7.86 ± 0,18 | 25.00 ± 1.37 | 0.39 ± 0.08 | 63.77 ± 2.35 | 50.00 ± 2.11 | 0.167 ± 0.019 |
| GWSP3 | 7.77 ± 0,14 | 7.637 ± 040 | 0.61 ± 0.16 | 12.39 ± 0.46 | 82.00 ± 3.47 | 0.300 ± 0.035 |
| OCOMWWGi | 5.49 ± 0,04 | 50.50 ± 2.74 | 1.57 ± 0.35 | 32.16 ± 1.37 | 3.90 ± 0.30 | – |
| OCOMWWG3 | 6.06 ± 0,05 | 37.30 ± 2.05 | 1.98 ± 0.40 | 18.84 ± 0.85 | 5.60 ± 0.43 | – |
| PGGWi | 6.50 ± 0,06 | 35.33 ± 1.89 | 0.75 ± 0.15 | 47.10 ± 1.42 | 35.56 ± 2.76 | 0.500 ± 0.057 |
| PGGW3 | 8.00 ± 0,10 | 25.00 ± 1.36 | 1.40 ± 0.29 | 17.85 ± 0.62 | 44.80 ± 3.39 | 0.210 ± 0.024 |
| PMGWi | 8.50 ± 0,17 | 44.86 ± 2,26 | 2.59 ± 0.53 | 17.32 ± 0.41 | 19.25 ± 1.41 | 1.780 ± 0.14 |
| PMGW3 | 9.20 ± 0,21 | 36.05 ± 1.96 | 2.87 ± 0.61 | 12.56 ± 0.38 | 35.30 ± 1.89 | 1.800 ± 0.20 |

*GWDL* grass waste and dead leaves, *GWSP* grass waste and sludge of phosphate, *OCOMWWG* olive cake olive oil mill wastewater and garbage, *PGGW* phosphogypsum and green waste, and *PMGW* poultry manure and green waste, *i* initial, 3 months of composting and 6 months, *TOC* total organic carbon, *TKN*: total Kjeldahl nitrogen, *P* available phosphorus the frequency of infection (F) of palm root system with AMF remained high ($F > 60\%$), and it was substantially affected by the application of compost (Table 9.4). Plants inoculated with AMF produced higher above- and belowground biomasses than the control plants. The single application of the organic amendments has no impact on the root growth of the palm. Interestingly, the combination AMF + green waste compost had significantly higher shoot and root biomasses, relative water content, and water potential than the no-treated control plants. Indeed, our results showed that the association AMF and compost GWDL increased 1.6× shoot and 1.9× root biomasses than the control.

Moreover, amended plants with AMF+compost showed a relative water content (81%) slightly higher to the control (77%) (Table 9.4). The mycorrhizal date palms and amended with compost had a higher leaf water potential (−16.83 bar) than control plants (−30.37 bar). Exposure of date palm to the single AMF (2.21 s/cm) or compost (2.22 s/cm) or combined (2.12 s/cm) led to a considerable decrease in stomatal resistance (R) than no-treated plants (2.93 s/cm).

##### 9.3.3.1.2 Date Palm (*P. dactylifera*) Treated with AMF and/or Compost Showed Increased Minerals

We assayed the nutrient contents in shoots of date palm leaves amended with AMF and/or compost since the degree of growth depends on their uptake and translocation. Shoot N, P, K, Ca, and Mg was significantly higher in plants treated with single or synergism effect of AMF and compost than control plants (Table 9.5). These results could at least partly be explained by the effective contribution of mycorrhizal association in improving nutrients of plants through the development of fungal hyphae, allowing good use of the soil minerals and their mobilization to the plants. The positive effects of applying compost GWDL on mineral nutrition was clearly observed. The values of the ionic content are higher in palms amended with AMF+compost than control plants.

#### 9.3.3.2 The Case Study of the Underlying Crops

##### 9.3.3.2.1 Impacts of Biofertilizers on AMF Infectivity and Crop Growth

The calculated frequency of mycorrhization in alfalfa, tomato, wheat, and corn roots exceeds 90% and that in the absence of compost (Table 9.6). On the contrary, the rate of mycorrhization decreases and remains below 66% for plants amended with compost.

##### 9.3.3.2.2 Effect of GWDL Compost and AMF on Alfalfa Biomass

After 2 months of culture, the application of GWDL at doses of 5% has a beneficial effect on improving the production of the shoot and root dry biomasses of alfalfa (*Medicago sativa*) than the control (Fig. 9.6 and Table 9.6). The combination compost+AMF showed the highest values of the shoot and root biomasses.

**Table 9.4** Impact of biofertilizers on growth, water status, and stomatal resistance of date palm after 4 months of culture

| Treatment/ combination | Frequency of mycorrhization (F) (%) | Aerial dry mass (g) | Root dry mass (g) | Relative water content (RWC) (%) | Water potential Ψh (bar) | Stomatal resistance R (s/cm) |
|---|---|---|---|---|---|---|
| Control plants | – | 0.407 ± 0.015 | 0.116 ± 0.016 | 77.21 ± 2.78 | −30.37 ± 0.70 | 2.93 ± 0.10 |
| Compost **GWDL** | – | 0.448 ± 0.015 | 0.088 ± 0.007 | 80.75 ± 2.32 | −24.07 ± 0.98 | 2.22 ± 0.13 |
| AMF | 74.29 ± 5.05 | 0.460 ± 0.020 | 0.128 ± 0.007 | 79.10 ± 2.08 | −27.57 ± 0.40 | 2.21 ± 0.15 |
| **GWDL** + AMF | 62.89 ± 3.77 | 0.670 ± 0.010 | 0.216 ± 0.020 | 81.65 ± 2.02 | −16.83 ± 1.44 | 2.12 ± 0.11 |

*GWDL* grass waste and dead leaves, *AMF* mycorrhizal consortium of Tafilalet

**Table 9.5** Effects of mycorrhization and/or compost on the mineral composition of date palm after 4 months of culture

| Element | Treatments | Content (mg/g DM) |
|---|---|---|
| N (mg/g DM) | Control | 12.14 ± 0.452 |
| | AMF | 18.64 ± 0.516 |
| | Compost GWDL | 20.54 ± 0.935 |
| | Compost GWDL + AMF | 26.33 ± 0.539 |
| P (mg/g DM) | Control | 3.73 ± 0.428 |
| | AMF | 3.62 ± 0.436 |
| | Compost GWDL | 5.44 ± 0.180 |
| | Compost GWDL + AMF | 7.98 ± 0.921 |
| K (mg/g DM) | Control | 1.49 ± 0.231 |
| | AMF | 1.79 ± 0.141 |
| | Compost GWDL | 1.85 ± 0.107 |
| | Compost GWDL + AMF | 2.01 ± 0.104 |
| Ca (mg/g DM) | Control | 0.55 ± 0.127 |
| | AMF | 0.86 ± 0.032 |
| | Compost GWDL | 1.03 ± 0.168 |
| | Compost GWDL + AMF | 1.15 ± 0.078 |
| Mg (mg/g DM) | Control | 1.49 ± 0.122 |
| | AMF | 1.77 ± 0.1417 |
| | Compost GWDL | 2.07 ± 0.065 |
| | Compost GWDL + AMF | 2.74 ± 0.097 |

*GWDL* grass waste and dead leaves, *AMF* mycorrhizal consortium of Tafilalet palm grove, *DM* dry matter

#### 9.3.3.2.3 Tomato Has Better Growth After the Compost (GWSP) and/or AMF Applications

We evaluated tomato growth parameters after the amendment with mycorrhizal consortium Tafilalet (AMF) and/or GWSP compost (a mixture of dandelions and phosphates washing sludge, at 10% dose) (Table 9.6). The production of the shoot and root biomasses has increased following the colonization with AMF and/or compost, being more relevant after the combination of both treatments. Indeed, the synergism effects of both biofertilizers have markedly improved plant growth compared with uninoculated control.

#### 9.3.3.2.4 Wheat Plants Showed Increased Biomass in Response to Compost (OCOMWWG) and/or AMF

The application of compost-based waste pomace olive and garbage (OCOMWWG) at a dose 10% or the inoculation of the roots by *Glomus irregularis* (GI) increased slightly the above- and belowground biomasses of wheat than control plants (Table 9.6). The dual application of GI and OCOMWWG compost has no positive effect on the SDM of wheat, but this combination has increased the root biomass than control plants.

**Table 9.6** Impacts of biofertilizers on the growth of underlying crops

| Plants | | Treatments | Frequency of mycorrhization (F) (%) | Aerial dry mass (g) | Root dry mass (g) |
|---|---|---|---|---|---|
| Legume | Alfalfa | Control plants | – | 0.006 ± 0.00 | 0.006 ± 0.004 |
| | | Compost GWDL | – | 0.379 ± 0.01 | 0.234 ± 0.029 |
| | | AMF | 100.00 ± 5.77 | 0.079 ± 0.03 | 0.075 ± 0.025 |
| | | Compost GWDL + AMF | 52.00 ± 7.23 | 0.666 ± 0.05 | 0.361 ± 0.051 |
| Vegetable crops | Tomato | Control plants | – | 0.060 ± 0.01 | 0.020 ± 0.003 |
| | | Compost GWSP | – | 0.130 ± 0.01 | 0.030 ± 0.003 |
| | | AMF | 91.55 ± 4,57 | 0.100 ± 0.00 | 0.030 ± 0.002 |
| | | Compost GWSP + AMF | 65.25 ± 5,57 | 0.350 ± 0.02 | 0.100 ± 0.007 |
| Cereals | Wheat | Control plants | – | 0.068 ± 0.01 | 0.004 ± 0.001 |
| | | Compost OCOMWWG | – | 0.082 ± 0.01 | 0.018 ± 0.002 |
| | | GI | 95.00 ± 3,84 | 0.077 ± 0.01 | 0.017 ± 0.001 |
| | | Compost OCOMWWG+ *GI* | 60.00 ± 3,85 | 0.065 ± 0.01 | 0.013 ± 0.003 |
| | Corn | Control plants | – | 0.438 ± 0.03 | 0.139 ± 0.013 |
| | | Compost OCOMWWG | – | 0.648 ± 0.06 | 0.137 ± 0.036 |
| | | GI | 98.04 ± 3,85 | 0.534 ± 0.04 | 0.241 ± 0.02 |
| | | Compost OCOMWWG + GI | 65.77 ± 5.05 | 0.456 ± 0.08 | 0.183 ± 0.013 |

*GWDL* grass waste and dead leaves, *GWSP* grass waste and sludge of phosphate, *OCOMWWG* olive cake, olive oil mill wastewater, and garbage, *AMF* mycorrhizal consortium of Tafilalet, *GI* *Glomus irregularis*

#### 9.3.3.2.5 Compost (OCOMWWG) and AMF Promote Aerial and Root Traits of Corn Plants

The mixed treatment of corn with compost (OCOMWWG) and AMF (*G. intraradices*) increased shoot and root dry matters. In the same line of results, the single inoculation of the corn plant with GI improved the 1.25× the aerial biomass and 1.71× the root DM. However, we observed no difference of the mentioned parameters between the efficiency of compost alone on the belowground biomass.

### 9.3.4 The Potential Effects of Biofertilizers to Improve Crops Yield in the Field

We evaluated alfalfa, green and red lettuces, leek, and wheat treated with AMF (*G. intraradices*)-rhizobia and/or compost (poultry manure and green waste) in the field to compare the yield trait with those of untreated control plants. Each crop was

**Fig. 9.6** Effect of mycorrhizal consortium of Tafilalet (AMF) and grass waste and dead leaf (GWDL) compost on alfalfa growth after 2 months of culture

**Fig. 9.7** Implementation and randomization of crops plantations treated or not with different biofertilizers (AMF, PGPR, and/or compost)

randomly arranged in the different blocks of the managed parcels (Fig. 9.7). We found a significant difference in yielding after biological and organic fertilizers uses in leguminous (alfalfa), vegetable crops (lettuce, leek), and cereals (wheat) (Table 9.7). The application of indigenous AMF-rhizobia and compost (GWDL) increased 2× the total fresh biomass of alfalfa and green lettuce than control plants.

Composts (PMGW) promoted 4.4× the yield in leek plants and 1.5× red lettuce. The synergism of composts enriched by PGPR and *G. irregularis* improved yields of red lettuce and leek 1.6× than control plants. The single application of PGPR or its combination with compost and *GI* increased significantly red lettuce yield.

We also examined the important positive effects of the combination of AMF-PGPR and rhizobia on wheat yield compared to untreated control plants.

**Table 9.7** Impacts of the tested biofertilizers on crops yield in the field

| | Plant | Treatment/combination | Yield (g/plant) |
|---|---|---|---|
| Legume | Alfalfa | Control plants | 11.63 ± 1.15 |
| | | Compost GWDL | 14.93 ± 3.22 |
| | | AMF | 17.46 ± 2.26 |
| | | Rhizobia | 16.53 ± 2.97 |
| | | Compost GWDL + AMF + rhizobia | 26.06 ± 3.25 |
| Vegetable crops | Green lettuce | Control plants | 357.10 ± 43.28 |
| | | Compost PGGW | 464.75 ± 65.44 |
| | | GI | 536.75 ± 64.49 |
| | | *Compost PGGW* + GI | 635.38 ± 90.27 |
| | Red lettuce | Control plants | 465.60 ± 32.95 |
| | | Compost PMGW | 685.15 ± 59.68 |
| | | GI | 557.15 ± 40.78 |
| | | PGPR | 739.10 ± 43.33 |
| | | Compost PMGW + GI + PGPR | 761.57 ± 35.07 |
| | Leek | Control plants | 5.27 ± 1.53 |
| | | Compost PMGW | 23.00 ± 5.49 |
| | | GI | 6.32 ± 0.57 |
| | | PGPR | 6.28 ± 0.66 |
| | | Compost PMGW + GI + PGPR | 9.00 ± 1.28 |
| Cereal | Wheat | Control plants | 4.52 ± 0.18 |
| | | AMF | 5.73 ± 1.03 |
| | | PGPR | 5.45 ± 0.39 |
| | | Rhizobia | 7.37 ± 1.12 |
| | | AMF+PGPR+rhizobia | 11.00 ± 1.16 |

$n = 12$ belonging to each of the 12 repeated blocks for the same treatment and the same culture
*GWDL* grass waste and dead leaves, *PGGW* phosphogypsum and green waste, *PMGW*, poultry manure and green waste, *AMF* mycorrhizal consortium of Tafilalet, *GI Glomus irregularis*, *PGPR* plant growth-promoting rhizobacteria

These results suggest the beneficial role of the tripartite association AMF-PGPR-compost to increase the yield of underlying crops. This efficacy depends on the plant cultivar, the nature, and dose of the compost and the mycorrhizal and bacterial strains used.

### 9.3.5 Assessment of Physicochemical Parameters in Soil Samples Collected from the Agricultural Areas

We evaluated physicochemical properties of the field soils used before and after the experimentation (Table 9.8). The percentages of sand (74.75%) and silt (13.55%) were higher compared to the other soil elements. The soil used for our field trial is classified as calcareous with a pH value of 8.12. The soil conductivity was ranging between 0 and 500 μs/cm, and the percentage of limestone was 5.04%. In addition,

**Table 9.8** Physicochemical properties of agricultural soil before and after the plantation

| Date | Analyses | pH | Conductivity (µs/cm) | Total carbon (%) | Organic matter (%) | Total nitrogen (mg/g) | Available phosphorus (ppm) |
|---|---|---|---|---|---|---|---|
| **Before the experiment** | Soil (start experiment) | 8.12 ± 0,06 | 138.30 ± 2.91 | 0.50 ± 0,04 | 0.87 ± 0,15 | 9.98 ± 0,95 | 57.42 ± 1,21 |
| **At harvest time** cereals (wheat) | Control | 8.05 ± 0,04 | 506.66 ± 5.82 | 1.22 ± 0,07 | 2.10 ± 0,13 | 18.20 ± 1,92 | 57.00 ± 1,61 |
| | AMF + PGPR+rhizobia | 7.38 ± 0,03 | 204.50 ± 3.65 | 2.30 ± 0,11 | 3.97 ± 0,19 | 30.80 ± 2,87 | 63.20 ± 4,33 |
| **At harvest time** legume (alfalfa) | Control | 8.36 ± 0,08 | 150.00 ± 2.29 | 0.79 ± 0,05 | 1.20 ± 0,10 | 23.98 ± 2,07 | 61.40 ± 1,05 |
| | Compost GWDL + AMF+rhizobia | 8.25 ± 0,05 | 210.00 ± 4.34 | 1.19 ± 0,03 | 2.05 ± 0.17 | 42.65 ± 3,77 | 74.77 ± 3,95 |
| **At harvest time** vegetable crop (leek) | Control | 7.52 ± 0,18 | 257.00 ± 6.45 | 0.64 ± 0,10 | 1.40 ± 0,18 | 8.07 ± 1,62 | 46.71 ± 1,65 |
| | Compost PMGW+ GI +PGPR | 7.55 ± 0,08 | 246.00 ± 5.54 | 0.95 ± 0,06 | 1.92 ± 0,11 | 9.93 ± 1,62 | 195.86 ± 7,22 |
| **At harvest time** vegetable crop (red and green lettuces) | Control | 7.49 ± 0,03 | 424,00 ± 3.32 | 1.11 ± 0,06 | 2.42 ± 0,10 | 9.93 ± 1,62 | 106.16 ± 6,77 |
| | Compost PMGW+ GI +PGPR | 7.72 ± 0,02 | 237.00 ± 6.55 | 1.64 ± 0,06 | 2.18 ± 0,10 | 9.23 ± 1,60 | 257.43 ± 12,18 |

*GWDL* grass waste and dead leaves, *PMGW* poultry manure and green waste, *AMF* mycorrhizal consortium of Tafilalet, *GI*, *G. irregularis*, *PGPR* plant growth-promoting rhizobacteria the contents of total carbon (0.50%) and organic matter (0.87%) reflected the poorly remineralized soil organic matter. Furthermore, the field soil before starting the treatments contains 9.98 mg/g of total nitrogen and 57.42 ppm of available phosphorus.

After the field trials, our results (Table 9.8) showed that the treatments applied (compost, AMF, and PGPR) have improved soil quality than the before starting. Moreover, our treatments have slightly decreased the value of pH, with the exception of leguminous. The electrical conductivity has increased throughout the cultivation, being the untreated cereals (506.66 μs/cm) and lettuces (424 μs/cm) the highest. Relative to untreated control conditions, the biofertilizers increased total organic matter and total carbon content especially the combination AMF-PGPR-compost. The application of tripartite combination was correlated positively with the organic matter (3.97%) and total carbon (2.30%) in wheat crops. The total nitrogen content in soil was also enhanced by the application of biofertilizers. The highest value of total nitrogen (42.65 mg/g) was observed in alfalfa plants grown under the amended condition of AMF-rhizobia-compost. This element remained similar to the value obtained at the initial in leek and lettuce independently of the treatment.

Interestingly, the application of biofertilizers improved the soil available phosphorus at the harvest of different cultures, being the highest values observed in the rhizosphere of lettuce 4.5× (257.4 ppm) and leek 3.41× (195.8 ppm) compared to control (57.4 ppm).

## 9.4 Discussion

The purpose of our study was to investigate the growth promotion effect of single and combined (1) compost-based crop residues and animal wastes, microorganisms, (2) AMF (native AMF, mycorrhizal consortium of Tafilalet, and exotic AMF, *G. intraradices*), and (3) rhizobia and indigenous PGPR isolated from soil of the research sites. All these key players were tested in the greenhouse and the field for their effect on biomass, yield, development and physiology, and nutrient levels in several crop tissues. The mycorrhizal potential of a soil depends on the number of spores present in the rhizosphere, their quality, and capacity for adaptation and infectious properties. For instance, the number of mycorrhizal propagules of Tafilalet palm groves rhizosphere (1627/100 g) is 7.5× higher than palm grove northeast of Marrakesh in Morocco (219 propagules/100 g) (Meddich et al. 2017; Meena et al. 2014); Varma et al. 2017b; Kumar et al. 2018a). Whereas the mycorrhizal potential of saline soils of the Marrakesh palm grove does not exceed 149 propagules per 100 g of soil (Meddich et al. 2015c; Meena et al. 2015c; Yadav et al. 2018a). Changes in physicochemical properties of rhizospheric soil such as soil pH, water potential and partial pressure of $O_2$, and plant exudation could affect the ability of PGPR strains to colonize the rhizosphere.

The infectivity parameters (F% and M%) were higher in Tafilalet palm groves soils than the reference strain *G. irregularis*. The consortium mycorrhizal isolated from Tafilalet oasis area and selected *G. irregularis* showed a great ability to infect palms roots and the underlying crops (wheat, corn, alfalfa, leek, lettuce, and tomato). These results suggest the presence of variability in the parameters of infectivity of AMF according to the host plants and the conditions of the medium. A signal exchange between the two partners AMF-plant could be established, and molecules contained in the root exudates influence the development of the arbuscular mycelia (Gianinazzi-Pearson et al. 1996; Meena et al. 2018b; Dadhich et al. 2015; Sofi et al. 2018). Subsequently, AMF mycelia colonize cortical cells and give rise to fungal arbuscules representing the preferred site of exchange between the fungus and the host plant (Gianinazzi-Pearson and Gianinazzi-Silvio 1988; Gianinazzi-Pearson et al. 1996). The frequency of mycorrhization of palm roots and underlying crops with Tafilalet consortium and *G. intraradices* decreased following the application of compost. This could be due, at least partly, to the richness of compost in mineral elements or the high water retention inhibiting, by asphyxiation, thus the development of the symbiotic association and undermining the aggressivity of the mycorrhizal isolates. In addition, plants subjected to these conditions can directly benefit from the organic and mineral amendments and the absorption of water without establishing the relationship with AMF. Similar results were reported by Meddich (2001) for clover and barley imposed to increasing concentrations of mineral elements, especially phosphorus.

It is important to note the importance of mixed inoculation of AMF and compost in improving the growth and mineral nutrition of date palm. Mycorrhizal and amended palms with compost showed higher levels of N, P, K, Ca, and Mg than control plants, suggesting the compost and bacterium's ability to increase crops absorption of minerals.

The increase of waste temperature during the aerobic process of composting owing to the metabolism of microorganisms to solubilize the organic compounds. Hachicha et al. (2009) reported that during composting, a temperature exceeding 60 °C and maintained for several days ensures the destruction of pathogenic microorganisms. Generally, four phases of temperature fluctuation exist in the composting process: mesophilic, thermophilic, cooling, and maturation. The decrease in temperature during the maturity phase owing to the depletion of easily biodegradable organic matter (Gea et al. 2003; Petiot and Guardia 2004; Meena et al. 2016b; Sihag et al. 2015; Verma et al. 2015a).

The increase in pH can be explained by the accumulation of ammonia and/or a loss of short-chain fatty acids and volatiles resulting from the microbial activity (Lim et al. 2012; Shak et al. 2014). The rapid decrease in C/N observed in our compost at the third month of composting phase could be explained by the significant reduction in the metabolizable organic carbon related to the biodegradation of organic matter. Compost with a C/N ratio below 20 is considered mature and can be used without any restrictions (Jimenez and Garcia 1989). A C/N ratio close to 10–15 is often considered as an index of humic material formation and stability of composts (Lim et al. 2014; Meena et al. 2015a; Yadav et al. 2017a; Kakraliya et al. 2018). The application of low dose (5%) of compost GWDL, with low levels of available phosphorus, with AMF isolated from Tafilalet palm grove, showed a beneficial effect to improve *P. dactylifera* growth parameters. Roca-Pérez et al. (2009)

reported that the addition of organic matter by adding compost improves soil structure, fertility, porosity, and water retention. Several soil properties, including structure and porosity, affect root growth (Roca-Pérez et al. 2009; Datta et al. 2017b; Ram and Meena 2014). The high content of organic matter in composts stimulates the biological and enzymatic activities of substrates and the bioavailability of nutrients by mineralization of the organic matter (Hofman and Dušek 2003; Crecchio et al. 2004; Meena et al. 2015b). Also, the humic substances might promote nutrient uptake and can determine the rhizogenic activity (Eyheraguibel et al. 2008).

Overall, the low dose (5% and 10%) of the tested compost with or without mycorrhizal fungi has beneficial impacts on improving the growth of crops species: alfalfa, tomato, wheat, and corn. Similar results were observed in lettuce and maize amended with low compost concentrations (Mrabet et al. 2011). In contrast, the negative effect of 100% compost dose application has decreased both aerial and root biomasses and nutrients uptake than control (Meddich et al. 2017; Dhakal et al. 2016; Kumar et al. 2017a). This finding owing to the high concentration of mineral and elements in the substrate leading to inadequate assimilation of nutrients. Similar results were found in corn plant biomass amended with high compost concentration (Abouelwafa 2009).

Indeed, our study showed that the interaction of low-dose compost GWDL (5%) and AMF has significantly stimulated shoot and root biomasses of alfalfa. Similar results were observed in tomato plants treated with the combination of the low dose 10% of compost GWSP and AMF. The single application of GI and OCOMWWG promoted the growth of wheat and corn, while the combination GI+OCOMWWG has positively affected the root growth. These results confirm the good functioning of mycorrhization under limiting conditions and soils with low organic matter and nutrients supply (Meddich et al. 2017).

The application of different doses of phosphorus or NPK chemical fertilizers to substrates for growing non-mycorrhizal plants of clover, barley, and date palm did not lead to better results to those obtained with AMF or compost application (Meddich 2001; Meddich et al. 2015d). The availability and mobilization of phosphorus element with AMF or composts could not be solely responsible for improving plant tolerance to water and salt stress. Other nutrients such as Ca, K, N, and Mg could contribute to these strategies. Also, a better distribution of the water circulation in the plant can explain, partially, this tolerance in presence of AMF.

Our genetic analyses revealed the expression of three genes of MIP family coding for the synthesis of aquaporins in mycorrhizal clover roots with the Aoufous complex of Tafilalet and *G. monosporus* under severe drought stress (30% FC) (Zeze et al. 2007, 2008; Meena et al. 2017a). In this study, *P. dactylifera* inoculated with AMF or amended with compost showed similar RWC than control plants, whereas mycorrhizal palms amended with compost showed higher leaf water potential than control plants. At the same time, mycorrhizal and treated plants with compost showed the lowest stomatal resistance compared to control plants. The low stomatal resistance in amended plants could improve the mesophilic $CO_2$ uptake (Brown and Bethelenfalvay 1987) conferring an increase in photosynthesis (Lawlor 1987).

To assess the efficacy of microorganisms and compost on the underlying culture used in arid and semiarid regions, we exposed several crop species to the application of AMF-rhizobia and compost (GWDL). Our field assessment results indicate yield enhancement of wheat and alfalfa. Our results corroborate findings in *Vicia faba* (Jia et al. 2004), beans (Amrani 2009) and *Vigna unguiculata* (Clautilde et al. 2011) inoculated with rhizobia and AMF. The interactions between plants and AMF and/or rhizobia bacteria by which all partners could benefit from the mutual association may improve the growth of the plants owing to the mechanisms of growth promotion developed by the microorganisms such as the fixation of nitrogen, mineral solubilization, water uptake, and phytohormone production (Finlay 2007; Jalili et al. 2009; Oufdou et al. 2014; Verma et al. 2015b).

We also assessed the yield traits of underlying crops under the exposition to agricultural, animal, or agro-industrial wastes. Compost of green waste associated with poultry manure (PMGW) has considerably increased the yield of leek and red lettuce. The combination of this compost enriched with PGPR and *G. irregularis* improved also the yield of red lettuce. Our results are comparable to those obtained by Koulibaly et al. (2015) showing an increment of 65% of the cotton plant yield after the addition of compost. Copetta et al. (2011) showed that the use of AMF and compost from green waste considerably improved the yield and quality of tomato fruit. Composts improve the different physicochemical and biological properties of soils (Toumpeli et al. 2013; Mehta et al. 2014; Meena et al. 2017c) and consequently increase the yield of plants (Motta and Maggiore 2013). They are able to improve the mineral and water status of plants (Gharib et al. 2008; Meddich et al. 2015c; Layek et al. 2018). In addition, compost enriched soil with organic matter and microorganisms. These components contribute to make available and store nutrients for plants, promote the biological activity as a source of energy for microorganisms, and help on the structure, physicochemical properties, and aeration of the soil. Furthermore, the composts are involved in the maintenance of sandy soils and colloidal particles to avoid the erosion phenomena by retaining the particles set in motion by the rain and absorbing the drops (Bodet and Carioli 2001).

PGPR have the ability to solubilize complex phosphate, assimilate nitrogen, and reduce stresses by modulating the expression of ACC deaminase (Jalili et al. 2009). They are able to modulate the growth and architecture of crop roots by releasing phytohormones (i.e., auxin, cytokinins, etc.) or other antimicrobial and/or antifungal substances for the control of the harmful effects of pathogens (Souza et al. 2015). The pathway of AIA synthesis by PGPR could also stimulate plant growth (Barnawal et al. 2012; Chen et al. 2013; Meena et al. 2017b).

It is notable that AMF can solubilize phosphate and mobilize other nutrients for the benefit of the plant (Jia et al. 2004; Clautilde et al. 2011; Tarraf et al. 2017). Furthermore, AMF have the ability to improve the water status of plants (Zeze et al. 2007, 2008; Baslam et al. 2013). They are capable of mobilizing macro- and microelements in soil and water level in plants and controlling pathogens.

The agricultural soil analysis carried out before the plantation was able to characterize the sandy loam texture, low in organic matter (0.87%), low electrical conductivity (138.3 µs/cm), and slightly alkaline pH (8.12) owing to the high limestone content (5.04%). Moreover, the phosphorus content available from the soil (57.42 ppm) is relatively low, which is in favor of the formation, development, and proper functioning of symbiosis between plants and microorganisms such as PGPR and AMF (Meddich et al. 2015d, 2017). The application of compost can be of great interest in improving the fertility of agricultural soils and consequently improving crops growth and yield (Meddich et al. 2016; Meena et al. 2018a).

The AMF applied in our study were infectious and adapted to all the studied crop species showing their higher frequency of mycorrhization in seedlings treated with the association with compost. AMF infectivity and root colonization rates are positively correlated to improve crop biomass and plant physiological and water parameters (Meddich et al. 2015a, b). Sghir et al. (2014) observed that mycorrhizal frequencies of date palm roots and arbuscular contents decreased significantly in palm trees inoculated with the combination AMF-*Trichoderma harzianum* than in plants inoculated with only AMF. However, the double inoculation makes a major contribution to the growth and root architecture of date palm (Sghir et al. 2014; Meena et al. 2018c) and high yielding of soybean (Egberongbe et al. 2010). This suggests the existence of compromises and positive and complementary impacts between the symbiotic microorganisms and their host plant. The physicochemical properties of the agricultural soil after the crop harvests showed that all treatments had a positive effect on the nutritional and physicochemical properties of the rhizosphere. In fact, the contents of organic matter, carbon, and available phosphorus improved by the composts and/or microorganisms compared to uninoculated control soil. According to Caravaca et al. (2002), the mycorrhizal inoculation of *Olea europaea* was very effective in improving soil quality. Other studies (Bhattacharyya and Jha 2012; Sharma et al. 2013) have found that the ability of microorganisms especially PGPR and rhizobia improved the quality of soil and the availability of nutrients through different mechanisms including solubilization of phosphate and potassium, symbiotic and free nitrogen fixation, and the production of siderophores.

Together the results of field trials suggest that indigenous biofertilizers can constitute a better alternative well adapted to the use of chemical fertilizers in arid and semiarid conditions and can fulfill diverse beneficial interactions in plants leading to promising solutions for sustainable and environment-friendly agriculture.

## 9.5 Future Perspectives

Healthy soil is vital to life on Earth to maintain or increase the global yield production by at least 70% to feed the anticipated $9.6 \times 10^9$ people by 2050. Yield losses are caused by the effects of climate change and by indirect effects such as increased inputs in crop production. To counteract these negative effects, various adaptation strategies have been suggested. Benefiting the soil in terms of quality or health is closely linked by the adoption of best management practices. These principles call for the integrated use of beneficial microorganisms and organic manures to meet global food security and sustainable agriculture demands. Thus, this study clearly pointed out how the natural microbial-mediated process can impact positively the soil and consequently growth and yield of crops adapted to harsh environmental conditions. Our approach of rhizoengineering based on the single or multi-inoculation of native or exotic microorganisms such as AMF and PGPR, together with the use of different compost-based growing media, influences the nutrient use of plants and the rhizosphere quality. An understanding of the mechanisms of action of these complex interactions of the compost and/or microbial-promoted increase of crop yield and health and soil fertility has yet to be explored. Further, future researches hinging soil aspects in addition to the primary focus of crop yields are needed. At same time, long-term agronomic experiments in different agro-ecological zones across the world to provide practical datasets pertinent to soil quality are time-demanding tasks.

## 9.6 Conclusions

In summary, our results demonstrate:

- The soils of Tafilalet palm grove showed higher mycorrhizal potential and infectivity capacity. The mycorrhizal fungi isolated from this grove and *G. irregularis* were infectious and increased the biomass and other physiological parameters of the date palm and its underlying crops (wheat, corn, tomato, lettuce, alfalfa, and leek).
- AMF symbiosis may enhance the osmotic adjustment in plants conferring the maintenance of higher leaf water status.
- The use of the composts has clearly promoted the growth of date palm and underlying crops tested. The combination of low doses of native composts and indigenous AMF significantly improved the growth of *P. dactylifera* and the crop species.
- The combination of *G. irregularis* and 10% OCOMWWG compost has no positive effect on the production of shoot dry matter of wheat and corn but increased substantially the root biomass than control plants.
- The application of the various combinations and biological treatments in field conditions resulted in significant differences than the control. The tripartite combination of AMF-PGPR-compost significantly increased crop yields in all crop species: leguminous, cereals, and vegetable crops. This efficacy depends on the plant, the nature and dose of the compost, and the mycorrhizal and bacterial isolates tested.
- The use of such effective organic and biological amendments could constitute a biotechnological tool to improve yield and plant adaptation to soil and environmental constraints.

In general, our study elucidated the positive impacts of biofertilizers composts-AMF-PGPR on the growth, yield, and development of date palms and cultures underlying with the adoption of innovative practices. The application of composts and/or microorganisms could improve soil fertility, preserve water resources, respect the environment, and ensure the development of sustainable organic agriculture. The transfer of this technology in the open field will have a positive impact on the oasis environment by generating socioeconomic and environmental benefits such as improving farmers' incomes, reducing poverty, and preserving natural resources.

**Acknowledgments** This work was partially supported by Socially Responsible Projects, Cadi Ayyad University UCAM/RSU 2018 Morocco to Abdelilah Meddich Ph.D., PPR2/2016/42, CNRST, Morocco to Khalid Oufdou Ph.D., and KAKENHI Grants-in-Aid for Scientific Research (A) (15H02486) from the Japan Society for the Promotion of Sciences and by a Grant for Promotion of KAAB Projects (Niigata University) from the Ministry of Education, Culture, Sports, Science, and Technology, Japan, to Toshiaki Mitsui, Ph.D.

We thank the Urban Municipality of Marrakesh (Morocco) for the use of composting platform.

This work was also partly supported by the r4d project "Application of organic bio-fertilizer technology to improve the sustainability of date palm production and cultivation" with the grand number IZ07Z0_160904 funded by the r4d program, the Swiss Program for Research on global Issues for Development, a partnership of the Agency for Development and Cooperation and the Swiss National Science Foundation. Colleagues who have contributed to our thinking to set up some experiments include Paul Mäder Ph.D., Sarah Symanczik Ph.D. (project fertile date palm, Fibl, Switzerland), and Mohamed Hafidi Ph.D. (Ecology and Environment Laboratory, Faculty of Science Semlalia, Marrakesh, Morocco).

## References

Abouelwafa R (2009) Biodégradation aérobie des boues de station d'épuration des rejets du raffinage des huiles brutes pour une valorisation agronomique. National Doctoral thesis, Faculty of Science Semlalia, Cadi Ayyad University, Marrakesh, Morocco

AFNOR (1975) Dosage de l'azote ammoniacal. NF-T 90-015. Association Française de Normalisation.

AFNOR (1999) Matières fertilisantes et supports de culture. Tome 1: Normalisation. Dans Recueil normes et réglementation. Association française de normalisation, Paris, pp 433–436

Al-Karaki GN (2013) Application of mycorrhizae in sustainable date palm cultivation. Emir J Food Agric 25:854–862

Al-Karaki GN, Othman Y, Al-Ajmi A (2007) Effects of Mycorrhizal fungi inoculation on landscape Turf establishment under arabian gulf region conditions. Arab Gulf J Sci Res 25:147–152

Al-Yahya'ei MN, Oehl F, Vallino M, Lumini E, Redecker D, Wiemken A, Bonfante P (2011) Unique arbuscular mycorrhizal fungal communities uncovered in date palm plantations and surrounding desert habitats of Southern Arabia. Mycorrhiza 21:195–209

Amrani A (2009) Effet de la double inoculation Rhizobium-Champignons mycorhiziens sur la croissance de la féverole et du haricot nain. Doctoral thesis, University of Oran Es-Sénia, 111p

Arancon NQ, Edwards CA, Babenko A, Cannon J, Galvis P, Metzger JD (2008) Influences of vermicomposts, produced by earthworms and microorganisms from cattle manure food waste and paper waste, on the germination, growth and flowering of petunias in the greenhouse. Appl Soil Ecol 39:91–99

Ashoka P, Meena RS, Kumar S, Yadav GS, Layek J (2017) Green nanotechnology is a key for eco-friendly agriculture. J Clean Prod 142:4440–4441

Aubert G (1978) Méthodes d'analyse des Sols. Edition C.R.D.P., Marseille, France, 360 p

Augé RM, Stodola JW, Ebel RC, Duan X (1995) Leaf elongation and water relations of mycorrhizal sorghum in response to partial soil drying: two *Glomus* species at varying phosphorus fertilization. J Exp Bot 46:297–307

Awad MA (2006) Water spray as a potential thinning agent for date palm flowers (*Phoenix dactylifera* L.) c.v. 'Lulu'. Sci Hortic 111:44–48

Azcon-Aguilar C, Barea JM (1996) Arbuscular mycorrhizas and biological control of soil-borne plant pathogens – an overview of the mechanisms involved. Mycorrhiza 6:457–464

Baize D (1988) Guides des analyses courantes en pédologie. INRA, Paris, p 172

Balick MJ, Beck HT (1990) Useful palms of the world: a synoptic bibliography. Columbia University Press, New York

Barnawal D, Bharti N, Maji D, Chanotiya CS, Kalra A (2012) 1-Aminocyclopropane-1-carboxylic-acid (ACC) deaminase-containing rhizobacteria protect Ocimum sanctum plants during waterlogging stress via reduced ethylene generation. Plant Physiol Biochem 58(2):27–35

Baslam M, Goicoechea N (2012) Water deficit improved the capacity of arbuscular mycorrhizal fungi (AMF) for inducing the accumulation of antioxidant compounds in lettuce leaves. Mycorrhiza 22:347–359

Baslam M, Esteban R, García-Plazaola JI, Goicoechea N (2013) Effectiveness of arbuscular mycorrhizal fungi (AMF) for inducing the accumulation of major carotenoids, chlorophylls and tocopherol in green and red leaf lettuces. Appl Microbiol Biotechnol 97(7):3119–3128

Bearden B, Petersen L (2000) Influence of arbuscular mycorrhizal fungi on soil structure and aggregate stability of a vertisol. Plant Soil 218:173–183

Bhattacharyya PN, Jha DK (2012) Plant growth-promoting rhizobacteria (PGPR): emergence in agriculture. World J Microbiol Biotechnol 28:1327–1350. https://doi.org/10.1007/s11274-011-0979-9

Blaha D, Prigent-Combaret C, Mirza MS, Moënne-Loccoz Y (2006) Phylogeny of the 1-aminocyclopropane-1-carboxylic acid deaminase-encoding gene acdS in phytobeneficial and pathogenic Proteobacteria and relation with strain biogeography. FEMS Microbiol Ecol 56:455–470

Bodet JM, Carioli M (2001) Modalités pratiques d'emploi des composts élaborés à partir de produits d'origine non agricole. Les nouveaux défis de la fertilisation raisonnée, Gemas, pp 183–193

Botes A, Zaid A (2002) The economic importance of date production and international trade. In: Zaid A (ed) Date palm cultivation. Food and Agricultural Organisation of the United Nations, Rome

Brown MS, Bethelenfalvay GJ (1987) Glycine *Glomus-Rhizobium* symbiosis. Plant Physiol 85:120–123

Buragohain S, Sharma B, Nath JD, Gogaoi N, Meena RS, Lal R (2017) Impact of ten years of biofertilizer use on soil quality and rice yield on an inceptisol in Assam, India. Soil Res. https://doi.org/10.1071/SR17001

Caravaca F, Barea JM, Figueroa D, Roldán A (2002) Assessing the effectiveness of mycorrhizal inoculation and soil compost addition for enhancing reafforestation with Olea europaea subsp. sylvestris through changes in soil biological and physical parameters. Appl Soil Ecol 20:107–118. https://doi.org/10.1016/S0929-1393(02)00015-X

CFC/COI (2008) Valorisation des sous produits de l'olivier sur les terres agricoles. Agri Maghreb 26(33)

Chen L, Dodd IC, Theobald JC, Belimov AA, Davies WJ (2013) The rhizobacterium Variovorax paradoxus 5C-2, containing ACC deaminase, promotes growth and development of Arabidopsis thaliana via an ethylene-dependent pathway. J Exp Bot 64(6):1565–1573

Clapp CE, Chen Y, Hayes MHB, Cheng HH (2001) Plant growth promoting activity of humic substances. In: Swift RS, Spark KM (eds) Understanding and managing organic matter in soils, sediments and waters. International Humic Substances Society, St. Paul, 243–255pp

Clautilde M, Enama A, Robert N (2011) Effet simultané de la dilution et de la combinaison du rhizobium et des mycorhizes sur la production foliaire et les propriétés physico-chimiques des jeunes feuilles de *Vigna unguiculata* L. Walp. J Appl Biosci 40:2668–2676

Copetta A, Bardi L, Bertolone E, Berta G (2011) Fruit production and quality of tomato plants (*Solanum lycopersicum* L.) are affected by green compost and arbuscular mycorrhizal fungi. Plant Biosyst 145:106–115

Couillerot O, Prigent-Combaret C, Caballero-Mellado J, Moënne-Loccoz Y (2009) Pseudomonas fluorescens and closely-related fluorescent pseudomonads as biocontrol agents of soil-borne phytopathogens. Lett Appl Microbiol 48:505–512

Crecchio C, Curci M, Pizzigallo MDR, Ricciuti P, Ruggiero P (2004) Effects of municipal solid waste compost amendments on soil enzyme activities and bacterial genetic diversity. Soil Biol Biochem 36:1595–1605

Dadhich RK, Meena RS (2014) Performance of Indian mustard (*Brassica juncea* L.) in response to foliar spray of thiourea and thioglycollic acid under different irrigation levels. Indian J Ecol 41(2):376–378

Dadhich RK, Meena RS, Reager ML, Kansotia BC (2015) Response of bio-regulators to yield and quality of Indian mustard (*Brassica juncea* L. Czernj. and Cosson) under different irrigation environments. J App Nat Sci 7(1):52–57

Datta R, Baraniya D, Wang YF, Kelkar A, Moulick A, Meena RS, Yadav GS, Ceccherini MT, Formanek P (2017a) Multi-function role as nutrient and scavenger off reeradical in soil. Sustain MDPI 9:402. https://doi.org/10.3390/su9081402

Datta R, Kelkar A, Baraniya D, Molaei A, Moulick A, Meena RS, Formanek P (2017b) Enzymatic degradation of lignin in soil: a review. Sustain MDPI 9:1163. https://doi.org/10.3390/su9071163, 1–18

Dhakal Y, Meena RS, De N, Verma SK, Singh A (2015) Growth, yield and nutrient content of mungbean (*Vigna radiata* L.) in response to INM in eastern Uttar Pradesh, India. Bangladesh J Bot 44(3):479–482

Dhakal Y, Meena RS, Kumar S (2016) Effect of INM on nodulation, yield, quality and available nutrient status in soil after harvest of green gram. Legum Res 39(4):590–594

Dimkpa CO, Weinand T, Asch F (2009) Plant-rhizobacteria interactions alleviate abiotic stress conditions. Plant Cell Environ 32:1682–1694

Egberongbe HO, Akintokun AK, Babalola OO, Bankole MO (2010) The effect of *Glomus mosseae* and *Trichoderma harzianum* on proximate analysis of soybean (*Glycine max* (L.) Merrill.) seed grown in sterilized and unsterilised soil. J Agric Ext Rural Dev 2:54–58

Ehsine M, Belkadhi MS, Chaieb M (2014) Seasonal and nocturnal activities of the Rhinoceros Borer (Coleoptera: Scarabaeidae) in the north saharan oases ecosystems. J Insect Sci 14:1–5

El Cadi A, Fakih Lanjri A, Lalilti A, Chouaibi N, Asskali A, Khaddor M (2014) Characterization of the lipid fraction of phosphogypsum: origin and assessment of the degree of transformation of organic pollutants. Mater Environ Sci ISSN 2028–2508(5):2223–2229

Eyheraguibel B, Silvestre J, Morard P (2008) Effects of humic substances derived from organic waste enhancement on the growth and mineral nutrition of maize. Bioresour Technol 99:4206–4212

FAO (2012) Greening the economy with agriculture

FAOSTAT (2018) Rome: food and agricultural organization corporate statistical database

Finlay RD (2007) Ecological aspects of mycorrhizal symbiosis with special emphasis on the functional diversity of interactions involving the extraradical mycelium. J Exp Bot 59(11):15–26

Fisher RA, Yates F (1970) Statistical tables for biological agriculture and medical research, 6th ed. Hafner Publ. Comp., Davien

Francou C (2003) Stabilisation de la matière organique au cours du compostage de déchets urbains: influence de la nature des déchets et du procédé de compostage – Recherche d'indicateurs pertinents, Doctoral thesis, National Agronomic Institute Paris-Grignon, France, 288p

Fritz M, Jakobsen I, Lyngkjær MF, Thordal-Christensen H, Pons-Kühnemann J (2006) Arbuscular mycorrhiza reduces susceptibility of tomato to Alternaria solani. Mycorrhiza 16(6):413–419

Gea MT, Artola A, Sánchez A (2003) Application of experimental design technique to the optimization of bench-scale composting conditions of municipal raw sludge. Compost Sci Util 11:321–329

Gharib FA, Moussa LA, Massoud ON (2008) Effect of compost and bio-fertilizers on growth, yield and essential oil of sweet marjoram (*Majorana hortensis*) plant. Int J Agric Biol 10(4):381–387

Gianinazzi-Pearson V, Gianinazzi S (1988) Morphological integration and functional compatibility between symbionts in vesicular arbuscular mycorrhizal associations. In: Scannerini S, Smith D, Bonfante-Fasolo P, Gianinazzi-Pearson V (eds) Cell to cell signals in plant, animal and microbial symbiosis. Springer-Verlag, Berlin, pp 73–84

Gianinazzi-Pearson V, Dumas-Gaudot E, Gollotte A, Tahiri-Alaoui A, Gianinazzi S (1996) Cellular and molecular defense-related root response to invasion by arbuscular mycorrhizal fungi. New Phytol 133:45–57

Giusquiani PL, Pagliai M, Gigliotti G, Businelli D, Benetti A (1995) Urban watse composts: effects on physical, chemical and biochemical soil properties. Environ Qual 24:175–182

Glick BR (2005) Modulation of plant ethylene levels by the bacterial enzyme ACC deaminase. FEMS Microbiol Lett 251:1–7

Glick BR (2012) Plant growth-promoting bacteria: mechanisms and applications. Scientifica:1–15. https://doi.org/10.6064/2012/963401

Gogoi N, Baruah KK and Meena RS (2018) Grain Legumes: impact on soil health and agroecosystem. Meena RS et al. (eds.), Legumes for soil health and sustainable management, Springer, Singapore. https://doi.org/10.1007/978-981-13-0253-4_16

Grover M, Ali SKZ, Sandhya V, Rasul A, Venkateswarlu B (2011) Role of microorganisms in adaptation of agricultural crops to abiotic stresses. World J Microbiol Biotechnol 27:1231–1240

Haas D, Défago G (2005) Biological control of soil-borne pathogens by fluorescent pseudomonads. Nat Rev Microbiol 3:307–319

Hachicha S, Sellami F, Cegarra J, Hachicha R, Drira N, Medhioub K, Ammar E (2009) Biological activity during co-composting of sludge issued from the OMW evaporation ponds with poultry manure- Physico-chemical characterization of the processed organic matter. J Hazard Mater 162:402–409

Hayman DS, Barea JM, Azcón R (1976) Vesicular-arbuscular mycorrhiza in southern Spain: its distribution in crops growing in soil of different fertility. Phytopathol Mediterr 15:1–6

Hofman J, Dušek L (2003) Biochemical analysis of soil organic matter and microbial biomass composition a pilot study. Eur J Soil Biol 39:217–224

Iglesias-Jiménez E, Pérez García V (1989) Evaluation of city refuse compost maturity: a review

Jaiti F, Kassami M, Meddich A, El Hadrami I (2008) Effect of Arbuscular Mycorrhization on the accumulation of hydroxycinnamic acid derivatives in date palm seedlings challenged with *Fusarium oxysporum* f. sp. *albedinis*. J Phytopathol 156:641–646

Jalili F, Khavazi K, Pazira E, Nejati A, Rahmani HA, Sadaghiani HR, Miransari M (2009) Isolation and characterization of ACC deaminase-producing fluorescent pseudomonads to alleviate salinity stress on canola (*Brassica napus* L.) growth. J Plant Physiol 166:667–674

Jia Y, Gray VM, Straker CJ (2004) The influence of Rhizobium and arbuscular mycorrhizal fungi on nitrogen and phosphorus accumulation by *Vicia faba*. Ann Bot 94:251–258

Jung SC, Ainhoa MM, Lopez-Raez JA, Pozo MJ (2012) Mycorrhiza-induced resistance and priming of plant defenses. J Chem Ecol 38:651–664

Kakraliya SK, Singh U, Bohra A, Choudhary KK, Kumar S, Meena RS, Jat ML (2018) Nitrogen and legumes: a meta-analysis. In: Meena RS et al (eds) Legumes for soil health and sustainable management. Springer, Singapore. https://doi.org/10.1007/978-981-13-0253-4_9

Kammoun M, Ghorbel I, Charfeddine S, Kamoun L, Gargouri-Bouzid R, Nouri-Ellouz Q (2017) The positive effect of phosphogypsum-supplemented composts on potato plant growth in the field and tuber yield. J Env Manage 200:475–483

Koske RE, Tessier B (1983) A convenient, permanent slide mounting medium. Mycol Soc Am Newsl 34:59

Koulibaly B, Dakuo D, Ouattara A, Traoré O, Lompo F, Zombré PN, YaoKouamé A (2015) Effets de l'association du compost et de la fumure minérale sur la productivité d'un système de culture à base de cotonnier et de maïs au Burkina Faso. Tropicultura 33(2):125–134

Kumar S, Meena RS, Pandey A, Seema (2017a) Soil acidity management and an economics response of lime and sulfur on sesame in an alley cropping system. Int J Current Microb App Sci 6(3):2566–2573

Kumar S, Meena RS, Yadav GS, Pandey A (2017b) Response of sesame (*Sesamum indicum* L.) to sulphur and lime application under soil acidity. Int J Plant Soil Sci 14(4):1–9

Kumar S, Meena RS, Lal R, Yadav GS, Mitran T, Meena BL, Dotaniya ML, EL-Sabagh A (2018a) Role of legumes in soil carbon sequestration. In: Meena RS et al (eds) Legumes for soil health and sustainable management. Springer, Singpore. https://doi.org/10.1007/978-981-13-0253-4_4

Kumar S, Meena RS, Bohra JS (2018b) Interactive effect of sowing dates and nutrient sources on dry matter accumulation of Indian mustard (*Brassica juncea* L.). J Oilseed Brassica 9(1):72–76

Laarousi E, Hicham L, El Yachioui M, Ouhssine M (2006) Characterization, processing and recycling of poultry evisceration waste of Kenitra, Morocco. Afric Sci 02(I):102–115. ISSN 1813-548X

Lawlor DW (1987) Stress metabolism: its implication in breeding programmes. In: Srivastava JP, Porceddu E, Acevedo E, Icarda VS (eds) Drought tolerance in winter cereals. Wiley, Chichester, pp 227–240

Layek J, Das A, Mitran T, Nath C, Meena RS, Singh GS, Shivakumar BG, Kumar S, Lal R (2018) Cereal+Legume intercropping: an option for improving productivity. In: Meena RS et al (eds) Legumes for soil health and sustainable management. Springer, Singapore. https://doi.org/10.1007/978-981-13-0253-4_11

Li Y, Luo W, Li G, Wang K, Gong X (2018) Performance of phosphogypsum and calcium magnesium phosphate fertilizer for nitrogen conservation in pig manure composting. Bioresour Technol 250:53–59. https://doi.org/10.1016/j.biortech.2017.07.172

Lim SL, Wu TY, Sim EYS, Lim PN, Clarke C (2012) Biotransformation of rice husk into organic fertilizer through vermicomposting. Ecol Eng 41:60–64

Lim SL, Wu TY, Clarke C (2014) Treatment and biotransformation of highly polluted agro-industrial wastewater from a palm oil mill into vermicompost using earthworms. J Agri F Chem 62(3):691–698

Linderman RG (1994) Role of VAM fungi in biocontrol. In: Pfleger FL, Linderman RG (eds) Mycorhizae and plant health. American Phytopathological Society, St. Paul, pp 1–25

Meddich A (2001) Rôle des endomycorhizes à vésicules et à arbuscules des palmeraies marocaines dans la tolérance des plantes de zones arides au stress hydrique. National Doctoral Thesis. Cadi Ayyad University, Faculty of Sciences Semlalia, Marrakesh, Morocco, 234p

Meddich A, Boumezzough A (2017) First detection of *Potosia opaca* larva attacks on *Phoenix dactylifera* and *P. canariensis* in Morocco: Focus on pests control strategies and soil quality of prospected palm groves. J Entomol Zool Stud 5(4):984–991

Meddich A, Oihabi A, Abbass Y, Bizid E (2000) Rôle des champignons mycorhiziens à arbuscules de zones arides dans la résistance du trèfle (*Trifolium alexandrinum* L.) au déficit hydrique. Agronomie 20:283–295

Meddich A, Oihabi A, Jaiti F, Bourzik W, Hafidi M (2015a) Role of arbuscular mycorrhizal fungi on vascular wilt and drought tolerance in date palm (*Phoenix dactylifera* L.). Can J Bot 93(6):369–377. https://doi.org/10.1139/cjb-2014-0249

Meddich A, Jaiti F, Bourzik W, El Asli A, Hafidi M (2015b) Use of mycorrhizal fungi as a strategy for improving the drought tolerance in date palm (*Phoenix dactylifera* L.). Sci Hortic 192:468–474

Meddich A, Mézy M, Erwan A, Toutain G (2015c) Optimisation de la croissance et du développement du palmier dattier en pépinière par l'utilisation d'amendements biologiques, organiques et chimiques. Eur Sci J 11(24):396-408. ISSN: 1857-7881 (Print) e-ISSN 1857-7431

Meddich A, Hafidi M, Ait El Mokhtar M, Boumezzough A (2015d) Characterization of physical and chemical parameters and mycorrhizal potential of saline soils of the North-East of Marrakesh palm grove. J Mater Env Sc 6(9):2469–2475

Meddich A, Elouaqoudi F, Khadra A, Bourzik W (2016) Valorization of green and industrial wastes by composting process. J Rev Composit Adv Mater 26(3–4):451–469. https://doi.org/10.3166/RCMA

Meddich A, Ait El Mokhtar M, Wahbi S, Boumezzough A (2017) Evaluation of the mycorrhizal potential in relation with the physico-chemical properties of soils in Moroccan palm groves (Marrakech and Tafilalet). Les oasis en Afrique du Nord: dynamiques territoriales et durabilité des systèmes de production agricole. Cah Agric 26:45012. https://doi.org/10.1051/cagri/2017044

Meena RS, Lal R (2018) Legumes and sustainable use of soils. In: Meena RS et al (eds) Legumes for soil health and sustainable management. Springer, Singapore. https://doi.org/10.1007/978-981-13-0253-4_1

Meena H, Meena RS (2017) Assessment of sowing environments and bio-regulators as adaptation choice for clusterbean productivity in response to current climatic scenario. Bangladesh J Botany 46(1):241–244

Meena RS, Yadav RS (2014) Phonological performance of groundnut varieties under sowing environments in hyper arid zone of Rajasthan, India. J App Nat Sci 6(2):344–348

Meena RS, Yadav RS (2015) Yield and profitability of groundnut (*Arachis hypogaea* L) as influenced by sowing dates and nutrient levels with different varieties. Legum Res 38(6):791–797

Meena RS, Yadav RS, Meena VS (2014) Response of groundnut (*Arachis hypogaea* L.) varieties to sowing dates and NP fertilizers under Western Dry Zone of India. Bangladesh J Bot 43(2):169–173

Meena RS, Dhakal Y, Bohra JS, Singh SP, Singh MK, Sanodiya P (2015a) Influence of bioinorganic combinations on yield, quality and economics of Mungbean. Am J Exp Agri 8(3):159–166

Meena RS, Meena VS, Meena SK, Verma JP (2015b) The needs of healthy soils for a healthy world. J Clean Prod 102:560–561

Meena RS, Meena VS, Meena SK, Verma JP (2015c) Towards the plant stress mitigate the agricultural productivity: a book review. J Clean Prod 102:552–553

Meena RS, Yadav RS, Meena H, Kumar S, Meena YK, Singh A (2015d) Towards the current need to enhance legume productivity and soil sustainability worldwide: a book review. J Clean Prod 104:513–515

Meena RS, Yadav RS, Reager ML, De N, Meena VS, Verma JP, Verma SK, Kansotia BC (2015e) Temperature use efficiency and yield of groundnut varieties in response to sowing dates and fertility levels in Western Dry Zone of India. Am J Exp Agri 7(3):170–177

Meena H, Meena RS, Singh B, Kumar S (2016a) Response of bio-regulators to morphology and yield of clusterbean [*Cyamopsis tetragonoloba* (L.) Taub.] under different sowing environments. J App Nat Sci 8(2):715–718

Meena RS, Bohra JS, Singh SP, Meena VS, Verma JP, Verma SK, Shiiag SK (2016b) Towards the prime response of manure to enhance nutrient use efficiency and soil sustainability a current need: A book review. J Clean Prod 112:1258–1260

Meena RS, Gogaoi N, Kumar S (2017a) Alarming issues on agricultural crop production and environmental stresses. J Clean Prod 142:3357–3359

Meena RS, Kumar S, Pandey A (2017b) Response of sulfur and lime levels on productivity, nutrient content and uptake of sesame under guava (*Psidium guajava* L.) based agri-horti system in an acidic soil of eastern Uttar Pradesh, India. J Crop and Weed 13(2):222–227

Meena RS, Meena PD, Yadav GS, Yadav SS (2017c) Phosphate solubilizing microorganisms, principles and application of microphos technology. J Clean Prod 145:157–158

Meena BL, Fagodiya RK, Prajapat K, Dotaniya ML, Kaledhonkar MJ, Sharma PC, Meena RS, Mitran T, Kumar S (2018a) Legume green manuring: an option for soil sustainability. In: Meena RS et al (eds) Legumes for soil health and sustainable management. Springer, Singapore. https://doi.org/10.1007/978-981-13-0253-4_12

Meena H, Meena RS, Lal R, Singh GS, Mitran T, Layek J, Patil SB, Kumar S, Verma T (2018b) Response of sowing dates and bio regulators on yield of clusterbean under current climate in alley cropping system in eastern U.P. Indian Leg Res 41(4):563–571

Meena RS, Kumar V, Yadav GS, Mitran T (2018c) Response and interaction of *Bradyrhizobium japonicum* and *Arbuscular mycorrhizal* fungi in the soybean rhizosphere: a review. Plant Growth Regul 84:207–223

Mehta CM, Palni U, Franke-Whittle IH, Ak S (2014) Compost: its role, mechanism and impact on reducing soil-borne plant diseases. Waste Manag 34:607–622

Mitran T, Meena RS, Lal R, Layek J, Kumar S, Datta R (2018) Role of soil phosphorus on legume production. In: Meena RS et al (eds) Legumes for soil health and sustainable management. Springer, Singapore. https://doi.org/10.1007/978-981-13-0253-4_15

Morton JB, Benny GL (1990) Revised classification of arbuscular mycorrhizal fungi (Zygomycetes): a new order, Glomales, two new suborders, Glomineae and Gigasporineae, and two new families, Acaulosporaceae and Gigasporaceae, with an emendation of Glomacae. Mycotaxon 37:471–491

Motta SR, Maggiore T (2013) Evaluation of nitrogen management in maize cultivation grows on soil amended with sewage sludge and urea. Eur J Agron 45:59–67

Mrabet L, Belghyti D, Loukili A, Attarassi B (2011) Etude de l'effet du compost des déchets ménagers sur l'amélioration du rendement de maïs et de la laitue. Afri Sci 7(2):74–84

Nardi S, Concheri G, Dell'Agnola G (1996) Biological activity of humus. In: Piccolo A, Science BV (eds) Humic substances in terrestrial ecosystems. Elsevier, Amsterdam, pp 361–406

Oihabi A (1991) Effet des endomycorhizes V.A sur la croissance et la nutrition minérale du palmier dattier. PhD thesis of State Doctorate. Cadi Ayyad University, Marrakesh, Maroc/University of Bourgogne Dijon France, 117p.

Oihabi A, Meddich A (1996) Effet des mycorhizes à arbuscules sur la croissance et la composition minérale du trèfle (*Trifolium alexandrinum*). Cah Agr 5:382–386

Olsen SR, Sommers LE (1982) Methods of soil analysis. Part 2. Chemical and microbiological properties of phosphorus. ASA Monograp 9:403–430

Oufdou K, Benidire L, Lyubenova L, Daoui K, Fatemi ZE, Schröder P (2014) Enzymes of the glutathioneascorbate cycle in leaves and roots of rhizobia-inoculated faba bean plants (*Vicia faba* L.) under salinity stress. Eur J Soil Biol 60:98–103

Pequerul A, Perez C, Madero P, Val J, Monge E (1993) A rapid wet digestion method for plant analysis. In Optimization of plant nutrition, Springer, Dordrecht, 3-6pp.

Petiot C, Guardia A (2004) Composting in a laboratory reactor: a review. Compost Sci Util 12:69–79

Phillips JM, Hayman DS (1970) Improved procedures for clearing roots and staining parasitic and vesicular-arbuscular mycorrhizal fungi for rapid assessment of infection. Trans Br Mycol Soc 55:158–161

Plenchette C, Perrin R, Duver P (1989) The concept of soil infectivity and a method for its determination as applied to endomycorrhizas. Can J Bot 67:112–115

Raklami A (2017). Impacts de la double inoculation rhizobactéries-mycorhizes sur la croissance et le rendement de la fève et du blé. Mémoire du Master "Microbiologie Appliquée". Cadi Ayyad University, Faculty of Science Semlalia, Marrakesh, Morocco, 54p.

Ram K, Meena RS (2014) Evaluation of pearl millet and mungbean intercropping systems in Arid Region of Rajasthan (India). Bangladesh J Bot 43(3):367–370

Rauthan B, Schnitzer M (1981) Effects of soil fulvic acid on the growth and nutrient content of cucumber (*Cucumis sativus*) plants. Plant Soil 63:491–495

Roca-Pérez L, Martinez C, Marcilla P, Boluda R (2009) Composting rice straw with sewage sludge and compost effects on soil-plant system. Chemosphere 75:781–787

Rodier J (1984) L'analyse de l'eau : eaux naturelles, eaux résiduaires, eau de mer. 7th ed. Dunod, Paris, 1365 p

Ruiz-Lozano JM, Gomez M, Azcon R (1995) Influence of different Glomus species on the time-course of physiological plant response of lettuce to progressive drought stress periods. Plant Sci 110:37–44

Saaidi M (1992) Comportement au champ de 32 cultivars de palmier dattier vis-à vis du bayoud: 25 années d'observations. Agronomie 12:359–370

Sade N, Vinocur BJ, Diber A, Shatil A, Ronen G, Nissan H et al (2009) Improving plant stress tolerance and yield production: is the tonoplast aquaporin SlTIP2;2 a key to isohydric to anisohydric conversion. New Phytol 181:651–661

Sarangi NR, Babu LK, Kumar A, Pradhan CR, Pati PK, Mishra JP (2016) Effect of dietary supplementation of prebiotic, probiotic, and synbiotic on growth performance and carcass characteristics of broiler chickens. Vet World 9:313–319

Schnitzer RM, Poapest PA (1967) Effects of a soil humic compound on root initiation. Nature 213:598–599

Scholander PF, Hammel HT, Bradstreet ED, Hemmingzen EA (1965) Sap pressure in vascular plants. Science 148:339–346

Seul BL, Chang HL, Ki Yuol J, Park KD, Dokyoung L, Pil JK (2009) Changes of soil organic carbon and its fractions in relation to soil physical properties in a long-term fertilized paddy. Soil Tillage Res 104:227–232

Sghir F, Chliyeh M, Touati J, Mouria B, Touhami OA, Filali-Maltouf A, El Modafar C, Moukhli A, Benkirane R, Douira A (2014) Effect of a dual inoculation with endomycorrhizae and *Trichoderma harzianum* on the growth of date palm seedlings. Int J Pure Appl Biosc 2(6):12–26

Shabbir G, Dakheel AJ, Brown GM, Rillig MC (2011) Potential of arbuscular mycorrhizal technology in date palm production. In: Jain SM, Al-Khayri JM, Johnson DV (eds) Date palm biotechnology. Springer Netherlands, Dordrecht, 449–476pp

Shak KPY, Wu TY, Lim SL, Lee CA (2014) Sustainable reuse of rice residues as feedstocks in vermicomposting for organic fertilizer production. Environ Sci poll Res 21(2):1349–1359

Sharma SB, Sayyed RZ, Trivedi MH, Gobi TA (2013) Phosphate solubilizing microbes: sustainable approach for managing phosphorus deficiency in agricultural soils. Springerplus 2:587. https://doi.org/10.1186/2193-1801-2-587

Sihag SK, Singh MK, Meena RS, Naga S, Bahadur SR, Gaurav YRS (2015) Influences of spacing on growth and yield potential of dry direct seeded rice (*Oryza sativa* L.) cultivars. Ecoscan 9(1-2):517–519

Sofi PA, Baba ZA, Hamid B, Meena RS (2018) Harnessing soil Rhizobacteria for improving drought resilience in legumes. In: Meena RS et al (eds) Legumes for soil health and sustainable management. Springer, Singapore. https://doi.org/10.1007/978-981-13-0253-4_8

Souza RD, Ambrosini A, Passaglia LMP (2015) Plant growth-promoting bacteria as inoculants in agricultural soils. Genet Mol Biol 38(4):401–419

Stevenson FJ (1994) Humus chemistry: genesis, composition, reactions, 2nd edn. Wiley-Interscience, New York

Strullu DG (1986) Micropropagation of chesnut and conditions of mycorrhizal synthesis *in vitro*. New Phytol 102:95–101

Subramanian KS, Charest C (1997) Nutritional, growth and reproductive responses of maize (Zea mays L.) to arbuscular mycorrhizal inoculation during and after drought stress at tasselling. Mycorrhiza 7:25–32

Symanczik S, Błaszkowski J, Al-Hashmi H, Al-Farsi K, Al-Ismaili I, Patzelt A, Boller T, Wiemken A, Al-Yahya'ei MN (2014) Arbuscular mycorrhizal fungi from the Arabian Peninsula as biotechnological tools for revegetation and date palm production in the desert. In: International Congress on Mycorrhiza, 14–17 Octobre, Marrakesh, Morocco

Takeda M, Nakamoto T, Miyazawa K, Murayama T, Okad H (2009) Phosphorus availability and soil biological activity in an Andosol under compost application and winter cover cropping. Appl Soil Ecol 42:86–95

Tao L, Lin G, Zhang X, Yongliang C, Shubin Z, Chen B (2014) Relative importance of an arbuscular mycorrhizal fungus (*Rhizophagus intraradices*) and root hairs in plant drought tolerance. Mycorrhiza. https://doi.org/10.1007/s00572-014-0578-3

Tardieu F (2005) Plant tolerance to water deficit: physical limits and possibilities for progress. CR Geoscience 337:57–67

Tarraf W, Ruta C, Tagarelli A, De Cillis F, De Mastro G (2017) Influence of arbuscular mycorrhizae on plant growth, essential oil production and phosphorus uptake of Salvia officinalis L. Ind Crop Prod 102:144–153

Tautorus TE, Townsley PM (1983) Biological control of olive green mold in agaricus bisprus cultivation. Appl Environ Microbiol 45(2):511–515

Tejada M, García-Martínez AM, Parrado J (2009) Effects of a vermicompost composted with beet vinasse on soil properties, soil losses and soil restoration. Catena 77:238–247

Thygesen K, Larsen J, Bodker L (2004) Arbuscular mycorrhizal fungi reduce development of pea root-rot caused by *Aphanomyces euteiches* using oospores as pathogen inoculum. Eur J Plant Pathol 110:411–419

Toumpeli A, Pavlatou-Ve AK, Kostopoulou SK, Mamlos AP, Siomos AS, Kalburtji KL (2013) Composting *Phragmites australis* Cav. Plant material and compost effects on soil and tomato (*Lycopersicon esculentum* Mill.) growth. J Environ Manag 128:243–251

Trouvelot A, Kouch J, Gianinazzi-Pearson V (1986) Mesure du taux de mycorhization VA d'un système radiculaire: Recherche de méthodes d'estimation ayant une signification fonctionnelle. Les mycorhizes: Physiologie et Génétique, 1er Séminaire Européen sur les mycorhizes, Dijon, Edition Gianinazzi. INRA, Paris, pp 217–221

Vacheron J, Desbrosses G, Bouffaud ML, Touraine B, Moënne-Loccoz Y, Muller D, Legendre L, Wisniewski-Dyé F, Combaret CP (2013) Plant growth promoting rhizobacteria and root system functioning. Front Plant Sci 4(356):1–19

Varma D, Meena RS, Kumar S (2017a) Response of mungbean to fertility and lime levels under soil acidity in an alley cropping system in Vindhyan Region, India. Int J Chem Stu 5(2):384–389

Varma D, Meena RS, Kumar S, Kumar E (2017b) Response of mungbean to NPK and lime under the conditions of Vindhyan Region of Uttar Pradesh. Legum Res 40(3):542–545

Vassilev N, Requena AR, Nieto LM, Nikolaeva I, Vassileva M (2009) Production of manganese peroxidase by *Phanerochaete chrisosporium* grown on medium containing agro-wastes/rock phosphate and biocontrol properties of the final product. Ind Crop Prod 30:28–32

Verma JP, Jaiswal DK, Meena VS, Meena RS (2015a) Current need of organic farming for enhancing sustainable agriculture. J Clean Prod 102:545–547

Verma JP, Meena VS, Kumar A, Meena RS (2015b) Issues and challenges about sustainable agriculture production for management of natural resources to sustain soil fertility and health: a book review. J Clean Prod 107:793–794

Verma SK, Singh SB, Prasad SK, Meena RN, Meena RS (2015c) Influence of irrigation regimes and weed management practices on water use and nutrient uptake in wheat (*Triticum aestivum* L. Emend. Fiori and Paol.). Bangladesh J Bot 44(3):437–442

Vicente-Sánchez J, Nicolás E, Pedrero F, Alarcón JJ, Maestre-Valero JF, Fernández F (2014) Arbuscular mycorrhizal symbiosis alleviates detrimental effects of saline reclaimed water in lettuce plants. Mycorrhiza 24:339–348

Xiao LS, Ai-Rong L, Yan C, Kai-Yun G, Lu Z, Yan YL (2014) Arbuscular mycorrhizal fungi: potential biocontrol agents against the damaging root hemiparasite Pedicularis kansuensis. Mycorrhiza 24:187–195

Yadav GS, Babu S, Meena RS, Debnath C, Saha P, Debbaram C, Datta M (2017a) Effects of godawariphosgold and single supper phosphate on groundnut (*Arachis hypogaea*) productivity, phosphorus uptake, phosphorus use efficiency and economics. Indian J Agric Sci 87(9):1165–1169

Yadav GS, Lal R, Meena RS, Babu S, Das A, Bhomik SN, Datta M, Layak J, Saha P (2017b) Conservation tillage and nutrient management effects on productivity and soil carbon sequestration under double cropping of rice in North Eastern Region of India. Ecol Indi. http://www.sciencedirect.com/science/article/pii/S1470160X17305617

Yadav GS, Lal R, Meena RS, Datta M, Babu S, Das LJ, Saha P (2017c) Energy budgeting for designing sustainable and environmentally clean/safer cropping systems for rainfed rice fallow lands in India. J Clean Prod 158:29

Yadav GS, Das A, Lal R, Babu S, Meena RS, Saha P, Singh R, Datta M (2018a) Energy budget and carbon footprint in a no-till and mulch based rice–mustard cropping system. J Clean Prod 191:144–157

Yadav GS, Das A, Lal R, Babu S, Meena RS, Patil SB, Saha P, Datta M (2018b) Conservation tillage and mulching effects on the adaptive capacity of direct-seeded upland rice (*Oryza sativa* L.) to alleviate weed and moisture stresses in the North Eastern Himalayan Region of India. Arch Agron Soil Sci. https://doi.org/10.1080/03650340.2018.1423555

Zamioudis C, Pieterse CMJ (2012) Modulation of host immunity by beneficial microbes. Mol Plant-Microbe Interact 25:139–150

Zeze A, Brou Y, Meddich A, Marty F (2007) Molecular characterisation of a mycorrhizal inoculant that enhances *Trifolium alexandrium* L. resistance under water stress conditions. Afr. J. Biotechnol. ISSN 1684-5315. 6(13):1524-1528.

Zeze A, Brou Y, Meddich A, Marty F (2008) Molecular identification of MIP genes expressed in the roots of an arbuscular mycorrhizal *Trifolium alexandrium* L. under water stress. Afr J Agric Res ISSN 1991-637X Acad J 3(1):078–083

Ziouti A (1998) Aspects biochimiques de l'interaction *Phoenix dactylifera* L.- *Fusarium oxysporum* f. sp. *albedinis*, rôle des composés phénoliques. 3rd cycle thesis Faculty of Science Semlalia, Cadi Ayyad University, Marrakesh, Morocco, 123 p

# Organic Fertilizers for Sustainable Soil and Environmental Management

# 10

B. C. Verma, P. Pramanik, and Debarati Bhaduri

**Abstract**

The modernization of agriculture along with the "Green Revolution" transforms the agriculture practices in a new dimension where the traditional knowledge and techniques were replaced by the new technology to increase the productivity to feed the growing population. This Green Revolution changed the country status from importer to self-sufficient. Traditional source of nutrients was replaced by the synthetic and chemical fertilizers. Undoubtedly the inorganic fertilizers are keys behind the increasing productivity to a greater scale. However, inappropriate use of these chemical/synthetic fertilizers, unscientific management, over-utilization, etc. lead to soil and environmental pollution as well as deterioration of the soil quality. Moreover, continuous use of these fertilizers leads to toxicity as well as deficiency of some major and minor nutrients. In the scenario of global climate change, the unscientific use of these chemical inputs are major threats to environment. To reduce or minimize these ill effects, it is high time to shift the agriculture system from inorganic to organic mode to sustain the soil and environments for a longer period. Side by side, the use of chemical fertilizers should be minimized or avoided depending upon the cropping condition and demand of the system. Organic farming system and combined system (organic and inorganic or INM) both can promote agriculture toward the reducing use of chemical fertilizers, and that system must be popularized. Organic as well as INM have several advantages over the convention (chemical-based) system in terms of soil quality, environmental pollution, crop productivity, as well as the quality of pro-

B. C. Verma (✉)
Central Rainfed Upland Rice Research Station (ICAR-NRRI), Hazaribagh, India

P. Pramanik
Division of Agricultural Physics, ICAR-Indian Agricultural Research Institute, New Delhi, India

D. Bhaduri
Crop Production Division, ICAR-National Rice Research Institute, Cuttack, India



R. S. Meena (ed.), *Nutrient Dynamics for Sustainable Crop Production*,
https://doi.org/10.1007/978-981-13-8660-2_10

duce. This chapter aims to focus on the use of organic fertilizers (alone or in combination) for better soil and environmental management. However, the organic system also has the several limitations that must be addressed, and proper management must be evaluated to promote the organic production system. The popularization of the technology and techniques is governed by different factors, so the organic farming practices will be adopted by the farmers only when the technology will reach to the farmers with the clear message. Organic farming or organic nutrient management not only reduces the input cost but also provides an opportunity to recycle the waste unused materials, crop and plant residues to reduce the soil, water, and environment pollution. The use of organic fertilizers will improve the soil carbon status and soil quality which help in improving, carbon sequestration. With the several advantages associated with organic nutrient management, still proper demonstration, awareness, and training are required to popularize among the farmers and to get the best benefit out of it.

**Keywords**
Agricultural sustainability · Greenhouse gas · Integrated nutrient management · Management of soil carbon · Organic fertilizer · Soil and water pollution · Soil health

## Abbreviations

| | |
|---|---|
| BD | Bulk density |
| C/N | Carbon and nitrogen ratio |
| CEC | Cation exchange capacity |
| CMI | Carbon management index |
| EC | Electrical conductivity |
| FYM | Farmyard manure |
| GHG | Greenhouse gas |
| GWP | Global warming potential |
| IARI | Indian Agricultural Research Institute |
| IFOAM | International Federation of Organic Agriculture Movements |
| IGP | Indo-Gangetic plain |
| INM | Integrated nutrient management |
| IPCC | Intergovernmental Panel on Climate Change |
| MOC | Mustard oil cake |
| Mt. | Million tons |
| NPK | Nitrogen, phosphorus, and potassium |
| OF | Organic farming |
| ONM | Organic nutrient management |
| PFPN | Partial factor productivity of applied nutrient |
| PR | Penetration resistance |

| | |
|---|---|
| RDF | Recommended dose of fertilizer |
| RDN | Recommended dose of nitrogen |
| SOC | Soil organic carbon |
| SOM | Soil organic matter |
| SWC | Soil water content |
| WUE | Water use efficiency |

## 10.1 Introduction

Before the modernization/mechanization of agriculture or the advent of synthetic fertilizers, pesticides, agrochemicals, etc., the agricultural practices are aimed to use locally available materials and natural resources to sustain the productivity, and the farmers had no formal knowledge regarding the nutrient management or crop management. The "Green Revolution" in the early 1970s transforms the way of agriculture, and traditional farming practices were abandoned in the wake of new scientific methods and inputs with the main objective to enhance the productivity to feed the growing populations and changed country scenario from food importer to self-sufficient (Rena 2004). Chemical fertilizers have a major role in bringing the Green Revolution to boost the productivity; these are mainly formulated in appropriate concentrations and combinations to supply essential plant nutrients to sustain the crop growth. However, major portion of applied nutrients is lost to the environment through physical, chemical, and biochemical processes and cannot be absorbed by plants, causing substantial economic and resource losses but also possess very serious threats to soil and environment (Saigusa 2000; Meena et al. 2015). Along with this low use efficiency, unscientific and indiscriminate use of these inputs (agrochemical) causes some major problems like degradation of natural (soil and water) resources, depletion of soil health and quality, soil and water pollution, buildup of pesticide residues, micronutrient deficiencies in soil, ecological imbalance, etc. (Bhattacharyya et al. 2015). Simultaneously, increasing demand of food leads to overexploitation of land resources under intensive agricultural production systems which cause adverse impact and reduce or stagnate the productivity of the system. These impart a serious concern to the sustainability of agricultural system. Gradually it was realized that the agriculture production system needs some alternative practices for sustainable development to conserve these natural resources. The use of appropriate combination of mineral fertilizers along with organic source of nutrients (organic fertilizers) like farmyard manure (FYM), organic manures, crop residues, compost, vermicompost, etc. has proved to be beneficial in improving and sustaining soil and environmental quality and health (Vanlauwe et al. 2001; Sanchez et al. 2004; Ashoka et al. 2017). Such observations transform the farming systems where chemical fertilizers have either been minimized or avoided. The rising cost of chemical fertilizers has further focused attention on cycling of plant nutrients through organic materials or use of organic fertilizers. This is the key ingredient of organic farming/organic agriculture production system. The organic farming is a unique combination of environmentally sound practices with low external inputs. It relies not only on the fertilizers of organic origin (compost, manure, green manure, bone meal, etc.) but places emphasis on techniques such as crop rotation, companion planting, etc. Organic agriculture promotes practices such as extended crop rotations and soil amendments including animal manure and compost. The International Federation of Organic Agriculture Movements (IFOAM) defined organic farming as "a production system that sustains the health of soils, ecosystems and people. It relies on ecological processes, biodiversity and cycles adapted to local conditions, rather than the use of inputs with adverse effects. Organic agriculture combines traditional and scientific methods to benefit the shared environment and promote fair relationships and good quality of life for all involved" (IFOAM 2009).

Agriculture is a backbone of the Indian economy where large population depends on it either directly or indirectly. The agriculture sector itself produces a vast amount of residues/by-products that must be utilized properly. The major crop residues/by-products are obtained from paddy, wheat, millet, sorghum, pigeon pea, castor, mustard, groundnut, maize, cotton, jute, sugarcane, tea, etc. (Sugumaran and Sheshadri 2009; Kumar et al. 2017). In India the total amount of residues obtained from agriculture is about 435 million tons (Mt) in a year, out of which 70% of residues are excess, which are not used properly due to several constraints (Meena and Yadav 2015). On an average, among the states, Uttar Pradesh produces maximum crop residues followed by Punjab and Maharashtra; however, among the crops cereals top in terms of residue production, followed by fibers, oilseeds, pulses, sugarcane, etc. (Vishram 2014). The negative effect of this surplus residue is that a major portion was burnt in the field which not only causes the air pollution but also nutrient loss. Burning is considered as an easiest option for farmers, which can manage residues and help in preparing land for further sowing. Intentional burning of crop residues is promoted, which might be due to some reasons that include fertility enhancement and pest management (insects, disease, weeds, etc.) by killing them or changing their environment (Vishram 2014; Meena and Lal 2018).

Burning of crop residues is considered to be useful as it increases soil fertility; however, burning actually has differential mixed effect on soil fertility. It may sometimes increase the availability of phosphorus and potassium and also increase the soil pH; however, it triggers loss of other important nutrients like nitrogen and sulfur (Vishram 2014; Meena et al. 2016a). Burning of crop residues is the important factor which causes air pollution by releasing different pollutants. In Punjab alone, about 70–80 Mt. of rice and wheat straw are burned annually (Punia et al. 2008), releasing approximately 140 million tons of carbon dioxide along with different greenhouse gases and air pollutants.

These surplus residues must be converted into compost (organic fertilizers) for overall benefit of soil and environment. The per capita solid waste generation in our country lies between 200 and 500 g, and most of them are biodegradable in nature. This provides an opportunity to convert these wastes into organic amendments and reuse to improve the soil health, quality, and productivity. However, lack of knowledge and awareness is the major limitation in this process, and these waste materials are thrown away in water bodies and streams, and burn causing soil, water, and environmental pollution (Saikia et al. 2017; Yadav et al. 2017b). In recent years, there has been a rising demand for organic food due to increased public concerns about the negative environmental and health impacts of agrochemicals (pesticides, growth regulators, and mineral fertilizers) used in crop production (Baranski et al. 2014). Hence, there is a potential scope of the organic fertilizer for sustaining soil and environment, and organic agriculture is now seen as a significant means for sustainable food production.

## 10.2 Effect of Organic Fertilizer on Soil Health, Quality, and Productivity

### 10.2.1 Crop Production and Productivity

To counter the negative or ill effects of inorganic/synthetic/chemical fertilizers or agrochemicals, there is an approach in hand called the organic production system where the materials from organic origin are being used. To supply the plant nutrients, in place of chemical fertilizers, organic fertilizers (fertilizers having biological origin) are recommended. It is obtained mainly from the plants (residues, crop and vegetable biomass, etc.), animals (cow dung, urine, litters, etc.), waste materials, etc. Organic production system is a holistic approach providing preventive rather than reactive at the systems level. It is a system of production that doesn't allow any use of synthetic fertilizers and promotes application of compost and manures, organic wastes, crop rotations, legumes, pest control, etc. through biological measures. It restricts the use of synthetically produced fertilizers, pesticides, growth regulators, etc. It is an ecological production management system that promotes and enhances biodiversity, biological cycles, and biological health of the soil (Kumar et al. 2018b). The code of organic production system is to feed the soil rather than the crops to maintain optimum soil health, thus making the soil capable of providing the necessary nutrients to the crop for its growth and development. Basic principles of organic farming or organic crop production are to make maximum and sustainable use of locally available resources and minimize the leaching of nutrients through rotation with deep-rooted crops, produce quality food of high nutritional value, and maintain the genetic diversity of the production system and its surroundings including the protection of wildlife habitats. All the required plant nutrients can be provided through organic manures, which can increase the crop productivity and also help in succeeding crop with their residual effects (Ghosh et al. 2004; Kumar et al. 2018a). To improve the soil organic matter which directly influences physical, chemical, and biological properties of the soil, various kinds of organic materials/fertilizers such as animal manures (FYM, compost, poultry litter, etc.) sewage and sludge, crop residues, etc. are applied which finally leads to improvement in crop productivity (Debosz et al. 2002; Meena et al. 2016b).Organic fertilizers have nearly all the vital plant nutrients required for plant growth and produce other non-nutrient benefits also by providing foods for soil microbes, different growth-promoting organic acids, improving soil structure, water holding capacity, etc.; however, its real impact is not generally understood because their value was principally assessed in terms of nitrogen or nutrients only. Better plant growth has been linked with organic amendments, and ensuing better plant growth and vigor can be related to greater root development, water and nutrient use efficiency, etc. Consecutively, better-established roots can cover a larger area and draw more amount of water entering in a soil through infiltration. Improved soil structure because of better aggregation in organically amended soil will help to open up soil pores and channels in fine-textured soil, and consequently it helps in better growth of the crops. Besides, well-aggregated soil helps to reduce soil surface crusting which is beneficial for seedling emergence of the plants. Moreover, organically amended soil covers the soil surfaces which reduce the evaporative losses and increase water infiltration and soil structural stability. A study conducted by Pramanik and Prasad (2015) reported that bulk density (BD), soil water content (SWC), and porosity for 0–15 cm soil depth in organic wheat field were 1.61 $Mgm^{-3}$, 19.52%, and 39.08%, while inconventional wheat field were 1.72 $Mgm^{-3}$, 9.46%, and 35.26%, respectively (Table 10.1). They also reported that up to 15 cm soil depth, penetration resistance (PR) was lower (925 kPa) in organic wheat system as compared to conventional wheat field (1144 kPa).

Organic fertilizers help in keeping balance in carbon and nitrogen (C/N) ratio of the soil and also enhance the soil fertility and productivity. Due to more biological activities in soil, the nutrients that are in the lower depths are made available to the plants. The positive effects of organic fertilization are well-documented in terms of crop production and productivity. Rasul et al. (2015) reported that the poultry manure treatment is the most efficient in achieving highest grain yield, biological yield, and grain protein content as compared to the other organic and inorganic treatments. Likewise, organic manure especially poultry droppings significantly increased *Jatropha curcas* growth when compared to the inorganic sources of fertilizers. Prabhakaran (2003) reported that application of recommended dose of nitrogen through organic manure increases the nutrient uptake and yield in tomato. Jannoura et al. (2014) also reported that use of yard waste compost, horse manure, etc. (carbon rich organic fertilizers) helps in increasing pea yields by stimulating soil microbial biomass indices. Organic fertilizers increase the quality and yield of agricultural crops in ways similar to inorganic fertilizers (Bulluck et al. 2002; Dadhich and Meena 2014; Sofi et al. 2018) and also do not cause soil and environment pollution.

**Table 10.1** Physico-chemical properties of organic and conventional wheat field (Pramanik and Prasad 2015)

| Treatment | Depth (cm) | SWC (%) | Bulk density ($Mg\ m^{-3}$) | Porosity (%) | EC ($dS\ m^{-1}$) | pH |
|---|---|---|---|---|---|---|
| Organic wheat | 0–15 | 19.52 | 1.61 | 39.08 | 0.35 | 7.7 |
| | 15–30 | 17.38 | 1.77 | 33.25 | 0.36 | 7.9 |
| Conventional wheat | 0–15 | 9.46 | 1.72 | 35.26 | 0.41 | 8.05 |
| | 15–30 | 21.37 | 1.67 | 37.03 | 0.43 | 8.15 |

### 10.2.2 Nutrient Management

#### 10.2.2.1 Organic Nutrient Management (ONM)

As organic crop production depends upon concept that all the input in crop production must be of organic nature/origin, nutrients should also be provided from the organic sources (plant and animal origin). A large number of plant biomass, residues, weed biomass, as well as agricultural and animal wastes can be transformed decomposed and applied in the soil which can release nutrients to the plants. The biomass when applied in the soil gets converted into different fractions, and it will form a complex with the soil matrix. The present biomass and its fractions will work as a sink and source for available nutrients, which help in crop growth. The organic matter fraction forms a matrix with the soil and holds the nutrients and releases as per the requirement, which is illustrated (for understanding) in the below diagram (Fig. 10.1). The compost can be enriched and well-prepared with the help of technical know how and expert supervision to get the higher nutrient content. Although the use of organic resources is good for the soil and environment, but nutrient supply from this source is limited, and thus it is difficult to supply the nutrient as per the crop demand to get the maximum productivity. The sole ONM is not popular among the farming communities of lowland rice system and also because the uncertainty of the yield performance of different rice varieties is univocal.

But considering the sustenance of soil health, the locally available and cost-friendly investment of organic manures is sometimes preferred. Very few experiments documented the statistically at par yield under the ONM treatments, while few others emphasized using organic manure for quality rice production only. An almost similar situation is noticed for other field crops; however, practicing organic farming is not so uncommon for high-value crops like fruits and vegetables. It was reported that application of 50% nitrogen through FYM and rest 50% through vermicompost + biofertilizer was found equally productive and recorded 29.5% higher

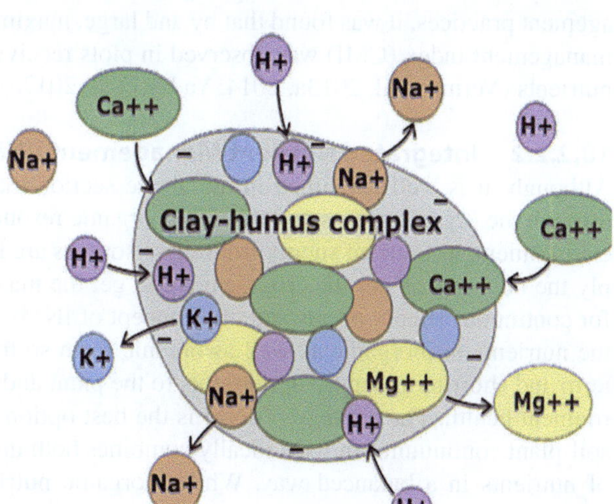


Fig. 10.1 Soil matrix as source and sink for nutrients (https://www.tankonyvtar.hu/en/tartalom/tamop425/0032_talajtan/ch05s04.html)

net returns and 9.9% higher energy use efficiency over the blanket farmer's practice (FYM at 1.0 t $ha^{-1}$) in rice-vegetable pea cropping system of eastern Himalayan soils (Singh et al. 2015; Varma et al. 2017). The combination of green manure and liquid organic manures improved the yield of japonica rice and almost all available major and micronutrient status of soil (Debbarma and Abraham 2015). A study from China reported that (PFAN) both physiological efficiency and partial factor productivity of applied nitrogen were inclined under sole organic farming in rice that relied on rapeseed cake, grass manure, and locally made Sanan organic fertilizer; more ever the cooking and eating quality was also found better for the tested varieties under organic farming over conventional farming; however, the yield, yield attributes, and dry matter accumulation were not impressive (Huang et al. 2016). While comparing the phenolic compounds and antioxidant activities of organically and conventionally grown japonica rice cultivars, Kesarwani et al. (2013) reported that increased phytochemical content was observed at the organically grown milled rice and evidenced to be a potent natural antioxidant. In another instance, the organically grown improved upland rice variety (IR 55419) and traditional variety (Speaker) had better tolerated the drought stress, and the yields were found consistently higher (Taylaran et al. 2013; Verma et al. 2015c). It was observed that under different crops and cropping systems, there is a scope for the nutrient management through organic fertilizers; however, the yield and quality are the important decisive factors. Apart from the nutrient-supplying capacity of the organic system as well as contribution towards the yield and quality of produce, it is well-accepted that the organic production system has more input of carbon as compared to all other systems. Carbon input and buildup in soil is very important in terms of soil carbon sequestration and soil health/quality. The carbon built up in soil is measured in terms of carbon management index (CMI) which measures the buildup of carbon in relation to stabilized as well as labile pools of carbon (Blair et al. 1995). It was observed that organic-based system showing the increase in CMI as compared to other system. Under different cropping systems with different set of nutrient management practices, it was found that by and large, maximum improvement in carbon management index (CMI) was observed in plots receiving 100% organic source of nutrients (Verma et al. 2013a, 2014; Yadav et al. 2017a).

#### 10.2.2.2 Integrated Nutrient Management (INM)

Although, it is well explained in the above section that nutrient can be provided through the organic means and the use of organic resources is good for the soil and environment, as nutrient supply from these sources are limited, it is difficult to supply the nutrients as per the crop demand to get the maximum productivity. Hence, for continuous supply of nutrient, the concept of INMs came, where we can supply the nutrients in inorganic as well as organic form so that it can take care of long-term and short-term supply of nutrients to the plant and can maintain soil and environment health. The strategy of INM is the best option in creating a balance in the soil-plant continuum. An INM ideally combines both inorganic and organic sources of nutrients in a balanced way. While inorganic nutrient forms readily supplied plant-essential nutrients and ensures better crop productivity, the organic forms of

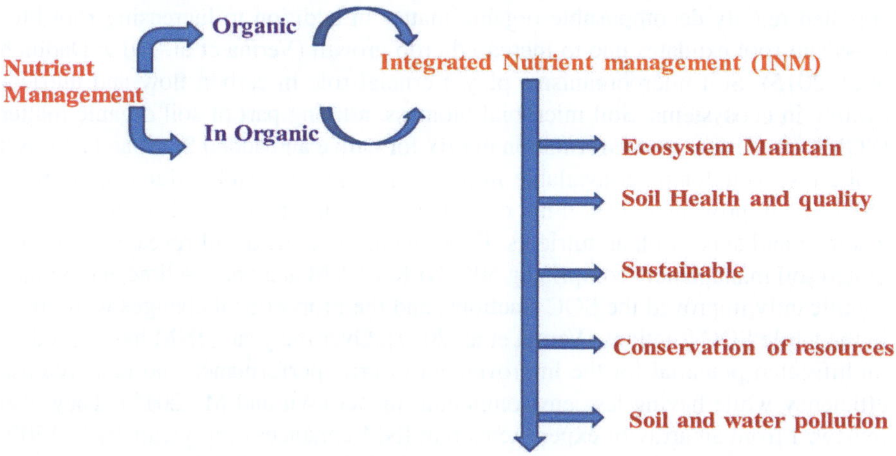


**Fig. 10.2** Concept of INM and its functions

nutrients on the other hand increase the nutrient use efficiency in soil and reduce the chances of soil and environment pollution. It also helps in proper utilization of natural resources and ecosystem balance. The concepts of INM and its functions are briefly explained with flowcharts (Fig. 10.2).

In the INM, flux of nutrients is very important and it should match with the crop demand. It was reported by the Verma et al. (2013b) that under the INM, urea-nitrogen can be substituted by the organic source of nutrients like FYM and *Sesbania* as much as 50% without hampering the nitrogen flux in soil. Long-term trials have revealed that application of manures with inorganic fertilizers results in higher yields over a longer period of time and also build up the soil carbon content. The practice of proper use of inorganic and organic sources of nutrients in appropriate proportion/combinations not only reduces the sole dependence and demand of chemical fertilizers but also builds up the soil fertility, soil health, and quality. Positive effects on soil fertility and productivity can be achieved by addition of organic matter and plant nutrients through effective management of crop residues, root biomass, stubbles, and weed biomass (Singh 2003; Kakraliya et al. 2018). It is well-documented that the soil fertility and crop productivity are increased by incorporation or retentions of organic manure or crop straw/residues on soil surface by benefiting the physical, chemical, and biological properties of soil positively. It increases hydraulic conductivity and reduces soil bulk density by altering soil structure, porosity, and soil aggregation. Mulching with plant/crop residues reduces the soil temperature fluctuations as it raises the minimum soil temperature in winter and decreases soil temperature during summer. When crop residues were evenly distributed over the soil surfaces as mulch than it reduces wind and water erosions also by influencing the hydraulic conductivity, surface runoff and soil moisture retention, etc. The management practices in which there is an addition of carbon in soil lead to increase the microbial biomass by accelerating the growth of microorganism. Organic amendments led to an increase in microbial biomass as these amendments supplied readily decomposable organic matter in addition to increasing root biomass and root exudates due to increased crop growth (Verma et al. 2010; Dadhich et al. 2015). Soil microorganisms play a crucial role in carbon flow and nutrient cycling in ecosystems. Soil microbial biomass, a living part of soil organic matter (SOM), constitutes a transformation matrix for native and added SOM and acts as a labile reservoir for plant-available nitrogen, phosphorus, and sulfur (Singh et al. 1989). Its improvement in soil not only increases nutrient availability but works as a source and sink of plant nutrients. Experiment under acid soil revealed that integrated soil management comprising 50% NPK + FYM at 5 t $ha^{-1}$ + lime at 0.5 t $ha^{-1}$ significantly improved the SOC fractions, and the proportional changes were more in the labile SOC fractions (Verma et al. 2017). Over the years, INM has proved its multifaceted potential for the improvement of crop performance and resource use efficiency while having less environmental impact (Wu and Ma 2015). They also reviewed from an array of experiments that INM enhances crop yields by 8–150% compared to conventional practices; moreover, it increases WUE and ensures economic benefit to farmers, maintaining soil health with fair grain quality. Mondal et al. (2016) observed the treatment(s) of 50% recommended dose of fertilizer (RDF) + 50% recommended dose of nitrogen (RDN) through mustard oil cake (MOC) or 75% RDF + 25% RDN through MOC + biofertilizer applied during hybrid rice cultivation in sandy loam soils of West Bengal, showed a good example of how best INM can assure higher grain and biomass yields, greater NPK removal, and higher partial factor productivity of applied nutrient (PFPN). In a long-term rice-wheat system-based experiment conducted at IARI, New Delhi, the highest rice yield was obtained in the INM treatment, 100% N (25% N substituted by FYM) (Bhaduri et al. 2014a). The fertilizer prescription equations (STCR-INM with a fixed dose of FYM at 5 t $ha^{-1}$) developed for rice was validated for alluvial soil (inceptisol) of Pratapgarh, eastern Uttar Pradesh, for achieving a yield target of 4.5 and 5.0 t $ha^{-1}$ with sustained soil fertility and able to give a good economic return to farmers over the existing farmers' practice and general recommended dose (Singh et al. 2017; Verma et al. 2015b). Similar beneficial effect with STCR-INM approach was achieved for wheat crop also in the Terai region of Uttarakhand and for productivity (Bhaduri and Gautam 2013) and soil nutrient recovery (Bhaduri and Gautam 2012), hence ensuring a more sustainable system. Another study reported from the coastal Sundarbans of West Bengal comprising the INM treatments of farmyard manure, green leaf manure, and vermicompost (75% RDF + 25% substitution) tested for the yields of rice-based vegetable systems (rice-tomato, rice-sunflower, and rice-chili) over conventional farmer's practice, and rice-sunflower system recorded maximum rice equivalent yield also overcoming the salinity stress (Mitran et al. 2017). The INM practiced in rice (fertilizer NPK in combination with 50% N through compost) resulted in the increase of organic matter content, available P, and available K content in soil, while the maximum available N was observed in other treatment, substitution of 50% N through sewage sludge (Gosal et al. 2017). The organic supplemented treatments (FYM, green manure, biofertilizers, crop residues, etc.) showed improvement on soil available N and P contents owing to faster mineralization of the organics and changing pH in rice-wheat system (Bhaduri et al.

2014b). These reports confirm that integrated nutrient management is the best way to provide nutrient according to the crop demand as well as to sustain the soil health and quality along with reducing the soil and environmental pollution. Hence, INM creates a scope for the organic fertilizers being used in agriculture.

### 10.2.3 Soil Carbon Storage/Carbon Sequestration

Soil organic carbon is the central key element which governs most of the soil properties. It regulates physical, chemical, as well as biological properties of soil. It is the key component to define the soil quality or soil health. Soil organic carbon (SOC) stocks contribute to the principal terrestrial carbon pool, even though these stocks decline after conversion of natural lands to agricultural lands (Guo and Gifford 2002). Agricultural land comprises a key proportion of the world's land surface, so SOC sequestration in agricultural ecosystems plays a vital role in managing carbon dynamics and mitigating climate change (Lal 2004a, b). Application of organic fertilizers (organic farming, OF) improves SOC stocks in agricultural lands as compared to traditional farming (García-Palacios 2018). Due to continuous intensive cropping systems, the long-term balances between input and output of carbon sources and sinks are disturbed (Kong et al. 2005), causing depletion of SOC in tropical and subtropical climatic situations. However, organic farming can improve the soil health and quality by improving the organic carbon content. Organically healthy soils can capture extra/surplus water through infiltration which thereafter increases the farm's resiliency to drought, substantial rainfall, and extreme weather events. A study conducted by Leifeld and Fuhrer (2010) revealed that SOC in organic farming (OF) system increased annually by 2.2% on average, while conventional farming did not change SOC significantly; hence, OF is associated with carbon sequestration. Organic mulch and crop residue play a vital role in sequestering SOC, where organic mulching is done through covering the soil surface by compost or farmyard manure (FYM) followed by applying dry organic matter above it. Organic mulching can improve the soil health by improving the growth of beneficial soil microbes and can improve the soil fertility status; besides it can sequester atmospheric carbon dioxide in soil. García-Palacios et al. (2016) reported that even with minimal dose of organic manure application, there is a significant improvement on soil respiration, carbon stocks, and sequestration rates. SOC stocks signify the net balance of long-term variation in soil C inputs and outputs (Crowther et al. 2016). The buildup of organic carbon depends upon the nature of organic matter as the different rate of organic matter decompositions causing different amounts of SOC losses from the soil contribute differential increase in SOC stocks. Soil organic matter or carbon decomposition is primarily controlled by the climatic conditions, morphological and chemical quality of plant residues, and soil characteristics (Parton et al. 2007; Cornwell et al. 2008; García-Palacios et al. 2016).The quality of plant residues available for the microbes determines the rate at which carbon will be sequestered in the soil or will be lost to the environment in different cropping systems (Faucon et al. 2017).The labile crop residues with more nitrogen content undergo rapid decomposition and greater carbon losses to the atmosphere (Cornwell et al. 2008; García-Palacios et al. 2016). According to the Microbial Efficiency-Matrix Stabilization framework, microbial biomass will increase after the degradation of labile litter and chemically fused with the soil matrix. This fusion and protection increase the stability of soil organic matter by protecting them from degradation (Cotrufo et al. 2013). So, the quality of organic residue added to the soils determines the carbon sequestration rate (Faucon et al. 2017) as it governs the soil organic carbon pools. Soil organic carbon pools are very important in determining the soil carbon sequestration. Organic materials with wider C/N ratio like FYM and crop residues had a more positive impact on the relatively stabilized fractions of soil organic carbon which may help in the carbon sequestration, while active fractions or labile form of carbon was more affected by the organic materials with narrower C/N ratio like green manure, etc. (Verma et al. 2014, 2015a). Active fractions of organic carbon pools are not stable, and it was more affected by the alternation in the agricultural management practices. Use of organic manures and compost enhances the organic carbon in soil as well as organic carbon pools. A study conducted by Pramanik and Prasad (2015) showed that SOC content was 0.71% in organically grown wheat and 0.43% in conventional wheat crop in the Bulandshahar district of Uttar Pradesh, India, after 10 years (Fig. 10.3).

Improvement in soil organic carbon content is the basic requirement to achieve the soil carbon sequestration under climate change scenario also. Nowadays, use of residues, compost, or other organic source has been partially replaced in some places by the application of biochar. This is the new dimension of utilization of residues in a productive way in agriculture. Biochar is a carbon-enriched substance produced by heating/burning of materials from organic origin in inert environment, or it is a recalcitrant organic carbon compound produced by incomplete combustion of biological materials (plant biomass, crop residues, waste materials, etc.) in the absence or with limited supply of oxygen at 300–1000 °C under the process of pyrolysis (Lehmann 2007).The properties (pH, EC, nutrient content, CEC, water holding capacity, surface area, etc.) of biochar depend on the type of feedstock and pyrolysis conditions under which the biochar has been prepared (Chan and Xu 2009; Singh et al. 2010).

The objective of biochar application is to improve soil properties and functions related to agronomic and environmental performance mainly related to enhanced water and nutrient retention and use efficiency (Lehmann and Joseph 2009; Woolf

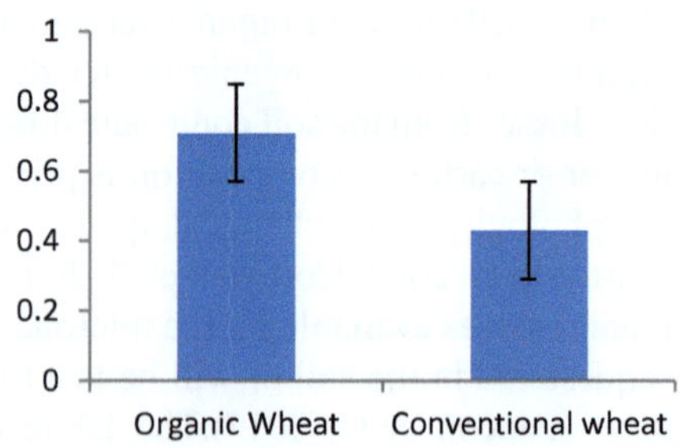


**Fig. 10.3** Variation of organic carbon content in organic and conventional farming under wheat in Bulandshahar district of Uttar Pradesh. (Pramanik and Prasad 2015)

et al. 2010). Biochar application is also encouraged to get the benefits of soil carbon sequestration as this is an important aspect in the global climate change. The biomass (crop residues, weed biomass, or product of organic origin), when converted to biochar, half of carbon goes with the bio-oil, biogas, etc.; however the remaining half of carbon is present in biochar, when this biochar will be applied in soil, which can improve the soil carbon content and help in soil carbon sequestration. This is well-illustrated in the below flowchart (Fig. 10.4).

Biochar can be used as a soil amendment to improve soil quality and health by influencing soil pH, CEC, structure, aggregation, and moisture retention and by reducing leaching of nutrients, soil acidity, irrigation, and fertilizer requirements (Peng et al. 2011; Zhang et al. 2012). Application of biochar with or without fertilizers was studied on maize under pot experiment, and it was found that biochar application has the positive effect on soil pH and biomass production (Mandal et al. 2015).

Biochar application has also positive effects on soil quality and crop productivity. Peng et al. (2011) has reported that, under pot experiment, application of biochar with or without fertilizers significantly alters the biomass production and found that with fertilizers, improvement is around two times as compared to without fertilizers (NPK). Several workers also reported that biochar applications affect positively to net crop productivity, grain yield and dry matter production, biomass production, etc. (Chan and Xu 2009; Major et al. 2010; Dhakal et al. 2015). Trail conducted on maize at IARI, New Delhi, farm reveled that application of biochar

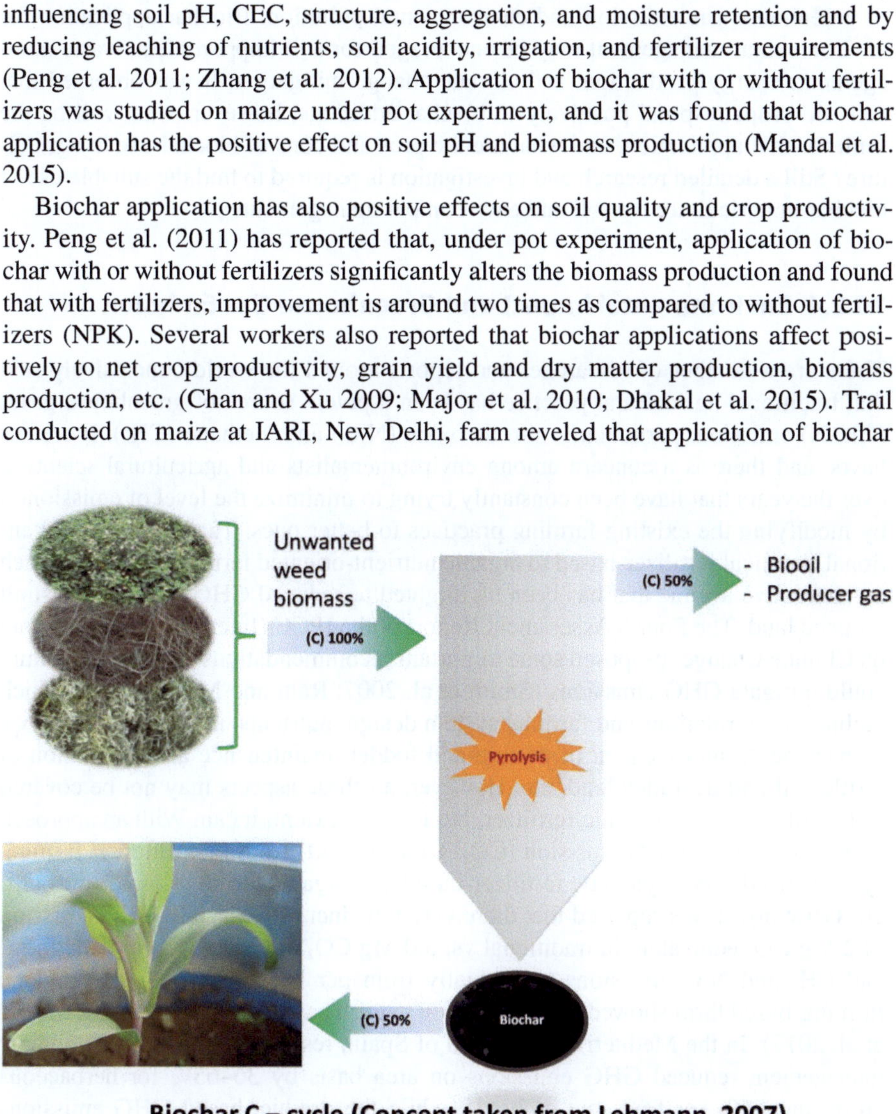


**Fig. 10.4** Carbon cycle under biochar application

with recommended dose of fertilizers increased the yield significantly and it was superior as compared to crop residue burning or incorporation. Major et al. (2010) reported that the biochar application decreases the exchangeable acidity and increases the availability of nutrients particularly calcium and magnesium. Application of biochar has positive effect that is related with the enhanced nutrient use efficiency, water holding capacity, microbial activity, etc. and shows positive significant role in crop growth. Biochar applications also stabilize carbon and absorb ionic solutes and hydrophobic organic pollutants which results in reduction of greenhouse gas emission and environmental pollution. Biochar application for environmental management may be encouraged for soil improvement, waste management, energy production, and climate change mitigation. It was observed that soil and crops respond positively to biochar additions; however, what will be the limit for that application, and what are the pros and cons of biochar use in agriculture? Still a detailed research and investigation is required to find the suitable use of biochar in agriculture for soil and environment management.

### 10.2.4 Environmental Impact and Greenhouse Gas Emission

There are a number of researches and reports around the world conclusively said that traditional agricultural practice has been a potent source of greenhouse gases (GHGs) emission, particularly nitrous oxide ($N_2O$) and methane ($CH_4$). It creates havoc and there is a concern among environmentalists and agricultural scientists over the years that have been constantly trying to minimize the level of emission(s) by modifying the existing farming practices to better ones. Turning from conventional chemical fertilizer-based to organic nutrient-oriented farming practice is such an alternative avenue that has been highlighted to reduced GHG emission per-unit cropped land. The Fourth Assessment Report of the IPCC (Intergovernmental Panel on Climate Change) proposed some important recommendations on how agriculture could mitigate GHG emissions (Smith et al. 2007; Ram and Meena 2014), which includes crop rotations and farming system design, nutrients, manure and livestock management, improvement of pasture and fodder, maintenance and restoration of fertile soil and degraded land, etc. However, all these aspects may not be covered fully with the use of organic fertilizer, but to some extent, it can. With an approach to assess the total GHG emission ($CO_2$, $CH_4$and $N_2O$) from two different farming systems (traditional synthetic fertilizer-based and organic farming-based) in southern Germany, it was reported that there was a distinct difference in GHG emission (4.2 Mg $CO_2$ equivalent in traditional vs. 3.0 Mg $CO_2$ equivalents in organic farm) and $CH_4$ and $N_2O$ emissions individually from per hectare land, where organic farming-based farm showed lower emission in each case (Flessa et al. 2002; Meena et al. 2017). In the Mediterranean region of Spain, researchers opined that organic management reduced GHG emissions on area basis by 36–65% for herbaceous crops and 56% for fruit tree orchards, while the product-based GHG emissions organic crops were also lower by 30% and 39%, respectively, with an exception of rice showing an increased $CH_4$ emission of 8% (Aguilera et al. 2015a, b). In three contrasting nutrient managements (INM, ONM, and NPK) of soybean-wheat rotation in Vertisols of central India, Lenka et al. (2017) found that the Global warming potential (GWP) per-unit grain yield was lowest under ONM over rest practices; GHG flux (in terms of cumulative $N_2O$ and $CO_2$ emissions) and annual GWP (in terms of $CH_4$ and $N_2O$ emissions) of ONM were better than INM practice followed.

In a more mechanistic approach, Skinner et al. (2014) concluded after a global meta-analysis that $N_2O$ emission from nonorganically managed soils is mainly influenced by total nitrogen inputs, whereas the same from organically managed soils is controlled by soil characteristics, while the lower $N_2O$ emissions from organically managed soils was reported over the nonorganic soils. Pathak (2015) opined that the application of organic manures and compost enhances the SOC pool and ensures better long-term soil C-sequestration, which may even persist for a century, over the use of equal amount of inorganic fertilizers. This increased SOC carries dual advantages: mitigation of $CO_2$ emission and enhancement of soil productivity.

Organic farming and reduced tillage techniques together showed interesting facts both in terms of C-storage in soil vis-à-vis reduced GHG emission; the long-term C-sequestration rates on arable land could be increased to a value of 500 kg C $ha^{-1}$ $year^{-1}$, which has a potential to mitigate 4.0 Gt $CO_2$-eq. $year^{-1}$ or cut down 65% of the agricultural GHG emission (Lal 2004a, b). Conservation agriculture, in the form of spreading of rice straw along with zero-tilled wheat cultivation in dominant rice-wheat systems of IGP, can be promoted for reducing the havoc of rice residue burning and lessen the air pollution since this particular management option reduced the $N_2O$ emission from soil (Gupta et al. 2016; Datta et al. 2017). Hence, it was well-explained that the organic fertilizer application is the requirement of the time for better soil and environmental management.

## 10.3 Factors Affecting the Use of Organic Fertilizers

Application, use and spread of any technology are governed by many factors, and it is difficult to manage single-handedly. According to Hadi et al. (2010), knowledge and understanding of certain new technology can be considered as an indicator of adoption level. Khanna (2001) viewed that higher level of education and experience can lead to higher adoption rates of new agricultural technologies. In acceptance of organic fertilizers in agriculture, several factors like household size, education level, and experience can negatively impact the adoption, whereas livestock numbers, extension contacts, access to information and media, and membership to farmer-based organizations positively influenced the decision to adopt organic fertilizer (Gelgo et al. 2016). As farm size and education are the most important factors for adopting organic fertilizers, awareness can be developed among the farmers regarding the importance of organic fertilizers through workshop, seminars, Krishi Vigyan mela, etc., and farmers should also be encouraged to operate large holding size. Access to information on different aspects of a new technology played a significant role in adoption of them. Various scientists also reported that environmental factors also play a very significant role in farmers' decision to adopt organic fertilizers. (Genius et al. 2006; Mzoughi 2001; Best 2010). The most important constraint are low livestock holding, lack of adequate labor, inadequate knowledge related to organic fertilizer adoption, high transaction costs, low skill, and capital.

Varma et al. (2017) found that a household head who is new and less experienced in agriculture, highly educated, had several exposures to extension, has bigger land holding, and is nearer to commercial organic fertilizer source is expected to adopt organic farming more as compared to households with contrasting characteristics. Likewise, the better information propagation through farmers' associations has encouraging impact on the farmer's decision to adopt advanced agricultural technologies. Better network among the farmers may access more information about several agricultural technologies. Ketema and Bauer (2011) revealed that a farmer with a big family is expected to adopt organic manures compared to synthetic fertilizer because he/she has sufficient labor for manure preparation and application. As this is comparatively less capital-intensive, farmers with low capital coupled with big households may swing from synthetic fertilizer to organic fertilizer. An increase in farm size may increase probability of manure application (Birungi 2007; Ketema and Bauer 2011). According to Diagne and Zeller (2001), farmers who have less fertile plots have positive perception toward adoption of the agricultural technologies such as organic fertilizer due to their expectation for better returns.

## 10.4 Limitation and Advantages of Organic Fertilizer

There were several enormous advantages of organic fertilizers mentioned; however, still we cannot assure the potential use of organic fertilizers. As the organic fertilizers can provide the nutrients slowly, the nutrient content in these materials are very low so huge amount of organic fertilizers are required to supply the nutrient as per the crop demand which creates a lot of limitation at the farmer's level. It became a labor-intensive operation to use the organic fertilizers. Nutrient supplying power through the organic fertilizers is slow, and it takes more time to provide the nutrients; even the organic fertilizers need several years to get the positive results. However, in commercial cultivation, farmers cannot wait for a longer period of time to get the positive effect of nutrient supply through organic fertilizers. Still there is some scope to get the quick response from the organic fertilizers if liquid formulation can be used like Vermiwash, seaweed extract, etc. Availability of these organic fertilizers is also not assured at the proper time to apply in the field, and the user could not wait to get them. Quality of organic fertilizers is very important to get the desired advantages, and there is uncertainty to get the quality product. It is very difficult to assure the quality of the organic fertilizers. In preparation of compost of good quality, it needs a proper condition and training, which is not always possible at the village levels. Compost preparation is a labor-intensive work, and getting labor during the composting is a serious concern for mixing of residues, filling the compost pit, etc. Proper benefit of organic fertilizers is dependent on the quality of the product as well as soil and environment conditions. It is highly dependent on soil temperature, moisture, aeration, and microbial diversity and density. Use of compost also has severe limitation as it is a potentially pathogenic as it bears seed of pathogen and negatively affects the soil and environment.

However, there are several limitations involved in proper use of organic fertilizer or organic source of nutrient; still practices have several advantages, which are briefly presented/summarized below. A large volume of waste obtained from the city, kitchen, as well as by-product of different industries, if not used properly. It became a challenge to handle, and the conversion of these residues into the valuable compost may definitively solve the problems. This prepared compost can be used as a source of nutrients in kitchen garden, pots, as well as in agricultural field along with fertilizers to support the crops by providing different nutrients after mineralization. This will be an excellent way to recycle the waste materials to productive nutrient source; otherwise, the nutrients present in that material will be locked as such and be slowly lost to the environment which has no use. These steps have put economic values to the waste materials and reduce the cost of production and finally lead to greater income from the farm. This mature compost is not only providing the nutrient but during the process of conversion/decomposition/mineralization, it produces several beneficial organic compounds like organic acids and fatty acids which positively affect the plant growth. It also releases vitamins, minerals for the betterment of the plants. It also provides organic matter to soil which helps in the buildup of the soil organic carbon and allows the microorganism to grow and multiply as it acts as a food source of these organisms. Application of organic fertilizers helps to maintain the carbon/nitrogen ratio in soil. Addition of compost or organic source of nutrient not only provides the major nutrients but provides almost all the nutrient which is required for the plant growth and also improve the soil quality and health by affecting the soil aggregation and soil structure formation which lead to an increase in the water holding capacity and infiltration and reduction of soil erosion. It also affects the soil temperature and moderates it at the optimum levels, which positively affects the germination and plant growth. With these numerous advantages of organic fertilizers, it reduces the dependency on fertilizers and conditioners. Apart of organic manure and compost, the use of green manure can improve the nitrogen status of soil and reduce the external nitrogen requirement as most of the green manure are leguminous in nature.

## 10.5 Scope and Future Prospects of Organic Fertilizer

A synthesis of a number of research papers, reviews, and meta-analyses supported well that the hypothesis of using organic fertilizer vis-à-vis organic farming systems is far more sustainable and environment-friendly than conventional farming systems (Reganold and Wachter 2016), with the following clear-cut benefits at a glance:

- Organic systems consistently have better soil carbon levels, hence sustaining good soil quality and health (specifically soil quality indicators) and less soil deterioration/erosion over the conventional systems.
- Organic farms usually support more plant and faunal diversity (insects, soil micro- and macro-fauna) and altogether a better diversified system. Most functional groups (herbivores, pollinators, predators and producers, i.e., plants) are more diverse in organic systems.
- With respect to nutrient leaching and greenhouse gas emissions, organic farming systems score better and reduce environmental pollution than conventional farming. Reports said that organic farms were found to have lower rate of nitrogen loss ($NO_3$ leaching, $N_2O$ emissions and $NH_3$ emissions).
- As organic agriculture uses virtually no synthetic fertilizers, there is a low to minimum risk associated with water pollution. Degradation in quality of freshwater and marine ecosystems is mainly linked to excessive use of chemical fertilizers (nitrogen and phosphorus) causing eutrophication (production of hypoxic zones) and becoming threats to aquatic lives.
- Organic systems are usually more energy-efficient (or, less energy-consumptive) than their conventional agriculture fields. European countries like Germany, Italy, Sweden, and Switzerland while partially trusted upon organic farming were found to use significantly less energy on a per-hectare basis.

## 10.6 Summary and Conclusion

The "Green Revolution" transforms the way of agriculture and farming practices (use of locally available materials and natural resources without any formal knowledge), and these old technologies were abandoned in the flow of new scientific methods and inputs with objective to produce more food for the growing populations. However, unscientific and indiscriminate use of these inputs (agrochemical) to achieve more productivity causes some major problems like degradation of natural (soil and water) resources, depletion of soil health and quality, soil and water pollution, buildup of pesticide residues, micronutrient deficiencies in soil, ecological imbalance, etc. In the context of soil and environment degradation and increasing cost of these inputs, it was realized that the agriculture production system needs some alternative practices for sustainable development to conserve these natural resources. The use of appropriate combination of mineral fertilizers along with organic source of nutrients (organic fertilizers) like farmyard manure (FYM), organic manures, crop residues, compost, vermicompost, etc. will be beneficial in improving and sustaining soil and environmental quality and health. The organic farming (organic fertilizer-based system) is a unique combination of environmentally sound practices with low external inputs. It relies not only on the fertilizers of organic origin (compost, manure, green manure, bone meal, etc.) but places emphasis on techniques such as crop rotation, green manuring, companion planting, etc. The use of materials of organic origin is not an issue in agriculture as this sector itself produces a vast amount of residues/by-product that must be utilized properly.

Only we required a sound idea and technology to reuse these residues for the betterment of soil and environment quality. The negative use of this surplus residue is that a major portion was burnt in the field which not only causes the pollution but also nutrient loss because burning is considered as an easiest option for farmers for fast management of crop/biomass residues to clear the agricultural field. Hence, there is a potential scope of the organic fertilizer for sustaining soil and environment, and organic agriculture is now seen as a significant means for sustainable food production. The code of organic production system is to feed the soil rather than the crops to maintain optimum soil health so that soil can provide necessary nutrients to the crop for its growth and development. Basic principles of organic farming or organic crop production are to make maximum and sustainable use of locally available resources and minimize the leaching of nutrients. Organic fertilizers have nearly all the important nutrients required for plant growth and produce other non-nutrient benefits also by providing foods for soil microbes and different growth-promoting organic acids and improving soil structure, water holding capacity, etc. Incorporation of crop residues or its retention on the soil surface has several positive impacts on physical, chemical, and biological properties of soil. It increases hydraulic conductivity and reduces soil bulk density by modifying soil structure, porosity, and aggregate stability. Apart from the nutrient-supplying power of the organic system as well as the yield and quality of produce, it is well-accepted that organic production system has more input of carbon as compared to all other systems. Carbon input and build up in soil is very impotent in terms of soil carbon sequestration and soil health/quality. Soil organic carbon is the central key element which governs most of the soil properties. It is the key component to define the soil quality or soil health. Improvement in soil organic carbon content is the basic requirement to achieve the soil carbon sequestration under climate change scenario also. Nowadays, the use of residues, compost, or other organic source has been partially replaced in some places by the application of biochar. This is the new dimension of utilization of residues in a productive way in agriculture. Biochar is intended to improve soil properties and functions relevant to agronomic and environmental performance mainly related to enhanced water and nutrient retention as well as improved soil structure and drainage. Biochar can be used as a soil amendment to improve soil quality by increasing soil pH, CEC, and moisture retention and by reducing leaching of nutrients, soil acidity, irrigation, and fertilizer requirements. Biochar application for environmental management can be motivated for soil improvement, waste management, energy production, and climate change mitigation. It was observed that soil and crops respond positively to biochar additions; however, still a detailed research and investigation is required to find the suitable use of biochar in agriculture for soil and environment management. Application, use, and spread of any technology are governed by many factors, and it is difficult to manage the single-handedly. Several factors like household size, education level, experience, livestock numbers, extension contacts, access to information media, and membership to farmer-based organizations affect to adopt organic fertilizer technology. The compost can be enriched and well-prepared with the help of technical knowledge and supervision to get the higher nutrient content. There were several enormous advantages of organic fertilizers mentioned; however, still we cannot assure the potential use of organic fertilizers. As the organic fertilizers can provide the nutrients slowly and the nutrient content in these materials are very low so huge amount of organic fertilizers are required to supply the nutrient as per the crop demand which creates a lot of limitation at the farmer's level. It became a labor-intensive operation to use the organic fertilizers. Quality of organic fertilizers is very important to get the desired advantages, and there is uncertainty to get the quality product. A large volume of waste obtained from the city, kitchen waste, as well as by-product of different industries, if not use properly, becomes a challenge to handle, and the conversion of these residues into the valuable compost may definitively solve the problems. This prepared compost can be used as a source of nutrients in kitchen garden, pots, as well as in agricultural field along with fertilizers to support the crops by providing different nutrients after mineralization. This will be an excellent way to recycle the waste materials to productive nutrient source; otherwise, the nutrients present in that materials will be locked as such and be slowly lost to the environment which has no use. So the use of organic fertilizer in agriculture for sustainable soil and environment management is very much required; however, the spread of this technology needs some improvement at the different levels of operation.

## References


Aguilera E, Guzmán G, Alonso A (2015a) Greenhouse gas emissions from conventional and organic cropping systems in Spain. I. Herbaceous crops. Agron Sustain Dev 35(2):713–724

Aguilera E, Guzmán G, Alonso A (2015b) Greenhouse gas emissions from conventional and organic cropping systems in Spain. II. Fruit tree orchards. Agron Sustain Dev 35(2):725–737

Ashoka P, Meena RS, Kumar S, Yadav GS, Layek J (2017) Green nanotechnology is a key for eco-friendly agriculture. J Clean Prod 142:4440–4441

Baranski M, Srednicka-Tober D, Volakakis N, Seal C, Sanderson R, Stewart GB, Benbrook C, Biavati B, Markellou E, Giotis C, Gnomadzka-Ostrawska J, Bembiaikowska D, Niggli U, Nicof D, Leifent C (2014) Higher antioxidant and lowercadmium concentrations and lower incidence ofpesticides residues in organically grown crops: asystematic literature review and meta analyses. Br J Nutr 112(5):794–811

Best H (2010) Environmental concern and the adoption of organic agriculture. Soc Nat Res Int J 23(5):451–468

Bhaduri D, Gautam P (2012) Balanced use of fertilizers and FYM to enhance nutrient recovery and productivity of wheat (*Triticumaestivum*cv UP-2382) in a Mollisol of Uttarakhand. Int J Agric Environ Biotechnol 5(4):435–439

Bhaduri D, Gautam P (2013) Optimization and validation of targeted yield equation-based fertilizer doses under INM for wheat in tarai region of Uttarakhand. India Indian Journal of Agricultural Research 47(1):16–25

Bhaduri D, Purakayastha TJ, Bhar LM, Patra AK, Sarkar B (2014a) Impact of integrated management on yield sustainability in relation to soil quality under a rice–wheat cropping system. Natl Acad Sci Lett 37(1):25–31

Bhaduri D, Purakayastha TJ, Patra AK, Singh M, Sarkar S (2014b) Combined effect of tillage-water-nutrient management under rice-wheat agro-ecosystem: a study on chemical indicators of soil quality. Agrochimica 58:63–76

Bhattacharyya R, Ghosh BN, Mishra PK, Mandal B, Rao CS, Sarkar D, Das K, Sankaranarayanan K, Lalitha M, MouliHati K, Franzluebbers AJ (2015) Soil degradation in India: challenges and potential solutions. Sustainability 7(4):3528–3570

Birungi P (2007) The linkages between land degradation, poverty and social capital in Uganda. Unpublished

Blair GJ, Lefroy RDB, Lisle L (1995) Soil carbon fractions based on their degree of oxidation, and the development of a carbon management index for agricultural systems. Aust J Soil Res 46:1459–1466

Bulluck LR, Brosius M, Evanylo GK, Ristaino JB (2002) Organic and synthetic fertility amendments influence soil microbial, physical and chemical properties on organic and conventional farms. Appl Soil Ecol 19(2):147–160

Chan KY, Xu Z (2009) Biochar: nutrient properties and their enhancement. In: Lehmann J, Joseph S (eds) Biochar for environmental management: science and technology. Earthscan, London, pp 67–84

Cornwell WK, Cornelissen JHC, Amatangelo K, Dorrepaal E, Eviner VT, Godoy O (2008) Plant species traits are the predominant control on litter decomposition rates within biomes worldwide. Ecol Lett 11:1065–1071

Cotrufo MF, Wallenstein MD, Boot CM, Denef K, Paul E (2013) The Microbial Efficiency-Matrix Stabilization (MEMS) framework integrates plant litter decomposition with soil organic matter stabilization: do labile plant inputs form stable soil organic matter? Glob Chang Biol 19:988–995

Crowther TW, Todd-Brown KEO, Rowe CW, Wieder ER, Carey JC, Machmuller MB, Bradford MA (2016) Quantifying global soil carbon losses in response to warming. Nature 540:104–108

Dadhich RK, Meena RS, Reager ML, Kansotia BC (2015) Response of bio-regulators to yield and quality of Indian mustard (*Brassica juncea* L. Czernj. And Cosson) under different irrigation environments. J App and Nat Sci 7(1):52–57

Datta R, Kelkar A, Baraniya D, Molaei A, Moulick A, Meena RS, Formanek P (2017) Enzymatic degradation of lignin in soil: a review. Sustain MDPI 9:1163. https://doi.org/10.3390/su9071163

Debbarma V, Abraham T (2015) Enhancement of soil fertility and productivity of organic rice (*Oryza sativa* (L.) sub sp. japonica). Res Environ Life Sci 8(4):687–690

Debosz K, Petersen SO, Kure LK, Ambus P (2002) Evaluating effects of sewage sludge and household compost on soil physical, chemical and microbiological properties. Appl Soil Ecol 19:237–248

Dhakal Y, Meena RS, De N, Verma SK, Singh A (2015) Growth, yield and nutrient content of mungbean (*Vigna radiata* L.) in response to INM in eastern Uttar Pradesh, India. Bangladesh J Bot 44(3):479–482

Diagne A, Zeller M (2001) Access to credit and its impact on welfare in Malawi. Research reports 116, International Food Policy Research Institute (IFPRI)

Faucon MP, Houben D, Lambers H (2017) Plant functional traits: soil and ecosystem services. Trends Plant Sci 22:385–394

Flessa H, Ruser R, Dörsch P, Kamp T, Jimenez MA, Munch JC, Beese F (2002) Integrated evaluation of greenhouse gas emissions ($CO_2$, $CH_4$, $N_2O$) from two farming systems in southern Germany. Agric Ecosyst Environ 91(1–3):175–189

García-Palacios P (2018) Data from: crop traits drive global soil carbon sequestration under organic farming. Figshare Digit Repos. https://doi.org/10.6084/m9.figshare.5659048.v1

García-Palacios P, McKie BG, Handa IT, Frainer A, Hättenschwiler S (2016) The importance of litter traits and decomposers for litter decomposition: a comparison of aquatic and terrestrial ecosystems within and across biomes. Funct Ecol 30:819–829

Gelgo B, Patience M, Zemedu ML (2016) Analysis of the impact of organic fertilizer use on smallholder farmers' income in Shashemene District, Ethiopia. Int J Agric Econ 1(4):117–124

Genius M, Pantzios CJ, Tzouveleka V (2006) Information acquisition and adoption of organic farming practices. J Agric Resour Econ 31(1):93–113

Ghosh PK, Ajay KK, Bandyopadhyay MC, Manna KG, MandalAK HKM (2004) Comparative effectiveness of cattle manure, poultry manure, phosphocompost and fertilizer-NPK on three cropping system in vertisols of semi-arid tropics. II. Drymatter yield, nodulation, chlorophyll content and enzyme activity. Bioresour Technol 95:85–93

Gosal SK, Gill GK, Sharma S, Walia SS (2017) Soil nutrient status and yield of rice as affected by long term integrated use of organic and inorganic fertilizers. J Plant Nutr. https://doi.org/10.1080/01904167.2017.1392570

Guo LB, Gifford RM (2002) Soil carbon stocks and land use change—a meta analysis. Glob Chang Biol 8:345–360

Gupta DK, Bhatia A, Kumar A, Das TK, Jain N, Tomer R, Malyan SK, Fagodiya RK, Dubey R, Pathak H (2016) Mitigation of greenhouse gas emission from rice–wheat system of the Indo-Gangetic plains: through tillage, irrigation and fertilizer management. Agric Ecosyst Environ 230:1–9

Hadi F, Abu DS, Sarkar SK (2010) Application of logit model in innovation adoption: a study on biotechnology academic researchers in Malaysia. Am Eurasian J Agric Environ Sci 9(3):282–287

Huang L, Jun YU, Jie YANG, Zhang R, Yanchao BAI, Chengming SUN, Zhuang H (2016) Relationships between yield, quality and nitrogen uptake and utilization of organically grown rice varieties. Pedosphere 26(1):85–97

International Federation of Organic AgricultureMovement (IFOAM) (2009) In: Godinho D (ed) Proceedings of the 1st IFOAM international conference on organic animal and plant breeding. IFOAM, Bonn

Jannoura R, Joergensen RG, Bruns C (2014) Organic fertilizer effects on growth, crop yield, and soil microbial biomass indices in sole and intercropped peas and oats under organic farming conditions. Eur J Agron 52:259–270

Kakraliya SK, Singh U, Bohra A, Choudhary KK, Kumar S, Meena RS, Jat ML (2018) Nitrogen and legumes: a Meta-analysis. In: Meena RS et al (eds) Legumes for soil health and sustainable management. Springer, Singapore. https://doi.org/10.1007/978-981-13-0253-4_9

Kesarwani A, Chiang PY, Chen SS, Su PC (2013) Antioxidant activity and total phenolic content of organically and conventionally grown rice cultivars under varying seasons. J Food Biochem 37(6):661–668

Ketema M, Bauer S (2011) Determinants of manure and fertilizer applications in Eastern Highlands of Ethiopia. Q J Int Agric 50(3):237–252

Khanna M (2001) Sequential adoption of site-specific technologies and its implications for nitrogen productivity: a double selectivity model. Am J Agric Econ 83(1):35–51

Kong AYY, Six J, Bryant DC, Denison RF, van Kessel C (2005) The relationship between carbon input, aggregation, and soil organic carbon stabilization in sustainable cropping systems. Soil Sci Soc Am J 69:1078–1085

Kumar S, Meena RS, Yadav GS, Pandey A (2017) Response of sesame (*Sesamum indicum* L.) to sulphur and lime application under soil acidity. Int J Plant and Soil Sci 14(4):1–9

Kumar S, Meena RS, Lal R, Yadav GS, Mitran T, Meena BL, Dotaniya ML, EL-Sabagh A (2018a) Role of legumes in soil carbon sequestration. In: Meena RS et al (eds) Legumes for soil health and sustainable management. Springer, Singapore. https://doi.org/10.1007/978-981-13-0253-4_4

Kumar S, Meena RS, Bohra JS (2018b) Interactive effect of sowing dates and nutrient sources on dry matter accumulation of Indian mustard (*Brassica juncea* L.). J Oilseed Brassica 9(1):72–76

Lal R (2004a) Soil carbon sequestration impacts on global climate change and food security. Science 304:1623–1627

Lal R (2004b) Soil carbon sequestration to mitigate climate change. Geoderma 123:1–22

Lehmann J (2007) A handful of carbon. Nature 447:143–144

Lehmann J, Joseph S (2009) Biochar for environmental management: Science and Technology. Earthscan, New York

Leifeld J, Fuhrer J (2010) Organic farming and soil carbon sequestration: what do we really know about the benefits? Ambio 39:585–599

Lenka S, Lenka NK, Singh AB, Singh B, Raghuwanshi J (2017) Global warming potential and greenhouse gas emission under different soil nutrient management practices in soybean–wheat system of Central India. Environ Sci Pollut Res 24(5):4603–4612

Major J, Rondon M, Molina D, Riha SJ, Lehmann J (2010) Maize yield and nutrition during 4 years after biochar application to a Colombian savanna oxisol. Plant Soil 333:117–128

Mandal S, Verma BC, Ramkrushna GI, Singh RK, Rajkhowa DJ (2015) Characterization of biochar obtained from weeds and its effect on soil properties of North Eastern Region of India. J Environ Biol 36(2):499–505

Meena RS, Lal R (2018) Legumes and sustainable use of soils. In: Meena RS et al (eds) Legumes for soil health and sustainable management. Springer, Singapore. https://doi.org/10.1007/978-981-13-0253-4_1

Meena RS, Yadav RS (2015) Yield and profitability of groundnut (*Arachis hypogaea* L.) as influenced by sowing dates and nutrient levels with different varieties. Legum Res 38(6):791–797

Meena RS, Yadav RS, Meena H, Kumar S, Meena YK, Singh A (2015) Towards the current need to enhance legume productivity and soil sustainability worldwide: a book review. J Clean Prod 104:513–515

Meena H, Meena RS, Singh B, Kumar S (2016a) Response of bio-regulators to morphology and yield of clusterbean [*Cyamopsis tetragonoloba* (L.) Taub.] under different sowing environments. J App and Nat Sci 8(2):715–718

Meena RS, Bohra JS, Singh SP, Meena VS, Verma JP, Verma SK, Shiiag SK (2016b) Towards the prime response of manure to enhance nutrient use efficiency and soil sustainability a current need: a book review. J Clean Prod 112:1258–1260

Meena RS, Gogaoi N, Kumar S (2017) Alarming issues on agricultural crop production and environmental stresses. J Clean Prod 142:3357–3359

Mitran T, Mani PK, Basak N, Biswas S, Mandal B (2017) Organic amendments influence on soil biological indices and yield in rice-based cropping system in coastal Sundarbans of India. Commun Soil Sci Plant Anal 48(2):170–185

Mondal S, Mallikarjun M, Ghosh M, Ghosh DC, Timsina J (2016) Influence of integrated nutrient management (INM) on nutrient use efficiency, soil fertility and productivity of hybrid rice. Arch Agron Soil Sci 62(11):1521–1529

Mzoughi N (2001) Farmers adoption of integrated crop protection and organic farming. Do moral and social cover and matter? Ecol Econ 70(8):1536–154

Parton W, Silver WL, Burke IC, Grassens L, Harmon ME, Currie WS (2007) Global-scale similarities in nitrogen release patterns during long-term decomposition. Science 315:361–364

Pathak H (2015) Greenhouse gas emission from Indian agriculture: trends, drivers and mitigation strategies. Proc Indian Natl Sci Acad 81(5):1133–1149

Peng X, Ye LL, Wang CH, Zhou H, Sun B (2011) Temperature- and duration-dependent rice straw-derived biochar: characteristics and its effects on soil properties of an Ultisol in southern China. Soil Tillage Res 112:159–166

Prabhakaran C (2003) Nutrient uptake and yield of tomato (*Lycoperscionesculentum*) with different organic manure. J Res Birsa Agric Uni 15(1):131–134

Pramanik P, Prasad S (2015) Impact of organic farming on soil physic- chemical properties and greenhouse gas emission. Journal of Agroecology and Natural Resource Management 2(4):258–263

Punia M, Nautiyal VP, Kant Y (2008) Identifying biomass burned patches of agricultural residue using satellite remote sensing data. Curr Sci 94:1185–1190

Ram K, Meena RS (2014) Evaluation of pearl millet and mungbean intercropping systems in Arid Region of Rajasthan (India). Bangladesh J Bot 43(3):367–370

Rasul GA, Ahmed ST, Ahmed MQ (2015) Influence of different organic fertilizers on growth and yield of wheat. American-Eurasian J Agric Environ Sci 15(6):1123–1126

Reganold JP, Wachter JM (2016) Organic agriculture in the twenty-first century. Nat Plants 2(2):15221

Rena R (2004) Green revolution: Indian agricultural experience – a paradigm for Eritrea. Eritrean Stud Rev 4(1):103–130

Saigusa M (2000) Broadcast application versus band application of polyolefin coated fertilizer on green peppers grown on andisol. J Plant Nutr 23:1485–1493

Saikia P, Savita Choudhury BU, Verma BC, Moirangthem P, Kumar M, Hazarika S (2017) Rural composting: clean environment, safe disposal of wastes, better soil health and crop productivity in Meghalaya. Indian Farmers' Digest 51(2):41–43

Sanchez JE, Harwood RR, Wilson TC, Kizilkaya K, Smeenk J, Parker E, Paul EA, Knezek BD, Robertson GP (2004) Managing soil carbon and nitrogen for productivity and environmental quality. Agron J 96:769–775

Singh DD (2003) Management of crop residue in summer rice and its effect on the soil properties and crop yield. Crop Res 25(1):191–193

Singh JS, Raghubanshi AS, Singh RS, Srivastava SC (1989) Microbial biomass acts as a source of plant nutrient in dry tropical forest and savanna. Nature 338:499–500

Singh B, Singh BP, Cowie AL (2010) Characterization and evaluation of biochars for their application as a soil amendment. Aust J Soil Res 48:516–525

Singh R, Babu S, Avasthe RK, Yadav GS, Rajkhowa DJ (2015) Influence of tillage and organic nutrient management practices on productivity, profitability and energetic of vegetable pea (*Pisumsativum* L.) in rice-vegetable pea sequence under hilly ecosystems of north-east India. Res Crops 16(4):683–688

Singh M, Singh YV, Singh SK, Dey P, Ram RL (2017) Validation of soil test and yield target-based fertilizer prescription model for rice on inceptisol of Eastern Zone of Uttar Pradesh. India Int J Curr Microbiol App Sci 6(2):406–415

Skinner C, Gattinger A, Muller A, Mäder P, Flieβbach A, Stolze M, Ruser R, Niggli U (2014) Greenhouse gas fluxes from agricultural soils under organic and non-organic management—a global meta-analysis. Sci Total Environ 468:553–563

Smith P, Martino D, Cai Z, Gwary D, Janzen H, Kumar P, McCarl B, Ogle S, O'Mara F, Rice C, Scholes B, Sirotenko O (2007) Agriculture. In: Metz B, Davidson OR, Bosch PR, Dave R, Meyer LA (eds) Climate change mitigation. Contribution of Working Group III to the fourth assessment report of the Intergovernmental Panel on Climate Change. Cambridge University Press, Cambridge/New York

Sofi PA, Baba ZA, Hamid B, Meena RS (2018) Harnessing soil rhizobacteria for improving drought resilience in legumes. In: Meena RS et al (eds) Legumes for soil health and sustainable management. Springer, Singapore. https://doi.org/10.1007/978-981-13-0253-4_8

Sugumaran P, Sheshadri S (2009) Evaluation of elected biomass for charcoal production. J Sci Ind Res 68:719–723

Taylaran RD, Salva Jr AG, Elmundo EM (2013) Organically-grown upland rice (*Oryza Sativa* L.) evaluation for drought adaptation. Philippine J Crop Sci. http://agris.fao.org/agris-search/search.do?recordID=PH2014000694

Vanlauwe B, Aihou K, Aman S, Iwuafor ENO, Tossah BK, Diels J, Sanginga N, Lyasse O, Merckx R, Deckers J (2001) Maize yield as affected by organic inputs and urea in the West African Moist Savanna. Agron J 93:1191–1199

Varma D, Meena RS, Kumar S (2017) Response of mungbean to fertility and lime levels under soil acidity in an alley cropping system in Vindhyan Region, India. Int J Chem Stu 5(2):384–389

Verma BC, Datta SP, Rattan RK, Singh AK (2010) Monitoring changes in soil organic carbon pools, nitrogen, phosphorus and sulfur under different agricultural management practices in the tropics. Environ Monit Assess 171:579–593

Verma BC, Datta SP, Rattan RK, Singh AK (2013a) Labile and stabilized fractions of soil organic carbon in some intensively cultivated alluvial soils. J Environ Biol 34:1069–1075

Verma BC, Datta SP, Rattan RK, Singh AK (2013b) Impact of temperature and moisture regimes on carbon and nitrogen mineralization in an alluvial soil amended with organics. Agrochimica 57(1):67–81

Verma BC, Datta SP, Rattan RK, Singh AK (2014) Impact of tillage, water and nutrient management practices on soil organic carbon pools in a seven year rice- wheat system. Agrochimica 58(4):11–15

Verma JP, Jaiswal DK, Meena VS, Meena RS (2015a) Current need of organic farming for enhancing sustainable agriculture. J Clean Prod 102:545–547

Verma JP, Meena VS, Kumar A, Meena RS (2015b) Issues and challenges about sustainable agriculture production for management of natural resources to sustain soil fertility and health: a book review. J Clean Prod 107:793–794

Verma SK, Singh SB, Prasad SK, Meena RN, Meena RS (2015c) Influence of irrigation regimes and weed management practices on water use and nutrient uptake in wheat (*Triticum aestivum* L. Emend. Fiori and Paol.). Bangladesh J Bot 44(3):437–442

Verma BC, Choudhury BU, Kumar M, Hazarika S, Ramesh T, Bordoloi LJ, Moirangthem P, Bhuyan D (2017) Soil organic carbon fractions and enzymes activities as affected by organic and inorganic amendments in an acid soil of Meghalaya. J Indian Soc Soil Sci 65(1):54–61

Vishram R (2014) Residues and tillage management towards mitigating climate change. In: Saikia et al (ed) Natural resource management for enhancement of adaptation and mitigation potential under changing climate. Published by ICAR Research Complex for NEH Region, Umiam, Meghalaya, India, pp 125–131

Woolf D, Amonette JE, Alayne F, Street-Perrott FA, Lehmann J, Joseph S (2010) Sustainable biochar to mitigate global climate change. Nat Commun 1(5):1–9

Wu W, Ma B (2015) Integrated nutrient management (INM) for sustaining crop productivity and reducing environmental impact: a review. Sci Total Environ 512:415–427

Yadav GS, Lal R, Meena RS, Babu S, Das A, Bhomik SN, Datta M, Layak J, Saha P (2017a) Conservation tillage and nutrient management effects on productivity and soil carbon sequestration under double cropping of rice in North Eastern Region of India. Ecol Indic. http://www.sciencedirect.com/science/article/pii/S1470160X17305617

Yadav GS, Lal R, Meena RS, Datta M, Babu S, Das LJ, Saha P (2017b) Energy budgeting for designing sustainable and environmentally clean/safer cropping systems for rainfed rice fallow lands in India. J Clean Prod 158:29–37

Zhang A, Bian R, Pan G, Cui L, Hussain Q, Li L, Zheng J, Zhang X, Han X, Yu X (2012) Effects of biochar amendment on soil quality, crop yield and greenhouse gas emission in a Chinese rice paddy: a field study of consecutive rice growing cycles. Field Crop Res 127:153–160

# Role of Nanotechnology for Enhanced Rice Production

11

Afifa Younas, Zubaida Yousaf, Nadia Riaz, Madiha Rashid, Zainab Razzaq, Maliha Tanveer, and Shiwen Huang

**Abstract**

Rice (*Oryza sativa* spp.) is a main cash crop all around the globe. It is grown under a wide range of environments. Food deficiency is a major issue in the world with growing global population. The current challenge in agriculture is food quality and quantity decline. In earlier times conventional farming techniques were used for rice cultivation. The major issue about conventional farming is to maintain the crop productivity, soil structure and fertility. Integrated farming, inorganic chemical fertilizers, ecological farming and Sri Lanka farming system are conventional practices which we mentioned in this chapter. These conventional farming practices raising rice crop showed decreased fertility of soil and increase the negative impact on environmental ecosystems. These conventional methods upgrade the risk of global warming and minimize the effective agricultural operations. To achieve required food production in the last few eras, nanotechnology has become one of the most promising techniques to revolutionize the conventional food science and technologies. Nanotechnology is the technology of the twenty-first century. This new discipline brings nano-agrochemicals, i.e. plant growth-promoting nanosystems (to enhance plant growth and production), nanopesticides and nanofertilizers. Nanotechnology offers the nanofood processing and advancement of nano-based food material, smart delivery of nutrients and bioactive materials. This chapter focused on

A. Younas
China National Rice Research Institute, Hangzhou, China

Department of Botany, Lahore College for Women University, Lahore, Pakistan

Z. Yousaf (✉) · N. Riaz · M. Rashid · Z. Razzaq · M. Tanveer
Department of Botany, Lahore College for Women University, Lahore, Pakistan

S. Huang (✉)
China National Rice Research Institute, Hangzhou, China
e-mail: huangshiwen@cass.cn





R. S. Meena (ed.), *Nutrient Dynamics for Sustainable Crop Production*,
https://doi.org/10.1007/978-981-13-8660-2_11

nano-agrochemicals, diagnosis of plant pathogen and nanofood-based technologies as advanced approaches of nanotechnology in the field of agriculture and food industry. In this chapter the potential uses and benefits of nanotechnology in precision agriculture are discussed. We also discussed the current and future uses of nanomaterials in agriculture, food safety and security and recommendations regarding to nanomaterial.

**Keywords**
Agriculture · Diagnosis of plant pathogen · Food quality · Nano-agrochemicals · Nanofoods · Nanofertilizers · Nanopesticides · Nanotechnology

## Abbreviations

| | |
|---|---|
| BARC | Bhabha Atomic Research Centre |
| EIS | Electrochemical impedance spectroscopy |
| ENMs | Engineered nanomaterials |
| FAO | Food and Agriculture Organization |
| FRET | Fluorescent resonance energy transfer |
| FYM | Farmyard manure |
| GNWA | Gold nanowire array |
| NGS | Next-generation sequence |
| NLCs | Nanostructure lipid carriers |
| PEG | Polyethylene glycol |
| PSA | Prostate-specific antigen |
| QD | Quantum dots |
| SERS | Surface-enhanced Raman scattering |
| SLNs | Solid lipid nanoparticles |
| SWNT | Single-walled nanotubes |
| VRE | Vancomycin-resistant enterococci |
| WFC | World Food Council |

## 11.1 Introduction

The practice and profits of nanotechnology in the agri-crops sector require substantial attention, especially in the synthesis of unique nano-agrochemicals like nanopesticides and nanofertilizers. The modern approaches in nanotechnology are acknowledged, and the most significant prospects awaiting the agricultural division from the latest scientific literature are addressed. In this chapter, discussion is on the significant use and recent application of nano-based technologies in the form of nanofertilizers, nanopesticides, nanosensors, plant protection and pathogen detection, food quality and safety measures. In novel nanoparticles for fertilizer covering, nanosensors have been reported for primary application in crop cultivation practices, food quality and packing techniques. Therefore, nanotechnology will change the agricultural sector for improved production and agri-food products. Nanotechnology plays an important role in effective nutrient utilization, plant pathogens and controlled released pesticides and fertilizers. Due to application of nanotechnology, we can overcome the scientific breaks between research, and fundamental questions have been addressed. Different subdivisions of nanotechnology like nanoscience and nanoengineering offer wide opportunities and provide a feasible alternative in agri and food processing sector by providing novel and advanced solutions (Kim et al. 2018; Ashoka et al. 2017).

Nanotechnology research activities are extremely important in agricultural sector in particular to global changes, population demand, climate change and limited availability of macro- and micronutrients of plants. Mei-Yan Wu in 2013 presented his work about the effect of nano-carbon and slow-released fertilizers on rice yield. In his results he mentioned that it is possible to use nano-carbon as coating agent for different types of combinations of fertilizers and it will be also helpful for reducing the water pollution especially Jingzhengda slow-released fertilizer and nano-carbon (JSCU+C) (Wu 2013; Meena and Lal 2018).

Nanotechnology and nanoparticles have a great potential as 'magic bullets' loaded with herbicides, fungicides, nutrients, fertilizers or nucleic acids, targeting specific plant tissues to release their charge to the desired part of the plant to achieve desired results.

There is a piece of work on rice with urea-HANP (*hydroxyapatite nanoparticle*) hybrids, and they showed 50% less consumption of urea in their results (Kottegoda et al. 2017). Due to use of nanofertilizers of urea, rice production was increased by approximately 7.9 tons/hectares which are higher than the normal urea fertilizers of 7.3 tons/hectares. HA is a bioceramic which provides calcium (Ca), phosphate (P) and other micronutrients. Urea coated with HA nanoparticles which slow down the release of nitrogen because of chemical bonding properties of HANPs and increase the uptake of urea into rice crop. NPs sized 10–200 nm in length and 15–20 width diameters, and it showed no penetration effect into rice crops (Kottegoda et al. 2017).

In this piece of work, we have evaluated different studies on rice production and consumption from view of supply and demand of different regions. Authors described performance of different rice cultivation methods and techniques. In this context, the present work discusses the key knowledge gaps due to orthodox methods and further high points for a promising approach of future agri-nanotechnology researches. This chapter is focused on modern strategies of nanotechnology used for management of agricultural land, fertilizers, pesticides, sensors, minimal use of synthetic or chemical pesticides and potential of nanoparticles in sustainable agriculture as modern approaches of nanotechnologies.

## 11.2 Rice as Functional Food Species

*Oryza sativa* L. commonly known as rice is the main and leading staple food crop eaten worldwide, providing basic nutrient intake to more than three billion people with worth 50–80% of their daily calories (Khush 2005). Rice has functional properties for both human and animals as it is a highly fibrous crop. Literature has also reported benefits of high fibre intake as compared to low fibre intake. In literature, 22% extruded moisture was reported from short- and long-grain rice flour at 70–120 °C, while the huge densities essentially stayed unchanged except water absorption and water solubility which increase with increasing extrusion temperature. Meanwhile, fat absorption indices decreased only at 55 °C. With increasing extrusion in temperature, cold paste viscosities progressively increased, whereas, the peak, breakdown, setback and final viscosities decreased. A substitution of 25% of extracted or 70 °C removed long-grain rice flour into a wheat flour-based cooked snack diminished its fat assimilation by 35–50% without influencing the general surface.

## 11.3 Rice Production All Over the World

Asia (East and West), the Middle East, Latin America and West Indies are cultivating rice on a large scale from centuries ago (FAO 2005). In the twenty-first century, a rapid increase in population is seen especially in Asia, the Middle East and African counties, and increasing population has uplifted its demand and supply to an estimated quantity of 2000 million metric tons by 2030 (Coats 2003; Bloom David 2011; FAO World Agriculture 2002). To achieve this target, new technologies are required to control and significantly improve this horrific scenario (Ainsworth 2008).

As the increasing population will shake off the current position of supply and demand of rice and therefore expected to become a major food crisis affecting the mankind and environment. The International Rice Research Institute in 2000 compare the current values of the population with food demand, and according to them, if this condition pursues for longer time period, the demand will rise to 800 million tons by 2025. In another report by the Food and Agriculture Organization (2000), currently the world population has expanded by 1.3%, which is slightly less than the growth rate between (1.9%) 1984 and 1994. With this tremendous population, the predicted rice production will be 424 million tons and demand 422 million tons in next 5 years (FAO 2005). This expectation from FAO was already proved by Yap; one of his researches highlighted the major contributing factors influencing negatively to rice supply and demand (Yap 1997). On overall scenario he made prediction that worldwide rice consumption would become 482 million tons till the year 2010 with almost 19 million tons and 463 million tons of consumption in industrial countries and developing countries, respectively. Keeping these future food crises in mind (Schwartz 1991), the research was pursued to clarify the relationship of future demand and supply of rice, which is well-connected with Asia.

In the year 2016, FAO production forecast confirmed the record outcome of the season. Global paddy production in 2016 was set to go beyond 751.9 million tons

(499.2 million tons, milled basis), which was 3.9 million tons and 1.6% more than the 2015 blue level. The 2016 season also stretched out well in Africa, where a record of 30.8 million tons was predicted to be gathered. More winning paddy costs in respect to contending crops likewise encouraged a production bounce back in the United States, yet the season demonstrated all the more difficult somewhere else. In Latin America and the Caribbean, a mix of sporadic climate and prospects of diminished edges discouraged yield in Argentina, Bolivia, Brazil, Ecuador, Guyana, Uruguay and Venezuela, overshading gains in Chile, Colombia, Cuba, the Dominican Republic, Mexico and Peru. In Oceania, constrained and expensive water supplies for water system also restricted production in Australia (FAO 2016).

Then coming to the next year 2017, FAO's preliminary forecast of global paddy production in 2017 was set at 758.9 million tons (503.8 million tons, milled basis). They said that the forecast would infer a 0.9% yearly development while recommending a likely strike in the rate of production growth next season. This could be particularly the case in Asia, which is probable to account for a significant part of the worldwide increase in production but where essential rice producers have seen returns diminished by large harvests of crops as of now undermined by harsh climate. Within the area, huge gains are expected to concern China (Mainland), India and Indonesia, where rice keeps on profiting from strong state incentives (FAO 2017). According to FAO rice production record from 2011 to 2022 (000 Tons) increases all over the world as shown in Table 11.1. Output was also expected to magnify in Bangladesh, the Democratic People's Republic of Korea, Malaysia, Myanmar, Nepal, Pakistan, the Philippines, Thailand, Turkey and Vietnam, more than reimbursing for compressions in Afghanistan, Cambodia, the Democratic People's Republic of Korea and Sri Lanka. In Africa, inconsistent downpours have imperfected the viewpoint for Madagascar and the United Republic of Tanzania, adding to a forecast of a shortage of rainfall in Egypt, as lands come back to cotton cultivation. However, provided no major setback is incurred, continued efforts to reduce reliance on imports could prompt further crosswise over West Africa, in this manner keeping yield in the continent close to the exceptional 2016 harvest. In Latin America and the Caribbean, constrictions postured by high costs of production and unpleasant rates have prohibited noteworthy territory recoveries in South America. However, crops have gained large profit by favourable climate, which is probable to manage retrieval in the area's yield. In the world elsewhere, Europe and the United States look headed towards making constrictions, in the midst of reduced edges, though yield in Australia is set to organize a strong retrieval because of bottomless water availabilities and lower water system costs.

## 11.4 Worldwide Supply and Demand for Rice

Looking at the current situation, in the next 35 years, the cereal and rice harvest area will uplift by 15% and 11% respectively. The harvested area of rice comprises about 23.5% of the cereal harvest area and accounts for 26.6% of worldwide cereal production and consumption (FAO 2018). Asia is cultivating rice on a vast scale due to

**Table 11.1** Rice productions over the entire world grain: world markets and trade US Department of Agriculture; Foreign Agricultural US Department of Agriculture; Foreign Agricultural Service Circular Series FG 12-11 December 2011 Service Circular Series FG 09-12 September 2012, June 2014, May 2016, August 2018

| | World rice production record from 2011 to 2022 (000 tons) | | | | | | | | |
|---|---|---|---|---|---|---|---|---|---|
| | 2011–2012 | 2012–2013 | 2013–2014 | 2014–2015 | 2015–2016 | 2016–2017 | 2017–2018 | 2018–2019 | 2019–2020 |
| United States | 7593 | 6035 | 6035 | 7106 | 6133 | 7117 | 5659 | 6763 | 6695 |
| Thailand | 20,262 | 20,300 | 20,300 | 18,750 | 15,800 | 19,200 | 20,370 | 21,200 | 21,200 |
| Vietnam | 36,001 | 36,929 | 36,947 | 28,166 | 27,584 | 27,400 | 28,943 | 29,069 | 29,069 |
| India | 95,300 | 100,000 | 100,000 | 105,485 | 104,408 | 109,698 | 110,000 | 109,000 | 109,000 |
| Pakistan | 4700 | 6650 | 6650 | 7003 | 6802 | 6849 | 7500 | 7400 | 7400 |
| Bangladesh | 32,900 | 33,000 | 33,000 | 34,500 | 34,500 | 34,578 | 32,650 | 34,700 | 34,700 |
| China | 137,000 | 141,000 | 140,500 | 144,560 | 145,770 | 144,953 | 145,989 | 142,200 | 142,200 |

**Table 11.2** Trade of rice all over the world grain: world markets and trade US Department of Agriculture; Foreign Agricultural US Department of Agriculture; Foreign Agricultural Service Circular Series FG 12-11 December 2011 Service Circular Series FG 09-12 September 2012, June 2014, May 2016, August 2018

| | World trade rice export record from 2011 to 2022 (000 tons) | | | | | | | | |
|---|---|---|---|---|---|---|---|---|---|
| | 2011–2012 | 2012–2013 | 2013–2014 | 2014–2015 | 2015–2016 | 2016–2017 | 2017–2018 | 2018–2019 | 2019–2020 |
| United States | 3247 | 3500 | 3350 | 3381 | 3355 | 3349 | 2950 | 3300 | 3320 |
| Thailand | 10,647 | 6500 | 8000 | 9779 | 9867 | 11,615 | 10,500 | 11,000 | 11,000 |
| Vietnam | 7000 | 7000 | 7000 | 6606 | 5088 | 6488 | 7000 | 7000 | 7000 |
| India | 4637 | 8000 | 6500 | 11,046 | 10,040 | 12,056 | 12,800 | 12,500 | 12,500 |
| Pakistan | 3000 | 3750 | 4000 | 4000 | 4100 | 3642 | 4300 | 4200 | 4300 |
| Ungurary | 975 | 850 | 850 | 718 | 996 | 1051 | 860 | 800 | 800 |
| China | 500 | 600 | 600 | 262 | 368 | 1173 | 1600 | 1800 | 1800 |

excessive field area, and that's why its demand and supply are going smoothly and steadily until now. Trade of rice all over the world record from 2011 to 2022 (000 tons) is shown in Table 11.2.

According to FAO and United Nations Procurement Division (UNPD) reports, crop and other commodity demand and production will be higher than 40% from 2009 to 2030, as we can say that the increase will be of 1.5% per year (FAO 2006; UNPD 2006). Using reference of World Bank (2008), it is also calculated that 50% increase will occur in demand of cereals between 2000 and 2030.

## 11.5 Rice Protects Asian Areas from Starvation

Asia and Africa are on the edge of starvation. Economic and Social Development and World Food Council (WFC) in 1974 in its 1st annual convention mainly discussed about the prevention of starvation in Asian and African states. Then later in their 10th annual convention held in 1984, they made conclusions and recommendations on exterminating starvation all over the world keeping emphasis on the reputation of food approaches to solve this serious food problem rooted mainly in Africa (Imoki 1984). This initiative showed fruitful results during those 10 years by eliminating Asian starvation especially in countries such as India and China facing large food issues and has now become self-sufficient.

## 11.6 Conventional Practices for Raising Rice Crop

Crop productivity depends upon several factors like physical and chemical properties of soil and water availability. Overall trends in Asia indicate declined rice yields. One of the big issues for farmers is to maintain productivity by conventional agricultural practices and low soil organic matter. Crop production and tenability depend on soil fertility structure. Soil fertility is a functional element describing physical, chemical and biological condition of the intrinsic or artificially maintained soil environment. These properties collectively influence nutrient supply in plants and successive crop production (Bhuiyan and Tuong 1999). The urge for headway of changes in farming practice has prompted the advancement of a twofold and threefold harvested rice pattern in Bangladesh, and comparable rice-based harvesting system is present all through Asia. There are apprehensions that in the year 2030, by following conventional methods particularly in case of rice, its yield will decline (Cassman and Pingali 1995).

However, this proof is not totally acceptable as unreliable confirmation from agriculturists proposes that efficiency components might change instead of yields, i.e. greater inputs (labour, fertilizer, pesticide, etc.) will be required to maintain the level of yield. The Bhabha Atomic Research Centre (BARC) proposals spot the soil nutrient status as the real cause of soil fertility. However, physical properties of soil are also important determinants along with nutrient status. As a well-refined soil is thought to have the best structure for most crops, permitting free permeation of surplus water and at the same time provide more pore area for roots growth (Fitz Patric 1986). While scientists also alarmed the decline in rice yield despite such rigorous farming in a number of Asian countries including Bangladesh (Cassman and Pingali 1995).

### 11.6.1 Integrated Farming Practices

New and advanced farming methods are now gaining attraction towards rural communities. Nayak et al. (2018) showed the vibrant and temporary characteristics of rice field ecosystems in combination with well-suited components such as fish and duck for integrated insect/pest management. This technique can enhance crop quality and overall efficiency through proper nutrient reutilizing. Anyhow, the natural mechanism underlying the rice-fish-duck system effectiveness was not studied in the past especially on soil and water interaction, microbial population and their work proficiency. So Nayak et al. (2018) study the ecological significance of organisms' interaction to conserve soil, nutrient and yield management and to enhance fertility in rice-based integrated farming systems. In this system fish-rice, duck-rice and rice-fish-duck and different physiological and chemical parameters of water, organic matter, alkalinity and soil nutrient level were higher than the conventional system. This addition of fertility was due to faecal matter mixing with soil by fish and duck in the paddy field. Integrated system provides higher yield and effectiveness in terms of rice equivalent yields, and input-to-output ratio of farming was higher than the conventional farming system. Thus, the integrated farming system augments total yield and revenue by increasing soil and water index through better nutrient cycling to produce maximum production of rice.

### 11.6.2 Inorganic Chemical Fertilizers and Rice Cultivation

The contribution of chemical fertilizers and the successive subsidy to ensure the affordability have influenced and encouraged farmers to use the organic matter within their farming systems (Khan et al. 1996). Past administrations emphasized the use of nitrogen fertilizers to farmers for speedy yield and low labour requirement. This is the major reason in declining of organic manure in crop production. Also soil addiction to chemical fertilizers increases soil hardening due to lack of OM (organic manure) inputs.

Therefore, concerns about potential nitrogen load from agriculture have led towards more usage of organic nitrogen sources and regulations reducing the use of nitrogen fertilizers, whereas only 1% of the world's cropland (about 16 Mha) is currently under certified organic production, and demand for organic food is expected to grow specifically in developed countries, and the organic agriculture might become a more common alternative to traditional agriculture in the next 30 years (FAO World Agriculture 2002). Though it is usually believed that organic agriculture offers a number of environmental benefits, the scientific basis for such an insight is weakly developed. Recent results have indicated drawbacks of controlling the fate of nitrogen from organic sources to be just as difficult as managing the fate of mineral nitrogen fertilizer (Poudel et al. 2002; Dadhich et al. 2015).

### 11.6.3 Ecological Farming

Ecological farming holds a strong grip on successive cropping system from ancient time. Proshika, a Bangladesh-based organization, has promoted ecological farming by using quick compost, made of a mixture of cow dung, rice bran and oil cake, in the ratio of 4:2:1 and recycling of plant remainders to soil as a substitute of chemical fertilizer and pesticides. They also encouraged farmers to use FYM (farmyard manure), household wastes, oils and green manures. Farmers following Proshika found change in their fields, i.e. softness, fertility and soil uniformity, with enhanced physical and biological properties. Keeping in view the concerns of scientists and farmers about the sustainability and profitability of rice-based farming systems purely based on mineral fertilization, this manuscript examines the soil fertility status of fields under ecological and current farming practices.

### 11.6.4 Sri Lanka's Farming Systems

Sri Lanka's cultural diverse system of cultivation has combined a wide range of ecological landscapes resulting in a wide variety of farming practices. This type of farming system has been used over thousands of years to incorporate a rich display of farming systems and cultivated crops including grains, rice, spices, vegetables and fruits which have introduced new varieties formally or informally.

The indigenous communities of Sri Lanka lived happily with self-sufficiency of food due to acclimatized agricultural practices. Due to uncertainty of rainfall and fragility of ecosystem, soil erosion damaged insect pest and wildlife. To control all these limiting factors, indigenous civilization of Sri Lanka generally known as hydraulic civilization adopted indigenous agricultural practices to support all these factors.

However, SRI method can be applied in combination of organic or inorganic fertilizer to increase crop quality and quantity (Bezner-Kerr et al. 2012). The advantages of application of SRI method on the conventional method include less seed requirement; up to 50% water savings; almost 50% reduction in the use of inorganic fertilizers, if coupled with 50% organic fertilizers or with a combination of organic fertilizer and biological fertilizer; 20% reduction in production costs; and increase in yield (Hutabarat 2011; Kumar et al. 2018).

## 11.7 Problems in Rice Production

1. Conventional farming system where it tends to decrease the maximum negative impact on environment also decreases fertility of soil. Systematic repetitions of rice crop rice on same land showed decreased plant nutrient supply and low crop productivity by altering physical and chemicals factors in field ecology.
2. Environmental ecological degradation also occurs by chemical fertilizers regarding ecosystem safety. Intensive human interference dramatically changes the natural habitats by influencing biogeochemical cycles and modifying biotic communities with concerns of loss of biodiversity and ecosystem. In modern agricultural practices, the conventional system of rice intensification is subjected to the application of heavy doses of different agrochemicals negatively impacting the quality of ecosystem.
3. In the mid-twentieth century, the beginning of the Green Revolution has alarmed a serious threat for the traditional agriculture system by inducing Western agricultural technology.
4. Then keeping in mind all consequences, it can be expected for future crop production to adopt the integrated farming system where risks of environmental degradation and global warming effects could be significantly minimized in order to get advantageous agricultural operational cost.

## 11.8 Improvement of Agro-Farming by Nanotechnology

### 11.8.1 Agrochemicals and Nanotechnology

For crop sustainability application of biosynthesized nanoparticles intends towards better development. Nanomaterials are carrier of agrochemicals also called nanobiosensors (Bhattacharyya 2009), with facilitated target transportation of basic supplements directly improving the development and yield of the product. They also perform as nano-biosensors for crop protection (Singh et al. 2015a; Kumar et al. 2017). Furthermore, much importance is given to the new improvement in plant science referring to nano-biotechnology that accentuates on agricultural practice, plant cultivation and so forth (Bhattacharyya et al. 2009, 2015; Sofi et al. 2018).

Nanoparticle synthesis occurs by bio-reduction of nanomaterials which is obtained from both in vivo and in vitro processes. Proteins, sugars, enzymes and phytochemicals, like phenolics, terpenoids, flavonoids and cofactors, generally act as reducing and stabilizing representatives for the synthesis of nanoparticles (Prasad et al. 2014). Synthesis of TiO2-NPs by using leaf extract of false daisy (*Eclipta prostrata* L.) noticeably indicates that nanomaterials, particularly, titanium hydroxide, may organize at room temperature (Rajakumar et al. 2012). The decrease was supported because of the presence of carboxyl group (-COOH) stretch and amine group (-NH), alongside other numerous secondary metabolites, existing in the plant extract. It can likewise be viewed that the Cu ions present in the plant support to condense the biosynthesis of CuO NPs (5–10 nm, spherical size) and they were observed to be lively against both Gram-negative and Gram-positive bacteria (Awwad et al. 2015; Acharyulu et al. 2014; Philip 2009; Vardhana and Kathiravan 2015; Meena and Meena 2017). Additionally, the nanoparticles display more noteworthy antibacterial properties against *Bacillus subtilis* as compared to ampicillin. $CeO_2$ NPs (5 nm, spherical shape) with antibacterial activities were magnificently synthesized from the leaf extract of flame lily (*Gloriosa superba* L.) (Kargara et al. 2015; Arumugama et al. 2015; Singh et al. 2015a). The utilization of nanoparticles for the discharge of anti-microbiological or drug molecules will be exceedingly testing assignment in close desire for the direct of all pathological pest plants. The conceivable payback of nanotechnology for agriculture and food needs to be unbiased alongside concern as the water, soil and environment are linked with this process (Bhattacharyya et al. 2010; Khot et al. 2012; Singh et al. 2015a, b).

### 11.8.2 Fertilizers and Nanotechnology

Over the past decade, the field of nanotechnology has had a substantial impact on all aspects of our society from electronic to medicine and has seen outstanding growth over this earlier era. However, applications of nanotechnology in the rural division are still relatively unused. Nanotechnology can provide elucidation of the major problems affected to agri-crops by conventional fertilizer management. These days, agriculture is fronting greater capability due to growing population and a fading arable land base and water resources. Fertilizers are applied to soil-crop systems for fulfilling the vital nutrient requirements of the plants either naturally or synthetically. In improving crop yield, commercial fertilizers play a pivotal role, yet natural inefficiencies and expectable fertilizer management can lead to massive economic and environmental outlay. Negative environmental impressions such as leached fertilizer in the form of nitrate due to lost farmland water, air and other processes into marine ecosystem (Johnson and Raun 2003; Meena et al. 2015).

The significant financial impact of insufficient fertilization also cannot be overlooked. For example, farmers can improve their financial performance (approximately $4.7 billion per annum) by improving their nitrogen use efficiency up to 20% worldwide (Raun and Jonhson 1999). If global food manufacture and demands are to be met in an ecologically and economically sustainable manner, new approaches and technologies must be investigated in agriculture. Clear predictions exist for impacting agriculture productivity through the use of nanotechnology. Nanofertilizers are one potential yield that could be a major development for agriculture; the large surface area and small size of the nanomaterials result in increased interaction and capable uptake of nutrients for crop fertilization (Derosa et al. 2010; Meena and Yadav 2015). In fertilizers' products, nanotechnology may improve release profiles and increase uptake efficiency, foremost to significant financial and eco-friendly benefits. While nanotechnology may provide as an opportunity for the improvement of fertilizers, they may also be a source of unease. The increased reactivity and faster dissolution kinetics are the result of increased surface area of nanomaterials (Chahal and Kumari 2012; Kakraliya et al. 2018). These factors might impair inefficiency problems if nanofertilizer formulations are more easily dissolved and leached into the environment.

The use of nanomaterials in fertilizers would establish a planned input of nanomaterials into the environmental exposure (Fig. 11.1). Plant, particularly farmed crops, could serve as a potential pathway of nanoparticle bioaccumulation in the food chain. Thus, it is vital that the risk and benefits of nanotechnology in fertilizers be critically evaluated. The main purpose of this piece of work is to provide the

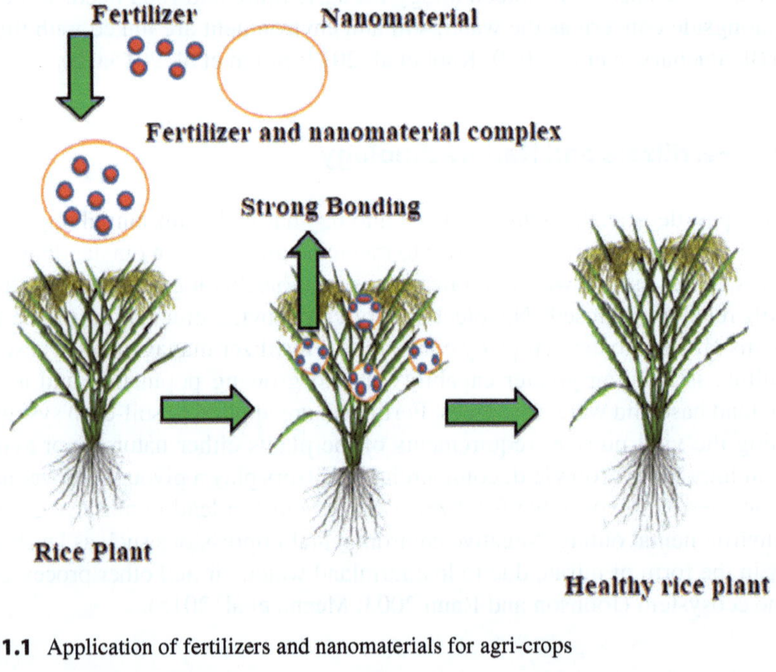


**Fig. 11.1** Application of fertilizers and nanomaterials for agri-crops

modern information about nanotechnology and their associations with area of agri-fertilizers and nutrition. Through this manuscript, we aim to provide an overview of the current state of nanotechnology and to emphasize opportunities for the interventions of nanotechnologies in the area of fertilizers and plant nutrition. Nanotechnology plays a significant role on rice production and disease management as shown in Table 11.3.

### 11.8.3 Pesticides and Nanotechnology

In the area of pesticides, competence of pesticides can be measured and assessed by their significant status during initial stages of plant growth as it provides assistances in reducing pest populace, thus having an effective control over pests for a longer phase. Therefore, the practice of dynamic constituents which can be functional at surface of the host collections is an economical and flexible way of monitoring pests. For biopesticides long-persistence nano-encapsulation has become a mandatory feature from adverse environmental conditions. Subsequently, nano-encapsulation of pesticides is one of the tremendous processes which can make it possible for time-controlled release or release upon the different environmental factors like temperature, humidity, photoperiod, etc. (Nair et al. 2010). Thus the advancement of nano-encapsulated pesticides is on the increasing development (OECD and Allianz 2008). Though its commercialization is still to take place in the future for the improvement of nano-formulations which deals with plant protection (insect pest management) and precludes build-up of wastes into the field or other ecological matrices (Manimaran 2015; Channabasava et al. 2015; Layek et al. 2018), formations of a nano-encapsulated pesticide have slow releasing properties, improved solubility, penetrability and consistency (Manimaran 2015; Channabasava et al. 2015; Ram and Meena 2014).

These assets are mainly achieved through either defending the encapsulated active ingredients from early poverty or increasing their insect pest control efficacy for a longer period (Fig. 11.2). Nano-encapsulated pesticide formulation is able to reduce the quantity of pesticides, and thus no human risk will be exposed. Moreover, it may reflect on as high quality of recyclable material for crop protection. However, be short of data of the mechanism of synthesis and there is no apposite information of a cost-benefit study of nano-encapsulation materials overdue their application process in relation to pesticide delivery. The scientific research brought about primary and critical information for technocrats for the use of nano-encapsulation techniques in relation to pesticide delivery (Nuruzzaman et al. 2016).

Nanoparticles also play a useful role for preparation of new formulations against insect/pest management (Barik et al. 2008; Gajbhiye et al. 2009). Nanotechnology is a proficient technology which can be useful in nanoparticle gene-mediated DNA transfer, desired DNA or chemicals into plant cells for safety and security of host plants against insect pests (Torney 2009).

Another type of nanoparticles, spongy hollow silica nanoparticles (PHSNs) laden with validamycin pesticide, can be worked as proficient transporter in

**Table 11.3** Role of nanotechnology improved rice production and disease management

| Form of disease scientific name | Nanotechnology type | Rice production/quality/output of experiment | References |
|---|---|---|---|
| *Ustilaginoidea virens* (false smut disease of rice) | 1. Aluminium nanoparticles<br>2. Silver nanoparticles<br>3. Titanium dioxide nanoparticles<br>4. Silicon carbide nanoparticles | Impact of fungicides and nanoparticles on *Ustilaginoidea virens* causing false smut disease of rice | Priya et al. (2018) |
| *Magnaporthe grisea* (blast disease) | Silver nanoparticles in concentrations 0, 25, 50, 100 and 200 ppm | Inhibition effects of silver nanoparticles against rice blast disease caused by *Magnaporthe grisea* | Elamawi Rabab and EL-Shafey (2013) |
| *Xanthomonas oryzae* pv. *oryzae* | Silicon dioxide nanospheres | Facile fabrication of rice husk-based silicon dioxide nanospheres loaded with silver nanoparticles as a rice antibacterial agent | Jianghu et al. (2016) |
| *Xanthomonas perforans* | | Nanotechnology in plant disease management DNA-directed silver nanoparticles on graphene oxide as an antibacterial against *Xanthomonas perforans* | Ocsoy et al. (2013) |
| | Foliar application of two silica sols | Foliar application of two silica sols reduced cadmium accumulation in rice grains. *Journal of Hazardous Materials* | Liu et al. (2009) |
| | Effects of nanoscale silica sol foliar application on arsenic uptake | Effects of nanoscale silica sol foliar application on arsenic uptake, distribution and oxidative damage defence in rice (*Oryza sativa* L.) under arsenic stress | Liu et al. (2014) |
| *Rhizoctonia solani*, *Pyricularia oryzae* and *Gibberella fujikuroi* | Nanomaterial treated water | The effects of nanomaterial treated water on the pathogens of rice diseases and fungicides | Li et al. 2005 |
| | Nano-carbon | Application effect of fertilizer added with nano-carbon on rice [J] | Liu et al. (2011) |
| | Silicon | Silicon supplying capacity of paddy soils and characteristics of silicon uptake by rice plants in cool regions in Japan | Sumida (1992) |

(continued)

**Table 11.3** (continued)

| Form of disease scientific name | Nanotechnology type | Rice production/quality/output of experiment | References |
|---|---|---|---|
| | Nano-silica fertilizers | The effects of nano-silica fertilizer on the number of stomata, chlorophyll content and growth of black rice (*Oryza sativa* L. Cv. Japonica) | Putri et al. (2017) |
| | | Response of upland rice (*Oryza sativa* L.) on the application of silicate and phosphate fertilizer on ultisol | Zulputra and Dan (2014) |
| | Potassium and phosphorus nanofertilizers | Effect of nanofertilizers on rice growth [J] | Zhang et al. (2010) |
| | Nanopesticides, nanofertilizers, nanoherbicide, biosensors with nanomaterials | Nanotechnology: the new perspective in precision agriculture | Duhana et al. (2017) |
| | Nanotechnology, silver nanoemulsion, biocontrol, nanofungicide | Agro-nanotechnology for plant fungal disease management: a review | Patel et al. (2014) |
| | Nanofertilizers, nanoherbicide, nanopesticides, nanosensors | Nanofertilizers and nanosensors – an attempt for developing smart agriculture | Rameshaiah et al. (2015) |
| | Nanoparticles, nanopesticides | Nanotechnology: scope and application in plant disease management | Khan and Rizvi (2014) |

agriculture for proficient delivery system for in-controlled delivery system release (Liu et al. 2006). Nanoemulsions are water and oil in combinations like pesticide formulations which are helpful and work against the pest population in agronomy (Wang et al. 2007). In the same way, lipid nanoparticles can be utilized for formulations of nanopesticides (Liu et al. 2006). Another example of nanoemulsions is nano-silica which can be utilized as nanopesticide.

In insects different kinds of cuticular lipids act as defensive system that prevents dehydration or death. A type of nano-silica insecticide when applied on plant surface can adhere to the cuticular lipids of insects and leads towards death. Barik et al. (2008) worked on charged nano-silica 3–5 nm; these can be effectively used for ectoparasite of animals and insects in agriculture. Polyethylene glycol nanoparticles (PEG nanoparticles) enclosed with garlic vital oil excellently used for red flour beetle (*Tribolium castaneum* Herbst,) insect these pest found in stored foodstuff. These nanoparticles can also commendable for red flour beetle. These types of insecticides can be slow and frequent release of active part of nanoparticles (Yang et al. 2009).

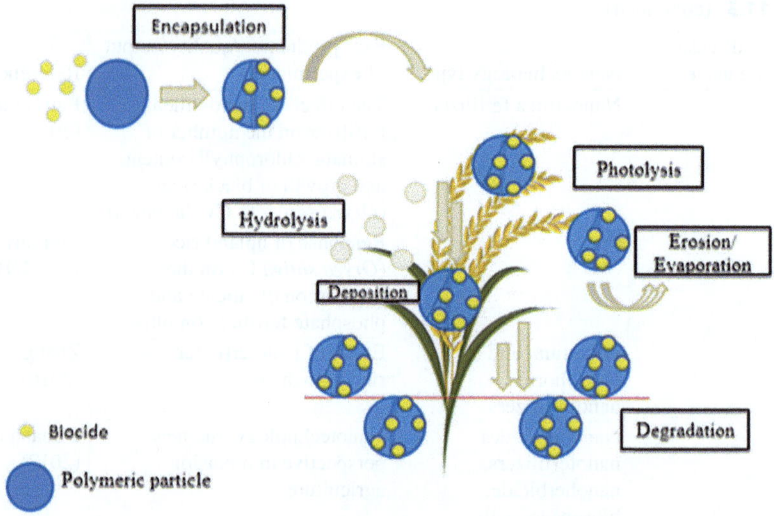


**Fig. 11.2** Presentation of pesticides and nanomaterial capsulation

Trends of nanoparticles can be varied like silver nanoparticles, aluminium oxide, zinc oxide and titanium dioxide in controlling of rice weevil and grasserie disease in silkworm (*Bombyx mori*) are caused by rice weevil (*Sitophilus oryzae* L.) and baculovirus BmNPV (*B. mori* nuclear polyhedrosis virus, respectively (Goswami et al. 2010). Alumina nano-insecticides worked against rice weevil and lesser grain borer (*Rhyzopertha dominica* F.); these pests stored in foodstuff move from one region to another by transportation (Teodoro et al. (2010)). If we used alumina nanostructured nanoparticles, then we found controlled mortality rate of pest in wheat. These nano-insecticides are more cheaper and favourable alternative for management of insects.

## 11.9 Pathogen Detection

Nanotechnology showed that surprised approach exhibits more benefits at scientific level with different types of nanoparticle properties (Fig. 11.3). Some nanoparticles act as antimicrobial agents like silver nanoparticles has developed more advanced and reasonable production. Silver nanoparticles showed different ways to express their actions against pathogens (microorganisms) (Young 2009); these types of nanoparticles can be used for numerous plant pests in a comparatively safer way as compared to commercially available fungicides. Due to their specific function, it can affect the pest biochemical process (Pal et al. 2007). These silver nanoparticles can also disturb the ATP synthesis related to proteins (Yamanka et al. 2005; Meena et al. 2018). In a nutshell, the detailed pathway of bio-molecule inhibition is yet to be understood.

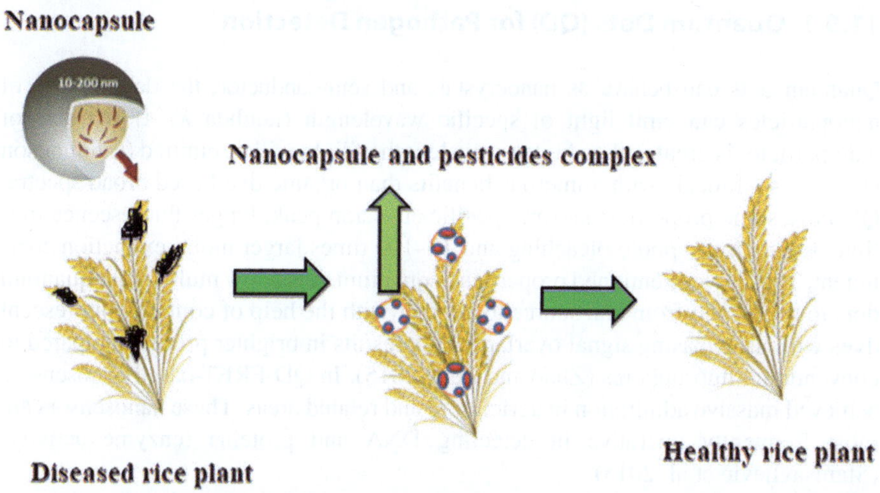


**Fig. 11.3** Pathogen detection by nanocapsules and pesticide complex (self-creation by authors)

Additionally, nanoparticles have been reflected as an alternate and efficient line in expressions of environmentally safer and economical (Kumar and Yadav 2009; Prasad et al. 2011; Swamy and Prasad 2012; Prasad and Swamy 2013). Nanoparticles ensure a pronounced effect on plant disease management as compared to chemical fungicides (Park et al. 2006). Zinc oxide (ZnO) and magnesium oxide (MgO) nanoparticles are most appropriate antibacterial and anti-odour agents (Shah and Towkeer 2010).

There are some factors which make it more efficient for antimicrobial activity like being easily dispersed, optical transparency and smoothness which make ZnO and MgO nanostructures an attractive antibacterial component in many products. These two nanoparticles have been suggested as an antimicrobial protective for wood or food products (Aruoja et al. 2009; Huang et al. 2005; Sharma et al. 2009; Varma et al. 2017). Some suitable nanoparticles deliver improved permeation through epidermis and allow controlled slow release of most active elements on specific target weeds. The practice of nanoparticle-based pesticides is more appropriate, and they are safer for plants and cause less ecological disturbance associated with conventional chemical pesticides (Barik et al. 2008).

In the past decade, nano-silver was most studied and consumed as powerful antibacterial and antimicrobial agent (Swamy and Prasad 2012; Prasad et al. 2012; Prasad and Swamy 2013). The main reason behind utilization of sliver is that it has high surface area, high fraction of surface area and high antimicrobial effect as compared to other substances (Suman et al. 2010). Nano-silver acts as antifungal agent against various plant pathogens. Nano-silver is the most effective inhibitor of fungal pathogen like most considerable inhibitions of plant pathogenic fungi observed on potato dextrose agar (PDA) and 100 ppm of AgNPs.

### 11.9.1 Quantum Dots (QD) for Pathogen Detection

Quantum dots can behave as nanocrystals and semiconductor; the definite size of nanoparticles can emit light of specific wavelength (lambda λ). If the size of nanoparticles is greater, then higher wavelength of light will be emitted (Edmundson et al. 2014). It deals with numerous benefits than organic dye-based broad spectra. QD have some properties: narrow specific emission peak, longer fluorescence lifetime, resistance to photo bleaching and 10–100 times larger molar extinction coefficient. The above-mentioned properties of quantum dots allow multicolour quantum dots to be excited from one source to another with the help of common fluorescent dyes without releasing signal overlap which results in brighter probes compared to conventional fluorophores (Zhao and Zeng 2015). In QD-FRET-based nanosensors achieved massive admiration in agriculture and related areas. These nanosensors are most frequently operative in detecting DNA and proteins (enzyme activity) (Stanisavljevic et al. 2015).

In 1989 Dameron et al. firstly worked on mycosynthesis of semiconductor nanomaterials on yeast, capable of producing cadmium sulphide (CdS) in reaction of cadmium salt stress. In another study different types of microorganisms have been used for the bio-formulation of CdS but partially focused on luminous properties (Yadav et al. 2015), and when *Fusarium oxysporum* was reacted with mixture of $CdCl_2$ and $TeCl_2$, a valuable myco-mediated fusion of highly illuminated CdTe quantum dot was obtained (Jain 2003; Kashyap et al. 2013; Alghuthaymi et al. 2015). In another research QD-based nanosensors capable of producing multiple enzymatic activities were exposed (Knudsen et al. 2013), and nowadays these CdTe quantum dot-based biosensors are used with specific antibody coatings against *Polymyxa betae*-specific glutathione-S-transferase (GST) protein (Safarpour et al. 2012). The combined affinity of antigen and antibody brought the CdTe quantum dots and rhodamine closer together.

Some immune sensors express high sensitivity by fluorescent resonance energy transfer (FRET), and they can screen plant samples with appropriate results within 30 min. Rad et al., in 2012, worked on quantum dots based on nano-biosensors which can detect phytoplasma witches' broom disease of lime (*Candidatus Phytoplasma aurantifolia*) in infected lime trees. These kinds of immunosensors expressed more specificity and defined detection of witches' broom. In another research they developed biosensors for the detection of specific sequence of DNA for basal stem rot of oil palm (*Ganoderma boninense*) (Bakhori et al. 2013).

Modified quantum dots which are 5–8 nm contained carboxylic groups connected with DNA probe (single-stranded DNA) through amide linkage. In QD conjugated with single-stranded DNA, these probes were labelled with Cy5 detection of specific sequence of basal stem rot of oil palm gene based on FRET signals. The manufactured biosensor sensitivity has been shown with a detection limit from $3.55 \times 10^{-9}$ M (Bakhori et al. 2013). This technique is also a proficient, simple, quick and sensitive method for detection of plant pathogen. Furthermore, quantum dots can be active using UV light, and illumination can be visualized with the naked eye; technology is directly practical into the field.

Research based on quantum dots is at initial stage for plant pathology and food toxins. So, in this area work must carry on for optimization of assay to attain efficient signals for low levels of pathogens in complex systems, whether they are food, plants or insects. The applications of functional quantum dots have various prospects like nanoparticle-based diagnostic system in agriculture and its related sector. This field is vast, and researchers are struggling to continue with latest technologies which can be used to shield agricultural harvests and food commodities from plant pathogens.

### 11.9.2 Nanofabrication Imaging

Nanotechnology offers unique opportunities to precisely tune and control the chemical and physical properties of contrast materials in order to overcome problems of toxicity, useful imaging time and specific tissue. Nie (2013) reported that mesoscopic nanoparticles (5–100 nm diameter) have large surface areas and are ideal for conjugating functional groups in multiple pathogen diagnosis assays. Electron beam and photolithography techniques are also used to fabricate topographies that mimic leaf surface features as well as the internal plumbing of plants, and then nano-imaging technologies are used to study how pathogen invade and colonize the leaf tissue (Mccandless 2005). Lithography was used to nanofabricate a pillared surface on silicon wafers. This lawn of miniature pillars (1.4 and 20 mm wide) was used to examine the movement across the surface by the fungus that mimicked some of the characteristics of the host plant.

Images of the red stalk rot of cereal (*Colletotrichum graminicola*) crawling across the nanofabricated surface assisted the researchers to determine that the fungus needs to make a minimum contact (at least 4.5 mm) former to initiation of appressoria formation. To develop disease resistant cultivars, the infection process and behaviour of *Xylella fastidiosa* causing Pierce's disease inside grapevine xylem were studied using nanofabrication methods (Meng et al. 2005). The application of carbon-coated magnetic nanoparticles and microscopy methods at different levels of resolution to visualize the transport and deposition of nanoparticles inside the plant host was reported by González-Melendi et al. (2008). Further, Szeghalmi et al. (2007) investigated nanostructured surface-enhanced Raman scattering (SERS) substrates for imaging applications at high spatial resolution (1 μm). They performed SERS imaging of dried fungal hyphae grown on commercially available nanostructured gold-coated substrates and concluded that this type of nanofabrication techniques offers a well-characterized and reproducible substrate for in situ or in vivo imaging studies of plant pathogen interactions. Rispail et al. (2014) evaluated the behaviour of quantum dots and superparamagnetic nanoparticles on *Fusarium* wilt (*Fusarium oxysporum*) and indicated integration of nanomaterials with the fungal hypha labelling the presence of the pathogenic fungus. Though they showed differential behaviour of nanomaterials with respect to internalization. This work represents the first study on the behaviour of quantum dots and superparamagnetic particles on fungal cells and creates the first and essential step to address the feasibility of new nanotechnology-based systems for early detection and eventual control of pathogenic fungi.

### 11.9.3 Nanopore System

This system is based on electric identification of DNA sequence and can performed with low quantification of samples more efficiently and effectively (Branton et al. 2008). This technique is used for agricultural point of view and applicable for genome of pathogen, gene function in addition to pathogen detection and estimation in agricultural crops. We can also identify the nucleotide by using nanopore system. Its mechanism is based on dimension of conductivity variation across a lipid membrane, whereas DNA can be dragged through a nanoscale pore by an electric current. Each nucleotide has specific conductivity and nucleotide identification by allowing passing through pore (Egan et al. 2012; Mitran et al. 2018).

Kumar et al. in 2012 introduced nanopore-based sequencing (nano-SBS) technique which can accurately distinguish four nucleotides, guanine (G), cytosine (C), arginine (A) and thymine (T), of DNA by detecting the different sized tags released from 5′-phosphate-modified nucleotides at the single molecule level for sequence determination. A most recent approach about nanopore technique, a portable DNA sequencing machine (MinION) was launched by UK-based Oxford Nanopore Technologies (Hayden 2015). This type of tool is able to sequence the single sense and antisense DNA strand (10 KB) and will make next-generation sequence (NGS).

### 11.9.4 Bio-barcode System

Nanomaterials are well-thought-out as very interesting sensitive materials. Keeping in mind their sensitivity, nowadays bio-barcode system is widely used with high sensitivity and selective detection approach system high sensitive and selective detection approach. Its technique depends on nanomaterial capability to transfer the target binding towards an enhanced detection system by recognition of selected target specificity of the elements (Charbgoo et al. 2016; Verma et al. 2015).

This is a hypersensitive approach as, for detection of proteins analytes have been entrenched. This system is magnetic microparticle probes-dependent associated with target-specific antibodies binding to a specific target called prostate-specific antigen (PSA) and nanoparticle probes that are encoded with DNA that is unique to the protein target of interest and antibodies that can sandwich the target captured by the microparticle probes. The magnetic separation of the complexed probes and target followed by dehybridization of the oligonucleotides on the nanoparticle probe surface allows the determination of the presence of the target protein by identifying the oligonucleotide sequence released from the nanoparticle probe. Because the nanoparticle probe carries with it a large number of oligonucleotides per protein binding event, there is substantial amplification and PSA can be detected at 30 attomolar concentration (Jwa-Min et al. 2003).

### 11.9.5 Nanosensors

The chemical or mechanical sensors used to detect chemical species and nanoparticles or to observe physical factors on nanoscale are known as nanosensors. Due to their small size, they least affect chemical or physical properties of species and provide such a mean of transduction or amplification which does not provide bulk structure of material flow. In 2004, Basu et al. (2004) tested anti-*E. coli*-bound gold nanowire arrays (GNWA) prepared on anodized porous alumina template for capturing *E. coli* 0157:H7. This bacterial antibody complex can alter the surface properties of the sensors, such as biomembrane. This alteration was detected using electrochemical impedance spectroscopy (EIS) to measure the amount of *E. coli*, and then the amount of bound E. coli was determined. Their initial results showed that GNWA biosensor can detect up to 50 *E. coli* cells with 1.78cm$^2$ sensor areas (Basu et al. 2004).

In another study carried out by Zhou et al. (2006), the attachment of single-walled nanotubes (SWNT) enhanced and reversed bacteria di-electrophoresis (DEP) mobility. Subsequently, the SWNT-bacteria groups accumulate rapidly (<5 min) into conducting bridges between two electrodes by positive alternating current DEP. Therefore, this approach revealed a detection doorstep of 104 CFU/ml of *E. coli* incorporation of functionalized SWNT will lead SWNT to play a more specific role as absorbers and transporters of pathogens in biosensors.

### 11.9.6 Metal Nanoparticles

Metallic nanomaterials themselves are known to have antibacterial effects (Fig. 11.4). Silver nanoparticles are found effective to both Gram-negative bacteria including *E. coli*, *P. aeruginosa* and *S. typhus* and Gram-positive *Staphylococcus* epidermis (Sondi and Salopek-Sondi 2004; Furno et al. 2004; Melaiye et al. 2005). Deposition of silver on nanoparticles of titanium dioxide significantly increases its bactericidal effects against *E. coli* (Kim et al. 2006; Gogoi et al. 2018). Silver nanoparticles in combination with amoxicillin resulted in a synergistic effect against *E. coli*, which was greater than when they applied it separately (Li et al. 2005). Vancomycin-capped gold nanoparticles exhibited enhanced activities against vancomycin-resistant enterococci (VRE) stains and Gram-negative *E. coli* (Gu et al. 2004).

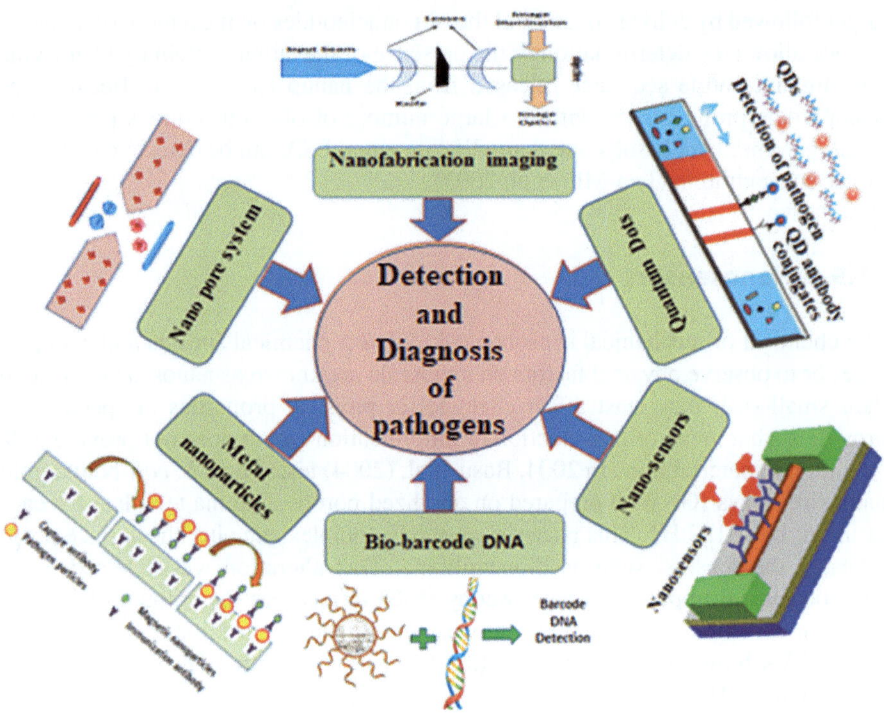


**Fig. 11.4** Application of nanotechnology: pathogens detection and diagnosis

## 11.10 Food Processing and Packaging by Nanotechnology

### 11.10.1 Nano-capsulation

This technique permits the safety of the sensitive bioactive food components from negative environmental surroundings, suppression of discordancy, solubilization or covering of unpleasant taste or aroma. The lipid-based carriers include nanoemulsions, nanoliposomes, solid lipid nanoparticles (SLNs) and novel generation of encapsulation system, namely, nanostructure lipid carriers (NLCs), which are more functionalized and stable concerning their production and physicochemical properties.

New technology in nanotechnology is nano-encapsulation for food and nutraceutical industries. It deals with encapsulation of bioactive compounds in an easy and compact way to serve as a collection of advanced techniques and systems in food and nutraceutical industries. The prospective of nano-encapsulation technologies also enhances their unusual applications in functional foods and nutraceutical systems (Fathi et al. 2012).

### 11.10.2 Smart Packing

Industries are using nanomaterials in food packaging because they are progressively being used in the food packaging industry due to their progressive efficient properties (Silvestre et al. 2011; Duncan 2011). It is also known as intelligent packing or smart packing which can sense any biochemical or microbial modification in the food, for example, detection of specific pathogen production in food or gases from food spoiling. Somewhere 'smart' packaging has also been developed to be used as a tracking device for safety or to warn about products imitation. Research is being done to explore more possibilities of 'smart' packaging surrounded by smart materials and stretchable nanoelectronic devices to regulate inside the environment of packaging and to alert about the food expiry.

Here we are discussing some more new advancement of nanotechnology that is the colour-changing property of plastic packaging due to nanoparticles which become active during food decay. It can be achieved either physically or chemically. In chemical mechanism, chemical indicator changes colour in the presence of certain gases given off during food oxidation. While the other mechanism is physical, nanoparticles embedded in the polymer layers change their visual properties depending upon their relative position in the matrix structure. This designed mechanism is used to produce a strong colour while the packaging stretches and giving off a clear sign of gas releasing decomposition of food.

## 11.11 Feasibility of Nanotechnology Towards Improved Rice Production Under Changing Climate

Parisi et al. (2014) described nanotechnology which contains up to 50% artificial and natural particle size ranging from 1 to100 nm (it is a billion part of 1 m). White and Elmer in 2016 hypothesized that nano-nutrients can stimulate the plant's own immune system. These nanoparticles are exclusive due to their small size, high volume-to-surface ratio, high solubility and increased reactivity with no chemical and genetic modifications into the plant biochemistry and surrounding environment. Christian and Prem (2018) examined numerous nanotechnologies for crops in Nigeria. Nanomaterials can address several crop plants like growth, yield, disease and resistance to environment such as drought stress. Nanotechnology might help farmers fight climate change, pests and disease – and boost yields (Genetic Literacy Project), November 16, 2017. A study conducted by the University of Leeds (2018) has publicized that global warming of only 2 °C will be damaging to crops (like rice, wheat, cotton, corn) in temperate and tropical regions, with less harvests from the 2030s onwards.

There is another study conducted by researchers Guerriero and Cai (2018) about HA nano-hybrid which is a bioceramic with nanoparticle combination which has maximum potential in different types of soils (different regions worldwide, e.g. temperate and tropical regions for rice crop), because nano-covering reduces the use of urea and makes possible an approach to more available global nitrogen issue commercially viable (Guerriero and Cai 2018; Yadav et al. 2017).

Nano-urea improved agronomic competence by 44.5% and grain yield 10.2%. Nanoparticles especially address excess fertilizer application and prevent early loss and release specific amount of demand nutrient (Wang et al. 2010). Some previous studies conducted in high-yielding rice cultivar Wandao153 was provided with ordinary urea and NMUrea (nanomaterial-coated urea) at N 0, 90, 135, 180, 225 and 270 kg/ha to examine the effects of NMUrea on rice yield and agronomic efficiency of nitrogen fertilizer (NAE). When they use the same N rates, tiller number, SPAD values and dry matter of plants were expressively higher in NMUrea treatments than without nanomaterial urea treatments.

They designed a model between grain yield and fertilizer rate; application could be reduced by 12.4–41.7% with nitrogen amount of 90–244.9 kg/ha. In this study the optimum fertilizer management was applied with application of nanomaterial urea at N 244.9 kg/ha, in which the highest grain yield (11174.7 kg/ha) and NAE (13.7 kg/kg) were obtained, being 9.25% and 4 kg/kg higher than in the ordinary urea treatment. With an N rate of 180 kg/ha, the grain yield and NAE were 10332.9 kg/ha and 18.5% in the NMUrea treatment, respectively, being 6.0% and 4.3 kg/kg higher than in the ordinary urea treatment (Wang et al. 2010).

Wu (2013) focused on nano-carbon slow-released fertilizer which reduced the quantity of fertilizer consumption and lessen environmental pollution. Nano-carbon increased the surface area (energy and chemical activity) and produced positive effect on rice yield nitrogen loss in shallow water of paddy soil. This experiment contained three treatments, i.e. control, Jingzhengda slow-released fertilizer+nano-carbon (JSCU+C) and Stanley slow-released compound fertilizer (SSRF+N-P) were used. In their findings they mention (JSCU+C) treatment fertilizer concentration 31.0%, and (SSCU+C) treatment 29.8% dropped. The time of nitrogen runoff loss due to rainfall was shortening 2.2 days and 1.8 days. Due to use of released fertilizer+nano-carbon, rice grain yield, and nitrogen use efficiency increased significantly. So Wu in 2013 recommended that nano-carbon can be used as coating material (nano-carbon) and slow-released fertilizer was a benefit for reducing water pollution.

Another study investigated the effect of nano-carbon nitrogen fertilizer on soil and rice cv. Changbai10 yield (Fan et al. 2012). They utilized different concentrations of nitrogen fertilizers and nano-carbon A1 (N 215 kg hm-2 + nano-carbon 1.194 kg hm-2), A2 (N 150.5 kg hm-2 + nano-carbon 0.836 kg hm-2), A3 (N 107.5 kg hm-2 + nano-carbon 0.597 kg hm-2), B1 (N 215 kg hm-2), B2 (N 150.5 kg hm-2 + nano-carbon 0.836 kg hm-2) and B3 (N 107.5 kg hm-2). The nitrogen accumulation of rice of A1, A2 and A3 treatments were 13.23%, 9.57% and 38.14% higher than B1, B2 and B3 treatments, respectively; total nitrogen residues of soil of A1, A2 and A3 treatments were 6.95%, 8.48% and 9.65% lower than B1, B2 and B3 treatments, respectively; dry biomass of aerial part of A1, A2 and A3 treatments were 58%, 15.7% and 19.3% higher than B1, B2 and B3 treatments. They indicated combined treatment of nano-carbon and nitrogen fertilizer most appropriate in their results. The application ratio of nitrogen fertilizer increased with combination of nano-carbon which can save the N fertilizer in production practice. So, the collective treatment is more appropriate and dissemination in soda saline-alkali soil in the agriculture.

## 11.12 Sustainable Approaches in Relation to Nanotechnology Towards Natural Resource Management

Nanomaterials can provide us a sustainable approach towards a broad spectrum. Engineered nanomaterials (ENMs) and ENM-enabled products have emerged as potentially high-performance substitutes to conventional materials and chemicals. There is a crucial need to incorporate environmental and human health objectives into ENM selection and design processes. So, an adapted framework based on the Ashby material selection strategy is presented as an enhanced selection and design process, which includes functional performance as well as environmental and human health considerations. The utility of this framework is demonstrated through two case studies, the design and selection of antimicrobial substances and conductive polymers, including ENMs, ENM-enabled products and their alternatives. Further, these case studies consider both the comparative efficacy and impacts at two scales: (i) a broad scale, where chemical/material classes are readily compared for primary decision-making, and (ii) within a chemical/material class, where physicochemical properties are manipulated to tailor the desired performance and environmental impact profile. Development and implementation of this framework can inform decision-making for the implementation of ENMs to facilitate promising applications and prevent unintended consequences.

## 11.13 Risk Management

Nanotechnology, which delivers a comprehensive zone, is also a border area of science and technology that has been evolved sharply recently and gives a wide application perspective of things, life, information, environment, energy and security at national level (Bai et al. 2009). But unfortunately, nanotechnology is also a two-edged sword; while it can be used in several areas to achieve unlimited benefits, many adverse effects affecting danger to human life, animals, plants and environment has also reported. Management and controlling establishments as well as environmental boards, non-governmental administrations and scientific establishments around the globe are concerning about the risk valuation of nanotechnology and have specified their suggestions, views and guidance (Vyom et al. 2012; COT-COM-COC 2005). In addition, nanomaterial size, shape, surface area, pH value, temperature and light density will affect its toxicity, while the same nanomaterials having the same chemical but different physical properties will affect differently in its toxicity.

In this chapter we mention another risk of nanotechnologies which is social risk of nanotechnology. Nanotechnology acts as advanced contaminant gaining more attention. According to researchers, specific properties of nanomaterial and their environmental behaviour are the main reason of ecological toxicity. Nanotechnologies also show extensive usage scenarios in the field of chemistry, chemical engineering, biomedical, composite materials, information technology, catalysts and other sides (Krrlik and Biffis 2001). Nanotechnology is widely used and necessarily discharged into the environment causing harm to the ecological system, whereas there are a huge number of eco-toxicological research reports reported to overcome this emerging pollution.

On the other hand, there are several reports on efficiency of various nanotechnologies which affect the whole ecosystem diversity. These are 31% $TiO_2$, nano-Co (18%), nano-ZnO (17%), nano-Ag (13%), single-walled carbon nanotubes (9%) and nano-CuO (9%) (Kahru and Dubourguier 2010) based on chemical and physical interactions of nanotechnology and environment (Yin et al. 2011).

The overall scenario demands that there should be some agency to undertake, collaborate, support and guide the researcher to provide better understanding and sharing of information about nanomaterial. That agency can detect the environmental disorder detection and analysis and environmental fate and classify the nanomaterial (chemical and physical identification) impending discharge, human vulnerability and environmental effect evaluation assessment (Henshaw and O'Carroll 2009).

At present, monitoring organization, industry group and health organization at public, nationwide and worldwide stages are allowing to control and monitor promising health threats from engineered nanoscale materials by using a full life cycle assessment method. Now, it has created a challenging situation to develop management frameworks for safe production, handling and disposal of engineered nanoscale materials supposed to be the large number of uncertain queries.

Countries like the United States and Europe have already taken action on precautionary measures into risk management agenda to accomplish determined uncertainties connected with several nanomaterials. Many practical agendas have been established that keep equally existing technical information, expert's suggestions and protective policies side by side to fill the current larger gaps in knowledge. To meet up the standards of any community, research institutes or universities, it is mandatory to have this framework of equilibrium to balance the value of research against cost of protective measures.

## 11.14 Future Perspective

Nanotechnology is capable of converting the current conventional agricultural practices into advanced techniques. It provides a promising defence system against many agriculture-related problems like enhancing insect/pest management caused by traditional methods and adversity of chemically used pesticides as well as development of improved crop varieties. Therefore, nanomaterial in several forms can be applied to efficiently manage insect/pest formulations. Nanoparticle facilitating gene transfer can be fruitful in producing such insect-/pest-resistant varieties. Hence, it can be concluded that nanotechnology can give us green eco-friendly formulation without altering nature as an alternative to commonly used harmful chemicals.

Through progressed knowledge of nanomaterials and the realization of their potential in the food industry, initiation of nanotech foods will provide solutions for persisting problems associated with foods and will offer long-term economic benefits. Internationally, nations will profit from increased food productivity with cost-effective revenues and innovative products with tunable properties to deliver smarter and healthier foods along with intelligent packaging systems with improved storage properties for healthier food protection. Nanomaterials in foods will have a huge impact on sustainability and will be accompanied by health and environmental benefits if regulated suitably.

However, every innovation created a challenge in evaluating the safety of complex novel nano-foods and nano-packaging to ensure that human and environmental concerns are not compromised as new products are released. Therefore, every introductory technology must be sufficiently slowed to allow potential risks for proper check and balance by regulatory and reliable measuring tools which are currently absent for nano-foods. Also, if nano-foods are to be implemented successively in our food cycle, the benefits of nanotech foods must be accompanied by greater transparency of the risks of such foods publicly to build consumer confidence. Public engagement acting in concert with public opinion is expected to perform the main role in the acceptance of nano-processed foods.

Evidently, there is a chance for this technology to leave a considerate influence upon energy, on environment and by enhancing the selection the economy and the environment by improving the screening procedures. Advance new visions for incorporating nanotechnologies into nano-biosensors need to be exposed, familiar of any possible threat to environment or mankind. With this focused effort on targeted goals related to agro-products by researchers and government, we believe on transformation of nanotechnology in the area of agriculture towards achieving the goal of sustainable agriculture in the globe.

## 11.15 Recommendation

First of all, to fulfil the desired demand and supply of rice, there is a need for a detailed and focused advanced research program on the established stabilizing food cost at national-level economies, as well as an emphasis on application of strategy, administration of food logistics organizations and tools to overcome corruption in these organizations. Secondly, there is need for larger rice reserves at four different global levels starting from private sector rice economy to minor trade in countries by the public sector then in large rice producing and consuming countries held publicly finally at international-level community. If we look at the above conditions, the basic and utmost recommendations are as follows:

1. Assessment of figures from general to specific element interacting with biological system.

   Mutual promotion and collaborative promotion and collaborative efforts by researchers to further understand and explore the innovative properties of science, as well as advanced methodologies to detect and measure nanoparticles on nanoscale
2. Collection, collation and interpretation of scientific information based on product review strategies
3. Construction of infrastructure setup to transfer knowledge inside and outside, looking for collection, manufacture and build upon information from individual studies of nanoscale materials
4. Establishment of agencies for ensuring consistent transfer and application of relevant knowledge for products containing nanoscale materials
5. Sufficiency of testing methods to test a product regardless of whether it is subject to premarket authorization or not
6. Evaluation of availability and reliability of current testing approaches to assess safety, effectiveness and excellence of nanoscale material products
7. Efforts to take part in promotion and development of description procedures and principles for nanoscale materials
8. Encourage and contribute in the expansion of models for the behaviour of nanoscale particles in vitro and in vivo

## 11.16 Conclusion

Nanotechnology has presented countless welfare in precision agriculture. Nanoparticles have exclusive characteristics like economically cheap and are required in lesser amount and can be easily manufactured from different sources (biological and synthetic sources). Due to limited resources of water and agricultural land, progress of agriculture can be probable by using different modern nanotechnologies for increasing efficiency with minimal destruction in climate. Nanotechnology has potential to update the agriculture by using nanofertilizers for growth and improved yield production of agri-crops, controlled released formulations of nanoparticles for micronutrients, nanopesticides for pest management and nanosensors for disease detection. Nanotechnology can deliver green, efficient and ecologically sound environment in agriculture. The modern nanotechnology tools can provide a promising future, and it can also assist to achieve sustainable agriculture sector.

## References


Acharyulu NPS, Dubey RS, Swaminadham V, Kalyani RL, Kollu P, Pammi SV (2014) Green synthesis of Cuo nanoparticles using *Phyllanthus amarus* leaf extract and their antibacterial activity against multidrug resistance bacteria. Int J Eng Res Technol 3:639–641

Ainsworth EA (2008) Rice production in a changing climate: a meta-analysis of responses to elevated carbon dioxide and elevated ozone concentration. Glob Chang Biol 14:1642–1650

Alghuthaymi MA, Almoammar H, Rai M, Said-Galiev E, Abd-Elsalam KA (2015) Myconanoparticles: synthesis and their role in phytopathogens management. Biotechnol Biotechnol Equip 29(2):221–236

Arumugama A, Karthikeyan C, Hameed ASH, Gopinath K, Gowri S, Karthika V (2015) Synthesis of cerium oxide nanoparticles using *Gloriosa superba* L. leaf extract and their structural, optical and antibacterial properties. Mater Sci Eng 49:408–415

Aruoja V, Dubourguier H, Kasamets C, Kahru KA (2009) Toxicity of nanoparticles of CuO, ZnO and TiO2 to microalgae, *Pseudokirchneriella subcapitata*. Sci Total Environ 407:1461–1468

Ashoka P, Meena RS, Kumar S, Yadav GS, Layek J (2017) Green nanotechnology is a key for eco-friendly agriculture. J Clean Prod 142:4440–4441

Awwad AM, Albiss BA, Salem NM (2015) Antibacterial activity of synthesized copper oxide nanoparticles using *Malva sylvestris* leaf extract. SMU Med J 2:91–100

Bai W, Zhang CC, Jiang WJ, Zhang ZY, Zhao YL (2009) Progress in studies on environmental behaviors and toxicological effects of nanomaterials. Asian J Ecotoxicol 2:174–182

Bakhori NM, Yusof NA, Abdullah AH, Hussein MZ (2013) Development of a fluorescence resonance energy transfer (FRET)-based DNA biosensor for detection of synthetic oligonucleotide of Ganoderma boninense. Biosen 3(4):419–428

Barik TK, Sahu B, Swain V (2008) Nanosilica-from medicine to pest control. Parasitolol Res 103:253–258

Basu M, Seggerson S, Henshaw J, Jiang J, del A Cordona R, Lefave C, Boyle PJ, Miller A, Pugia M, Basu S (2004) Nano-biosensor development for bacterial detection during human kidney infection: use of glycoconjugate-specific antibody-bound gold NanoWire arrays (GNWA). Glycoconj J 21(8–9):487–496

Bezner-Kerr R, McGuire KL, Nigh R, Rocheleau D, Soluri J, Perfecto I (2012) Effects of industrial agriculture on climate change and the mitigation potential of small-scale agro-ecological farms. Anim Sci Rev 2011:69

Bhattacharyya A (2009) Nanoparticles from drug delivery to insect pest control. Akshar 1(1):1–7

Bhattacharyya A, Bhaumik A, Nandi M, Viraktamath S, Kumar R, Nayak A, Chowdhury MU, Mandal S (2009) Nano—a new frontier in present century. Adv Life Sci 3:18–23

Bhattacharyya A, Bhaumik A, Usha Rani P, Mandal S, Epidi TT (2010) Nano-particles—a recent approach to insect pest control. Afr J Biotechnol 9(24):3489–3493

Bhattacharyya A, Antoney PU, NaikaH R, Reddy SJ, Adeyemi MM, Omkar (2015) Nanotechnology and butterflies: a mini review. J Appl Biosci 41:27–32

Bhuiyan SI, Tuong TP (1999) Increasing water-use efficiency in rice production: farm-level perspectives. Agric Water Manag 40(1):117–122

Bloom David E (2011) 7 Billion and counting. Science 333:562. https://doi.org/10.1126/science.1209290

Branton D, Deamer DW, Marziali A, Bayley H, Benner SA, Butler T, Di VM, Garaj S, Hibbs A, Huang X, Jovanovich SB, Krstic PS, Lindsay S, Ling XS, Mastrangelo CH, Meller A, Oliver JS, Pershin YV, Ramsey JM, Riehn R, Soni GV, Tabard-Cossa V, Wanunu M, Wiggin M, Schloss JA (2008) The potential and challenges of nanopore sequencing. Nat Biotechnol 26(10):1146–1153

Cassman KG, Pingali PL (1995) Intensification of irrigated rice systems: learning from the past to meet future challenges. Geo J 35(3):299–305

Chahal H, Kumari N (2012) Consumer perceived value: the development of a multiple item scale in hospitals in the Indian context. Int J Pharm Health Mark 6:167–190. https://doi.org/10.1108/17506121211243086

Channabasava A, Lakshman HC, Jorquera MA (2015) Effect of fungicides on association of arbuscular mycorrhiza fungus Rhizophagus fasciculatus and growth of Proso millet (*Panicum miliaceum* L.). J Soil Sci Plant Nutr 15:35–45

Charbgoo F, Soltani F, Taghdisi SM, Abnous K, Ramezani M (2016) Nanoparticles application in high sensitive aptasensor design. Trends Anal Chem 85:85–97

Christian OD, Prem SB (2018) Nanofertilizers: new products for the industry? J Agric Food Chem 66(26):6462–6473. https://doi.org/10.1021/acs.jafc.7b02150

Coats B (2003) Global rice production. In: Smith CW, Dilday RH (eds) Rice origin, history, technology and production. Wiley, Hoboken, pp 247–470

COT-COM-COC (2005) Joint statement on nanomaterial toxicology. http://cot.food.gov.uk/pdfs/cotstatements2005nanomats.pdf

Dadhich RK, Meena RS, Reager ML, Kansotia BC (2015) Response of bio-regulators to yield and quality of Indian mustard (*Brassica juncea* L. Czernj. and Cosson) under different irrigation environments. J App Nat Sci 7(1):52–57

Dameron CT, Reeser RN, Mehra RK, Kortan AR, Carroll PJ, Steigerwaldm ML, Brus LE, Winge DR (1989) Biosynthesis of cadmium sulphide quantum semiconductor crystallites. Nature 338(6216):596–597

Derosa M, Monreal C, Schnitzer M, Walsh R, Sultan Y (2010) Nanotechnology in fertilizers. Nat Nanotech 5:91

Duhana JS, Kumara R, Kumara N, Kaura P, Nehrab K, Duhanc S (2017) Nanotechnology: the new perspective in precision agriculture. Biotech Rep 15:11–23

Duncan TV (2011) Applications of nanotechnology in food packaging and food safety: barrier materials, antimicrobials and sensors. J Colloid Interface Sci 363(1):1–24

Economic and Social Development – World food council (WFC) (1974) http://www.nationsencyclopedia.com/United-Nations/Economic-and-Social-Development-WORLD-FOOD-COUNCIL-WFC.html#ixzz5Jv7uVQrL

Edmundson MC, Capeness M, Horsfall L (2014) Exploring the potential of metallic nanoparticales within synthetic biology. New Nanotech 31(6):572–578

Egan AN, Schlueter J, Spooner DM (2012) Applications of next-generation sequencing in plant biology. Am J Bot 99(2):175–185

Elamawi Rabab MA, EL-Shafey RAS (2013) Inhibition effects of silver nanoparticles against rice blast disease caused by magnaporthe grisea. Egypt J Agric Res 91(4):1271–1283

Fan L, Wang Y, Shao X, Geng Y, Wang Z, Ma Y, Liu J (2012) Effects of combined nitrogen fertilizer and nano-carbon application on yield and nitrogen use of rice grown on saline-alkali soil. J Food Agric Environ 10:558–562

FAO (2000) World agriculture towards 2005, New York

FAO (2005) A record-breaking year for the world rice economy. Food and Agriculture Organization of the United Nation http://www.fao.org/newsroom/en/news/2006/1000269/index.html

FAO (2006) World agriculture towards 2030/2050. FAO, Rome

FAO (2016) Food outlook. Biannual report on global food market. Food and Agriculture Organization of the United Nations. ISSN: 0251-1959

FAO (2018) FAO Cereal supply and demand brief. Cereal production and stock forecasts revised up, but prospects for 2018/19 still point to much tighter supplies than in 2017/18. http://www.fao.org/worldfoodsituation/csdb/en/

FAO World Agriculture (2002) Towards 2015/2030 summary report. Food and Agriculture Organization of the United Nations, Rome

Fathi M, Mozafari MR, Mohebbi M (2012) Nanoencapsulation of food ingredients using lipid based delivery systems. Trend Food Sci Technol 23(1):13–27

Fitz Patric EA (1986) An introduction to soil science, 2nd edn. Longman, Harlow, xiii + 255 pp; distributed in USA by John Wiley. Price £7.95 (paperback). ISBN 0 582 30128 9 (Longman); 0 470 20670 5 (Wiley)

Food and Agriculture Organization of united states (FAO) (2017) Rice market monitor. (RMM), vol XX(1). http://www.fao.org/economic/RMM

Furno F, Morley KS, Wong B, Sharp BL, Arnold PL, Howdle SM, Bayston R, Brown PD, Winship PD, Reid HJ (2004) Silver nanoparticles and polymeric medical devices: a new approach to prevention of infection? J Antimicro Chemother 54(6):1019–1024. https://doi.org/10.1093/jac/dkh478

Gajbhiye M, Kesharwani J, Ingle A, Gade A, Rai M (2009) Fungus mediated synthesis of silver nanoparticles and its activity against pathogenic fungi in combination of fluconazole. Nanomed 5(4):282–286

Genetic Literacy Project. November 16, 2017

Gogoi N, Baruah KK, Meena RS (2018) Grain legumes: impact on soil health and agroecosystem. In: Meena et al (eds) Legumes for soil health and sustainable management. Springer. https://doi.org/10.1007/978-981-13-0253-4_16

González-Melendi P, Fernández-Pacheco R, Coronado MJ, Corredor E, Testillano PS, Risueño MC, Marquina C, Ibarra MR, Rubiales D, Pérez-de-Luque A (2008) Nanoparticles as smart treatment-delivery systems in plants: assessment of different techniques of microscopy for their visualization in plant tissues. Ann Bot 101(1):187–195

Goswami A, Roy I, Sengupta S, Debnath N (2010) Novel applications of solid and liquid formulations of nanoparticles against insect pests and pathogens. Thin Solid Films 519:1252–1257

Gu H, Ho PL, Tong E, Wang L, Xu B (2004) Presenting vancomycin on nanoparticles to enhance antimicrobial activities. Nano Lett 3:1261–1263

Guerriero G, Cai G (2018) Interaction of nano-sized nutrients with plant biomass: a review. In: Phytotoxicity of nanoparticles. Springer, Cham, pp 135–149

Hayden EC (2015) Pint-sized DNA sequencer impresses first users. Nature 521:15–16

Henshaw DL, O'Carroll MJ (2009) Scientific Committee on Emerging and Newly Identified Health Risks (SCENIHR) assessment of products of nanotechnologies. European Commission, Brussels

Huang L, Dian-Qing L, Yan-Jun W, Min David G, Xue ED (2005) Controllable preparation of nano-MgO and investigation of its bactericidal properties. J Inorg Biochem 99:986–993

Hutabarat TR (2011) Populasi mikrob tanah emisi metan dan produksi padi dengan kombinasi pemupukan pada budidaya padi SRI (System of rice intensification) [Thesis] Bogor: Bogor Agricultural University (2011)

Imoki A (1984) Rice production and consumption of the world. The World International problem report. IRRI. 2000. "Bigger Harvest a Cleaner Planet." http://www.irri.org/publications/annual/pdfs/ ar2000/biggerharvests.pdf

Jain K (2003) Nanodiagnostics: application of nanotechnology (NT) in molecular diagnostics. Expert Rev Mol Diagn 3(2):153–161

Jianghu C, You L, Desong Y, Yingliang L (2016) Facile fabrication of rice husk based silicon dioxide nanospheres loaded with silver nanoparticles as a rice antibacterial agent. Sci Rep 6:21423. https://doi.org/10.1038/srep21423

Johnson GV, Raun WR (2003) Nitrogen response index as a guide to fertilizer management. J Plant Nutr 26:249–262

Jwa-Min N, Shad TC, Chad AM (2003) Nanoparticle-based bio-bar codes for the ultrasensitive detection of proteins. 301(5641):1884–1886. https://doi.org/10.1126/science.1088755

Kahru A, Dubourguier HC (2010) From ecotoxicology to nano ecotoxicology. Toxicology 2–3:105–119

Kakraliya SK, Singh U, Bohra A, Choudhary KK, Kumar S, Meena RS, Jat ML (2018) Nitrogen and legumes: a meta-analysis. In: Meena RS et al (eds) Legumes for soil health and sustainable management. Springer. https://doi.org/10.1007/978-981-13-0253-4_9

Kargara H, Ghasemi F, Darroudid M (2015) Bioorganic polymer-based synthesis of cerium oxide nanoparticles and their cell viability assays. Ceram Int 41:1589–1594

Kashyap PL, Kumar S, Srivastava AK, Sharma AK (2013) Myconanotechnology in agriculture: a perspective. World J Microbiol Biotechnol 29(2):191–207

Khan MR, Rizvi TF (2014) Nanotechnology: scope and application in plant disease management. Plant Path J 13(3):214–231

Khan PK, Musa M, Shahzad MA, Aadal NK, Nasim M (1996) Wheat performance on green manured and fertilized fields. Pak J Soil Sci 2:84–86

Khot LR, Sankaran S, Maja JM, Ehsani R, Schuster EW (2012) Applications of nanomaterials in agricultural production and crop protection: a review. Crop Prot 35:64–70

Khush GS (2005) What it will take to feed 5.0 billion rice consumers in 2030. Plant Mol Biol 59:16

Kim M, Lee JH, Koh H, Lee SY, Jang C, Chung CJ, Sung JH, Blenis J, Chung J (2006) Inhibition of ERK-MAP kinase signaling by RSK during Drosophila development. EMBO J 25(13):3056–3067

Kim DY, Kadam A, Shinde S, Saratale RG, Patra J, Ghodake G (2018) Recent developments in nanotechnology transforming the agricultural sector: a transition replete with opportunities. J Sci Food Agric 98(3):849–864

Knudsen BR, Jepsen ML, Ho Y-P (2013) Quantum dot-based nanosensors for diagnosis via enzyme activity measurement. Expert Rev Mol Diagn 13(4):367–375. https://doi.org/10.1586/erm.13.17

Kottegoda N, Sandaruwan C, Priyadarshana G, Siriwardhana A, Rathnayake UA, Berugoda Arachchige DM, Kumarasinghe AR, Dahanayake D, Karunaratne V, Amaratunga GA (2017) Urea-hydroxyapatite nanohybrids for slow release of nitrogen. ACS Nano 11(2):1214–1221. https://doi.org/10.1021/acsnano.6b07781

Krrlik M, Biffis A (2001) Catalysis by metal nanoparticles supported on functional organic polymers. J Mol Catal A Chem 1:113–138. PII:S1381-1169(01)00313-2

Kumar V, Yadav SK (2009) Plant-mediated synthesis of silver and gold nanoparticles and their applications. J Chem Technol Biotechnol 84:151–157

Kumar S, Meena RS, Pandey A, Seema (2017) Soil acidity management and an economics response of lime and sulfur on sesame in an alley cropping system. Int J Curr Microb Appl Sci 6(3):2566–2573

Kumar S, Meena RS, Lal R, Yadav GS, Mitran T, Meena BL, Dotaniya ML, EL-Sabagh A (2018) Role of legumes in soil carbon sequestration. In: Meena RS et al (eds) Legumes for soil health and sustainable management. Springer. https://doi.org/10.1007/978-981-13-0253-4_4

Layek J, Das A, Mitran T, Nath C, Meena RS, Singh GS, Shivakumar BG, Kumar S, Lal R (2018) Cereal+legume intercropping: an option for improving productivity. In: Meena RS et al (eds) Legumes for soil health and sustainable management. Springer. https://doi.org/10.1007/978-981-13-0253-4_11

Li P, Li J, Wu C, Wu Q, Li J (2005) Synergistic antibacterial effect of lactum antibiotic combined with silver nanoparticle. Nanotech 16:1912–1917. https://doi.org/10.1088/0957-4484/16/9/082

Liu F, Wen LX, Li ZZ, Yu W, Sun HY, Chen JF (2006) Porous hollow silica nanoparticles as controlled delivery system for water soluble pesticide. Mat Res Bull 41:2268–2275

Liu C, Li F, Luo C, Liu X, Wang S, Liu T, Li X (2009) Foliar application of two silica sols reduced cadmium accumulation in rice grains. J Hazard Mater 161:1466–1472

Liu J, Ma Y, Zhang ZM, Liu WZ, Guo ZD (2011) Application effect of fertilizer added with nano carbon on rice [J]. Phosphate Comp Ferti 6:028

Liu C, Wei L, Zhang S, Xu X, Li F (2014) Effects of nanoscale silica sol foliar application on arsenic uptake, distribution and oxidative damage defense in rice (*Oryza sativa* L.) under arsenic stress. RSC Adv 4:57227–57234

Manimaran M (2015) A review on nanotechnology and its implications in agriculture and food industry. Asian J Plant Sci Res 5:13–15

Mccandless L (2005) Nanotechnology offers new insights into plant pathology. College of Agriculture and Life Sciences News, Cornell University, Ithaca, pp 17–18

Meena RS, Lal R (2018) Legumes and sustainable use of soils. In: Meena RS et al (eds) Legumes for soil health and sustainable management. Springer. https://doi.org/10.1007/978-981-13-0253-4_1

Meena H, Meena RS (2017) Assessment of sowing environments and bio-regulators as adaptation choice for clusterbean productivity in response to current climatic scenario. Bangladesh J Botany 46(1):241–244

Meena RS, Yadav RS (2015) Yield and profitability of groundnut *(Arachis hypogaea* L) as influenced by sowing dates and nutrient levels with different varieties. Legum Res 38(6):791–797

Meena RS, Dhakal Y, Bohra JS, Singh SP, Singh MK, Sanodiya P (2015) Influence of bioinorganic combinations on yield, quality and economics of mungbean. Am J Exp Agric 8(3):159–166

Meena BL, Fagodiya RK, Prajapat K, Dotaniya ML, Kaledhonkar MJ, Sharma PC, Meena RS, Mitran T, Kumar S (2018) Legume green manuring: an option for soil sustainability. In: Meena et al. (ed) Legumes for soil health and sustainable management. Springer. https://doi.org/10.1007/978-981-13-0253-4_12

Melaiye A, Sun Z, Hindi K, Milsted A, Ely D, Reneker DH, Tessier CA, Youngs WJ (2005) Silver(I)–imidazole cyclophane *gem*-diol complexes encapsulated by electrospun tecophilic nanofibers: formation of nanosilver particles and antimicrobial activity. J Am Chem Soc 127(7):2285–2291. https://doi.org/10.1021/ja040226s

Meng Y, Li Y, Galvani CD, Turner JN, Burr TJ, Hoch HC (2005) Upstream migration of *Xylella fastidiosa* via pilus-driven twitching motility. J Bacteriol 187:5560–5567

Mitran T, Meena RS, Lal R, Layek J, Kumar S, Datta R (2018) Role of soil phosphorus on legume production. In: Meena et al (eds) Legumes for soil health and sustainable management. Springer. https://doi.org/10.1007/978-981-13-0253-4_15

Nair SK, Wang N, Turuspekov Y et al (2010) Cleistogamous flowering in barley arises from the suppression of microRNA-guided *HvAP2* mRNA cleavage. Proc Nat Acad Sci U S A 107:490–495

Nayak P, Nayak AK, Panda B, Lal B, Gautam P, Poonam A, Shahid M, Tripathi R, Kumar U, Mohapatra SD, Jambhulkar N (2018) Ecological mechanism and diversity in rice based integrated farming system. Ecol Indic 91:359–375. https://doi.org/10.1016/j.ecolind.2018.04.025

Nie L (2013) Biomedical nanotechnology for optical molecular imaging, diagnostics, and therapeutics. JSM Nanotechnol Nanomed 1:1–2

Nuruzzaman M, Rahman MM, Liu Y, Naidu R (2016) Nanoencapsulation, nano-guard for pesticides: a new window for safe application. J Agric Food Chem 64:1447–1483

Ocsoy I, Paret ML, Ocsoy MA, Kunwar S, Chen T, You M, Tan W (2013) Nanotechnology in plant disease management: DNA-directed silver nanoparticles on graphene oxide as an antibacterial against *Xanthomonas perforans*. ACS Nano 7(10):8972–8980

OECD and Allianz (2008) Report in cooperation with the OECD international futures programme. http://www.oecd.org/dataoecd/32/1/44108334

Pal S, Tak YK, Song JM (2007) Does the antibacterial activity of silver nanoparticles depend on the shape of the nanoparticle? A study of Gram negative bacterium *Escherichia coli*. Appl Environ Microbiol 73:1712–1720

Parisi C, Vigani M, Rodríguez-Cerezo E (2014) Proceedings of a workshop on "Nanotechnology for the agricultural sector: from research to the field". Publications Office of the European Union, Luxembourg

Park HJ, Kim SH, Kim HJ, Choi SH (2006) A new composition of nanosized silica-silver for control of various plant diseases. Plant Pathol J 22:25–34

Patel N, Desai P, Patel N, Jha A, Gautam HK (2014) Agronanotechnology for plant fungal disease management: a review. Int J Curr Microbio Appl Sci 3(10):71–84

Philip D (2009) Biosynthesis of Au, Ag and Au-Ag nanoparticles using edible mushroom extract. Spectrochim Acta A 73:374–381

Poudel DD, Horwath WR, Lanini WT, Temple SR, Bruggen AHC (2002) Comparison of soil N availability and leaching potential, crop yields and weeds in organic, low-input and conventional farming systems in northern California. Agric Ecosys Environ 90:125–137

Prasad R, Swamy VS (2013) Antibacterial activity of silver nanoparticles synthesized by bark extract of *Syzygium cumini*. J Nanopart 2013:1. https://doi.org/10.1155/2013/431218

Prasad KS, Pathak D, Patel A, Dalwadi P, Prasad R, Patel P, Kaliaperumal SK (2011) Biogenic synthesis of silver nanoparticles using Nicotiana tobaccum leaf extract and study of their antibacterial effect. Afr J Biotechnol 9(54):8122–8130

Prasad R, Swamy VS, Varma A (2012) Biogenic synthesis of silver nanoparticles from the leaf extract of *Syzygium cumini* (L.) and its antibacterial activity. Int J Pharm Bio Sci 3(4):745–752

Prasad R, Kumar V, Prasad KS (2014) Nanotechnology in sustainable agriculture: present concerns and future aspects. Afr J Biotechnol 13:705–713

Priya B, Amarendra K, Sanjeev K, Azad CS (2018) Impact of fungicides and nanoparticles on Ustilaginoidea virens causing false smut disease of rice. J Pharmacogn Phytochem 7(1):1541–1544

Putri FM, Suedy SWA, Darmanti S (2017) The effects of nano-silica fertilizer on the number of stomata, chlorophyll content and growth of black rice (*Oryza sativa* L. Cv. Japonica) "www.ejournal.undip.ac.id2017[Online]. Available: http://www.ejournal.undip.ac.id/indek.php/baf/index. Accessed 1 Feb 2017

Rad F, Mohsenifar A, Tabatabaei M, Safarnejad MR et al (2012) Detection of Candidatus Phytoplasma aurantifolia with a quantum dots FRET-based biosensor. J Plant Pathol 94(3):525–534

Rajakumar G, Rahuman AA, Priyamvada B, Khanna VG, Kumar DK, Sujin PJ (2012) Eclipta *prostrata* leaf aqueous extract mediated synthesis of titanium dioxide nanoparticles. Mater Lett 68:115–117

Ram K, Meena RS (2014) Evaluation of pearl millet and mungbean intercropping systems in Arid Region of Rajasthan (India). Bangladesh J Bot 43(3):367–370

Rameshaiah GN, Pallavi J, Shabnam S (2015) Nano fertilizers and nano sensors – an attempt for developing smart agriculture. Int J Eng Res Gen Sci 3(1):314–320. ISSN 2091-2730

Raun WR, Jonhson GV (1999) Improving nitrogen use efficiency for cereal production. Agron J 91:357–363

Rispail N, Matteis LD, Santos R, Miguel AS, Custardoy L, Testillano PS, Risueño MC, Pérez-de-Luque A, Maycock C, Fevereiro P, Oliva A, Fernández-Pacheco R, Ibarra MR, Fuente JM, Marquina C, Rubiales D, Prats E (2014) Quantum dot and superparamagnetic nanoparticle interaction with pathogenic fungi: internalization and toxicity profile. CS Appl Mater Interfaces 6(12):9100–9110. https://doi.org/10.1021/am501029g

Safarpour H, Safarnejad MR, Tabatabaei M, Mohsenifar A et al (2012) Development of a quantum dots FRET-based biosensor for efficient detection of Polymyxa betae. Can J Plant Pathol 34(4):507–515

Schwartz SJ (1991) World rise trade: prospects and issue for the nineties. Rice situation and outlook report. USDA Economic Research Service.

Shah MA, Towkeer A (2010) Principles of nanosciences and nanotechnology. Narosa Publishing House, New Delhi

Sharma VK, Yngard RA, Lin Y (2009) Silver nanoparticles: green synthesis and their antimicrobial activities. Adv Colloid Interf Sci 145:83–96

Silvestre C, Duraccio D, Sossio C (2011) Food packaging based on polymer nanomaterials. Prog Poly Sci 36(1):1766–1782

Singh A, Singh NB, Hussain I, Singh H, Singh SC (2015a) Plant-nanoparticle interaction: an approach to improve agricultural practices and plant productivity. Int J Pharm Sci Invent 4:25–40

Singh S, Singh BK, Yadav SM, Gupta AK (2015b) Applications of nanotechnology in agricultural and their role in disease management. Res Nanosci Nanotechnol 5:1–5

Sofi PA, Baba ZA, Hamid B, Meena RS (2018) Harnessing soil rhizobacteria for improving drought resilience in legumes. In: Meena RS et al (eds) Legumes for soil health and sustainable management. Springer. https://doi.org/10.1007/978-981-13-0253-4_8

Sondi I, Salopek-Sondi B (2004) Silver nanoparticles as antimicrobial agent: a case study on *E. coli* as a model for Gram-negative bacteria. J Colloid Interface Sci 275(1):177–182

Stanisavljevic M, Son K, Vaculovicova M, Kizeka R, Adama V (2015) Quantum dots-fluorescence resonance energy transfer-based nanosensors and their application. Biosens Bioelectron 74(15):562–574

Suman PR, Jain VK, Varma A (2010) Role of nanomaterials in symbiotic fungus growth enhancement. Curr Sci 99:1189–1191

Sumida H (1992) Silicon supplying capacity of paddy soils and characteristics of silicon uptake by rice plants in cool regions in Japan. Bulletin Tohoku Agric 85:1–46

Swamy VS, Prasad R (2012) Green synthesis of silver nanoparticles from the leaf extract of Santalum album and its antimicrobial activity. J Optoelectron Biomed Mater 4(3):53–59

Szeghalmi A, Kaminskyj S, Rösch P, Popp J, Gough KM (2007) Time fluctuations and imaging in the SERS spectra of fungal hypha grown on nanostructured substrates. J Phys Chem B 111(44):12916–12924

Teodoro S, Micaela B, David KW (2010) Novel use of nano-structured alumina as an insecticide. Pest Manag Sci 66(6):577–579

Torney F (2009) Nanoparticle mediated plant transformation. Emerging technologies in plant science research. Interdepartmental Plant Physiology Major Fall Seminar Series Phys, p 696.

UNPD (2006) United Nations Population Division. In: World population projections, the 2006 revision, New York

Vardhana J, Kathiravan G (2015) Biosynthesis of silver nanoparticles by endophytic fungi *Pestaloptiopsis pauciseta* isolated from the leaves of Psidium guajava Linn. Int J Pharm Sci Rev Res 31:29–31

Varma D, Meena RS, Kumar S (2017) Response of mungbean to fertility and lime levels under soil acidity in an alley cropping system in Vindhyan Region, India. Int J Chem Stu 5(2):384–389

Verma JP, Jaiswal DK, Meena VS, Meena RS (2015) Current need of organic farming for enhancing sustainable agriculture. J Clean Prod 102:545–547

Vyom S, Ashutosh K, Alok D (2012) Nanomaterials: exposure, effects and toxicity assessment. Review. Proceedings of the National Academy of Sciences, India Section B: Biological Sciences©. The National Academy of Sciences, India. https://doi.org/10.1007/s40011-012-0072-7

Wang L, Li Z, Zhang G, Dong J, Eastoe J (2007) Oil-in-water nanoemulsions for pesticideformulations. J Colloid Interface Sci 314:230–235

Wang XY, Ma GH, Di H, Tian XH, Wang Y (2010) Effects of NMUrea on rice yield and agronomic nitrogen efficiency. Plant Nutr Fertil Sci:1479–1485

World Bank (2008) Annual world development report. World Bank, New York

Wu M-Y. (2013) Effects of incorporation of nano-carbon into slow-released fertilizer on rice yield and nitrogen loss in surface water of paddy soil. Third international conference on intelligent system design and engineering applications, Hong Kong, pp 676–681. https://doi.org/10.1109/ISDEA.2012.161

Yadav A, Kon K, Kratosova G, Duran N et al (2015) Fungi as an efficient mycosystem for the synthesis of metal nanoparticles: progress and key aspects of research. Biotechnol Lett 37:2099–2120. https://doi.org/10.1007/s10529-015-1901-6

Yadav GS, Lal R, Meena RS, Babu S, Das A, Bhomik SN, Datta M, Layak J, Saha P (2017) Conservation tillage and nutrient management effects on productivity and soil carbon sequestration under double cropping of rice in North Eastern Region of India. Ecol Indic. http://www.sciencedirect.com/science/article/pii/S1470160X17305617

Yamanka M, Hara K, Kudo J (2005) Bactericidal actions of silver ions solution on *Escherichia coli* studying by energy filtering transmission electron microscopy and proteomic analysis. Appl Environ Microbiol 71:7589–7593

Yang FL, Li XG, Zhu F, Lei CL (2009) Structural characterization of nanoparticles loaded with garlic essential oil and their insecticidal activity against Tribolium castaneum (Herbst) (Coleoptera:Tenebrionidae). J Agric Food Chem 57(21):1015610162

Yap CL (1997) Major issue of concern for the world rice economy in medium term. FAO. International Rice Commission. Newsletter 46

Yin L, Cheng Y, Espinasse B, Colman BP, Auffan M, Wiesner M, Rose J, Liu J, Bernhardt ES (2011) More than the ions: the effects of silver nanoparticles on Lolium multiflorum. Environ Sci Technol 45(6):2360–2367. https://doi.org/10.1021/es103995x. Epub 2011 Feb 22

Young KJ (2009) Antifungal activity of silver ions and nanoparticles on phytopathogenic fungi. Plant Dis 93(10):1037–1043

Zhang Z, Xi-fu F, Lei SS, Xue-li C, Yu-juan L, Ming Z (2010) Effect of nano-fertilizers on rice growth [J]. Heilongjiang Agric Sci 8:19

Zhao M-X, Zeng E-Z (2015) Application of functional quantum dot nanoparticles as fluorescence probes in cell labeling and tumor diagnostic imaging. Nanoscale Res Lett 10:171. https://doi.org/10.1186/s11671-015-0873-8

Zhou R, Wang P, Chang HC (2006) Bacteria capture, concentration and detection by alternating current dielectrophoresis and self-assembly of dispersed single-wall carbon nanotubes. Electrophoresis 27:1376–1385

Zulputra W, Dan N (2014) Response of upland rice (*Oryza sativa* L.) on the application of silicate and phosphate fertilizer on ultisol. J Agroteknologi 4(2):1–10

MIX
Papier aus verantwortungsvollen Quellen
Paper from responsible sources
FSC® C105338

If you have any concerns about our products,
you can contact us on
**ProductSafety@springernature.com**

In case Publisher is established outside the EU,
the EU authorized representative is:
**Springer Nature Customer Service Center GmbH**
**Europaplatz 3, 69115 Heidelberg, Germany**

Printed by Libri Plureos GmbH
in Hamburg, Germany